The Neglected Ape

The Neglected Ape

Edited by

Ronald D. Nadler
Yerkes Regional Primate Research Center
Emory University
Atlanta, Georgia

Birute F. M. Galdikas
Orangutan Foundation International
Los Angeles, California

Lori K. Sheeran
California State University
Fullerton, California

and

Norm Rosen
University of Southern California
Los Angeles, California

Plenum Press • New York and London

Library of Congress Cataloging in Publication Data

The Neglected ape / edited by Ronald D. Nadler ... [et al.].
p. cm.
Papers from the International Orangutan Conference: The Neglected Ape, held March 5–7, 1994, in Fullerton, California and Papers from the Orangutan Population and Habitat Viability Analysis Workshop, held January 18–20, 1993, in Medan, North Sumatra, Indonesia.
Includes bibliographical references and index.
ISBN 0-306-45213-8
1. Orangutan—Congresses. 2. Wildlife conservation—Congresses. 3. Endangered species—Congresses. I. Nadler, Ronald D. II. International Orangutan Conference: The Neglected Ape (1994: Fullerton, Calif.).
QL737.P96N44 1996 95-53964
639.9′798842—dc20 CIP

Cover photograph by R. D. Nadler taken from "Socioendocrine interactions in great ape reproduction," by C. E. Graham and R. D. Nadler. In: *Socioendocrinology of Primate Reproduction* (T. E. Ziegler and F. B. Bercovitch, eds.).

Papers from the International Orangutan Conference: The Neglected Ape,
held March 5 – 7, 1994, in Fullerton, California
and
Papers from the Orangutan Population and Habitat Viability Analysis Workshop,
held January 18 – 20, 1993, in Medan, North Sumatra, Indonesia

ISBN 0-306-45213-8

A Division of Plenum Publishing Corporation
233 Spring Street, New York, N. Y. 10013

10 9 8 7 6 5 4 3 2 1

Printed in the United States of America

PREFACE

The orangutan is the most highly endangered species of great ape. Orangutans are threatened by deforestation, poaching, the illegal pet trade, and the isolation and fragmentation of dwindling wild populations. Their conservation is impeded by certain aspects of their ecology (e.g., a rain forest habitat) and certain features of their life history (e.g., an eight- to twelve-year interbirth interval). Added to the U.S. Endangered Species List in 1970, the orangutan is now clearly on the road to extinction. The number of wild orangutans in Borneo and Sumatra is currently estimated to have decreased to between 12,300 and 20,571 individuals. Only 2% of original orangutan habitat is protected and some of these areas are now being destroyed. Clearly, attention to ecology, demography, censusing, rehabilitation, and conservation is essential if the orangutan is to survive in the wild beyond the next century. The protection of orangutans is a complex, multifaceted problem, involving such pressing issues as human poverty, overpopulation, and the economic development of Southeast Asia. Although the orangutan has been placed in Appendix I of the Convention on International Trade in Endangered Species of Wild Fauna and Flora (CITES), more orangutans were sold illegally in Taiwan between 1990 and 1993 than are housed in all the world's zoos.

In the past, scientific and public attention has centered on the African apes. For this reason, the sole Asian great ape, the orangutan, has been called the "neglected ape." Over the last several years, however, this situation has been changing. In 1991, the government of Indonesia sponsored the International Great Ape Conference, which was held at Jakarta, Pangkalan Bun and Camp Leakey, Kalimantan Tengah. As the last official event for "Visit Indonesia Year 1991," this conference heralded a new era in awareness of the orangutan and its plight as an endangered species. Following the International Great Ape Conference, the first Population and Habitat Viability Analysis Workshop for orangutans was held in Medan, North Sumatra, approximately a year later. The workshop concentrated on the status of wild orangutans in Sumatra and Borneo, emphasizing the Sumatran population. Three working groups established at the workshop made contributions to the *Indonesian Orangutan Action Plan*, a plan designed to ensure the survival of orangutans in their natural habitats. The following year, an international conference entitled, "Orangutans: The Neglected Ape" was held at California State University, Fullerton. This conference was sponsored by the Department of Anthropology and the Anthropology Student Association, California State University, Fullerton, the Zoological Society of San Diego, and the Atlanta/Fulton County Zoo, Inc. (Zoo Atlanta). The conference brought together some 200 academic scientists, zoo personnel, field biologists, government officials, and concerned lay people from Europe, Southeast Asia, and North America. Keynote addresses were presented by J. A. R. A. M. van Hooff, J. Sugardjito, B. M. F. Galdikas, and T. L. Maple, all renowned for their contributions to our understanding of orangutan biology and behavior. Over a three-day period, approximately 56 presentations were delivered in 11 sessions, including conservation, census and demography, genetics, training and husbandry, social behavior, physical anthropology, cognition, veterinary medicine, enrichment and design, reproduction, and ecology.

As part of this new era of enlightenment, heightened media attention has broadened the public's exposure to orangutans. Two books (*Orangutans in Borneo* and *Reflections of Eden*) have recently been published, both aimed predominantly at the general public. Articles concerning orangutans in newspapers and popular magazines have multiplied over the past few years, while numerous orangutan documentaries have been featured on television.

This volume, a direct outgrowth of the Fullerton conference and the Medan workshop, is comprised of chapters that continue and extend this era in which the orangutan has now begun to receive the special attention and concern it requires. We selected authors to include in this volume who address the more important issues related to the survival of orangutans, their propagation in the wild and in captivity, and their unique contributions to the science of primatology. Some of the chapters review the accomplishments in specific research areas, but most present entirely new contributions of scientific data and analysis.

Section One of our edited volume provides an overview of the orangutan species, its past and current status and its predicament, as seen through the eyes of philosophers, conservationists, and a psychologist.

Section Two is devoted to orangutan conservation and recovery plans. Since the 1970s, attempts have been made to rehabilitate orangutans confiscated from animal dealers and private individuals. Rehabilitation demands considerable dedication because the ex-captives are often in poor physical and emotional condition — starved, abused, and unfamiliar with the natural lifestyle of wild orangutans of similar age. The number of orangutans being confiscated has not decreased and, therefore, rehabilitation efforts must be continued and improved. Furthermore, destruction of orangutan habitat has continued to increase from the 1980s through the present time, boding ill for the occupants.

Section Three in this volume is devoted to demography. More sophisticated methods than previously available are being applied to determine the current and projected number and distribution of wild orangutans. These techniques include Geographic Information Systems, which by displaying current forest distribution, vegetation type, and degree of fragmentation, allow for more accurate estimates of orangutan population sizes and distributions. The Population and Habitat Viability Analysis VORTEX model for orangutans projects population composition based on current demographic trends. Both of these techniques assist conservationists in more accurately identifying problem areas and in concentrating on key issues. Most importantly, the information so acquired provides forest officials with a more accurate appraisal of the possible consequences to the orangutan of converting its forests to farms, roads, and lumber.

In Section Four, research results are presented from the two locations where long-term field studies have been conducted, Tanjung Puting and Gunung Leuser. This section includes a theoretical discussion of evolutionary contributions to the unusual social organization of orangutans among the primates and two papers that document orangutan cognition in free-ranging wild-born ex-captives.

Section Five is devoted to scientific research on the social and sexual behavior, reproduction and physical development of orangutans and to their maintenance and propagation in captivity. The issues considered here enhance our scientific understanding of orangutan behavior and physiology and contribute experimental evidence for hypotheses derived from naturalistic observations in the wild. Hypotheses regarding reproductive strategies and mate choice, in particular, are susceptible to testing in a laboratory setting with experimental control over extraneous activities. Modern approaches to reproduction and captive management of orangutans are described by investigators in the forefront of their areas of specialization. The data can be applied to the furtherance of captive breeding programs, the improvement of captive facilities and the interpretation of data gathered from wild orangutans.

When all is said and done, we cannot be said to fully understand orangutans until we have studied and described their behavioral, ecological, and evolutionary complexities. We edited this volume to highlight current research and knowledge of the orangutan as a species in all its magnificent variability. We hope that this endeavor will help galvanize increased efforts on behalf of what was once the "neglected ape."

We thank all the authors who contributed chapters to this volume, Barbara McElduff for secretarial assistance and the Yerkes Regional Primate Research Center of Emory University for support (National Institutes of Health Grant No. RR-00165 from the National Center for Research Resources).

The Editors

CONTENTS

Section Three – Orangutan Demography and Population and Habitat Viability Analysis

Section Four – Social and Cognitive Bahavior

Section Five – Basic Science and Captive Maintenance

SECTION ONE – ORANGUTAN: MAN OF THE FOREST

Introduction

Section One recounts the diverse ways in which orangutans have been perceived throughout history and how those perceptions have influenced the status of orangutans in their natural habitat. Through a brief history of philosophy, Kaplan and Rogers illustrate how the orangutan has been hostage to the ambivalent concepts of nature. The pendulum has swung from glorifying the orangutan as the penultimate *noble savage* to vilifying it as a monster. Public affiliation with orangutans is problematic. Although humans identify with orangutans, they are not sufficiently empathetic to protect the species from what Rijksen refers to as "imminent extinction." Perhaps the commonest attitude is the one expressed in Mrs. Pryer's Diary of 1894, cited by Kaplan and Rogers. Mrs. Pryer explains, with some sympathy for the "poor creature," that the death of a crop-raiding female orangutan was necessary because, deprived of her habitat, the orangutan was "devouring the sugar cane." Human interests and expediency take precedence over compassion for another animal. In some sense, our fear of this "primordial, devilish creature" exemplifies our worst nightmares of human depravity.

Currently, the orangutan is a victim of global economic forces. These forces take precedence over all other considerations, human or otherwise, including the general good-will that modern humans might have theoretically toward orangutans. Rijksen laments the fact that the international conservation community did not come to the orangutan's aid during the 1980s. He argues that because conservationists were cautious, slow, and apparently interested mainly in compiling information about the plight of the orangutan rather than dealing with it, they completely neglected the deteriorating condition of the wild population and of its habitat.

The 1980s were, paradoxically, a time of considerable conservation activity concerning both the orangutan and its rain forest habitat. International conservation organizations such as the Orangutan Foundation International, with affiliates in Taiwan and Indonesia, were established during this time precisely to deal with the increasing problems of orangutan conservation. The wealth and influence of even the largest and most powerful public conservation organizations, however, are minuscule in comparison to the wealth of governments and large, multinational corporations. The global conservation picture throughout the 1980s, moreover, was so fraught with problem areas that the major conservation groups were forced to set priorities. There were so many species and ecosystems to protect that accomplishing even minimal goals required conservation organizations to focus on particular campaigns at the expense of others. Considerable funding, for example, was devoted to the well-publicized crises affecting elephants, dolphins, and mountain gorillas. Rijksen now makes a strong appeal for immediate action on behalf of the orangutan. He vehemently asserts that enough background information has been gathered pertinent to the conservation of this species; it is time for *action* on its behalf. He ends on a hopeful note, moreover, encouraged by the most recent steps in a positive direction taken by the government of Indonesia.

Eudey summarizes the extent to which the countries of Southeast Asia are already participating in the global economy with annual economic growth figures of 8% or 9%. These growth rates make Southeast Asia an important emerging component of the fastest growing economic region in the world, the Pacific Rim. As was true in the industrialization of today's "developed" countries, the modernization of Southeast Asia has had a profound, often negative, impact on endangered wildlife and their habitats. Eudey chronicles several economic and conservation trends in this region of the world.

Taking an optimistic approach, Rose sees great value for the conservation of orangutans in "Profound Interspecies Events (PIEs)." PIEs are peak experiences individuals describe as being revelatory and self-defining which derive from interactions with animals in general and orangutans in particular. In addition to being psychologically illuminating, PIEs have the potential to provide an emotional impetus for the conservation movement if they can be presented to the public in ways that evoke the inspiration felt by the individuals who experienced them initially. Rose reminds us that despite the justifiable pessimism we sometimes feel, humanity, at its core, has an affinity for the natural world. This affinity may help avert the extinction of animals such as the orangutan if primatologists will collaborate to promote it.

1

OF HUMAN FEAR AND INDIFFERENCE

The Plight of the Orangutan

G. Kaplan[1] and L. Rogers[2]

[1] Aboriginal and Multicultural Studies
University of New England
Armidale, NSW, 2351, Australia
[2] Department of Physiology
University of New England
Armidale, NSW, 2351, Australia

INTRODUCTION

To speak of western attitudes toward the orangutan in past and present is to evoke at once the ambivalent concept of "nature". Humankind has continued to make decisions on who and what is "in" nature and what ranking order and importance is to be assigned to nature and the living things deemed to be part of it. This paper proposes that we can glean from a study of shifting attitudes toward nature why there have been such shifts in the way orangutans were and are being perceived and treated. First, we briefly introduce some major thinkers concerned with nature and natural law in the pre-modern era. These are addressed again below because their views had profound implications for the way we view any species deemed to be part of nature. In the second section, the paper compresses popular writing and attitudes toward orangutans into a framework of the broad social and economic trends of modern times.

NATURAL LAW AND SOCIAL JUSTICE

The concept of nature has occupied philosophers, politicians and law makers throughout recorded history, clearly much longer than concepts about the apes. Empedocles, well before Aristotle, used the concept of the "survival of the fittest", but Aristotle disputed this. Medieval scholastic doctrine made the transition from nature as a universe to "nature" as a divine creation. According to this view, everything had a purpose because it was made by God. There also had to be a divine or "natural" law which, in turn, guided human society. The medieval scholar, Thomas Aquinas, argued, and this position can still be found in some attitudes and arguments of this century, that human-made laws had to reflect natural law:

The Neglected Ape, Edited by Ronald D. Nadler et al.
Plenum Press, New York, 1995

> "Every law framed by man bears the character of a law exactly to that extent to which it is derived from the law of nature. But if on any point it is in conflict with the law of nature, it at once ceases to be a law; it is a mere perversion of law." (cit. Tawney, 1969)

Bertrand Russell (1945) felt that John Locke's second *Treatise on Government* (1690) did not alter Aquinas' position. However, unlike the scholastic doctrine, he believed that a state of nature existed as a preceding historical stage in development, superseded as it were, by a human contract which instituted government. At the same time, Locke introduced a utopian element in his definition of the state of nature when he said:

> "Men living together according to reason, without a common superior on earth, with the authority to judge between them is properly the state of nature." (cit. Russell, 1945).

Evidently, Locke saw the rule of reason as compatible or even identical with natural law. One can see from this statement how much Locke's view of human "nature" and of natural law differs from that of Hobbes, whose writing preceded his by several decades. Hobbes was disinclined to think human beings were capable of the benign anarchism Locke envisaged, but he thought they were rather predisposed to a cruel and warring anarchism of "every man against every man". Natural law was thus defined by an intrinsic lawlessness or mutual destruction. It was necessary, therefore, to overcome the state of nature. In his work *Leviathan* (1651), Hobbes created what he called an "artificial animal", a commonwealth or state, greater in stature than nature and therefore, protecting and defending man. "Nature", he wrote, "the art whereby God hath made and governs the world, is by the art of man, as in many other things, so in this also imitated, that it can make an artificial animal". Without such a commonwealth, man would be thrown back into a "natural state", leading Hobbes to conclude in his famous words that in such a state of nature "the life of man [would be] solitary, poor, nasty, brutish and short" (Hobbes, 1651). It is worth remembering this conclusion when discussing attitudes toward orangutans and the other great apes.

In the 18th Century the French, English and Scottish philosophers developed the confidence that everything was knowable and could be ascertained and explained by reason. It is in the 18th Century that the term "common sense" was coined (Thomas Reid), that Linnaeus classified the natural world systematically into classes and categories and that the French philosophers, such as Diderot, Voltaire, D'Alembert and Rousseau worked on the largest enlightenment project, the encyclopedia.

Rousseau is undoubtedly one of the most important thinkers in terms of this discussion about the later attitudes toward apes. He, for his part, helped form a tradition of less Machiavellian disposition. In an intellectual climate that eventually culminated in the French Revolution (1789), the Academy of Dijon in 1749 advertised a prize-winning competition for an essay, asking "Has the progress of the arts and sciences tended to the purification or to the corruption of morality?". The question fell on fertile ground and in a sense started the career of Jean Jacques Rousseau, who answered it and won the prize. In a flash of inspiration, he rejected Hobbes' view of man and argued that "man is naturally good" (Hendel, 1937). He proceeded to perceive the corruption and artificiality of social/"man"-made systems and institutions and rejected reason as the one and only source of human endeavor.

More importantly, Rousseau's ideas stimulated a shift in focus from a God-given universe to a secular examination of people and society. He argued that humanity had drifted far away from its origin in nature. We changed from natural beings to social citizens and, as social citizens we were corrupted and put people in chains and morally depleted our institutions. It was time to listen to one's emotions and turn back to a harmonious "natural" state. But how was one to (re)discover the nature of man and how could one follow one's own "nature"?. What is it that must be learned and unlearned? In his work *Emile*, *The Social Contract* and *Discourse on Inequality*, he developed the ideas to answer these profound questions.

There were two ways for rediscovery: journeys to other places and cultures, in the hope of discovering people and life forms which were as yet not corrupted and "over-civilized", and a journey into self. As Bronowski and Mazlish (1960) so succinctly put it, Rousseau "rejected anthropology in favor of introspective psychology". In a romantic move, Rousseau suggested retiring to the woods to "behold your history... not in books written by your fellow creatures, who are liars, but in nature, which never lies".

However, Rousseau accepted the concept of his age of the "noble savage", acquired precisely from the experiences of travelers to the New World. Steeped in misconceptions and prejudices, the "noble savage" of "primitive" society was nevertheless proposed as the paradigm for the natural life which Rousseau could then fully utilize for his own purposes. Rousseau read widely in the travel literature of his age (cf Hendel, 1937). Of the native Americans, for example, he believed

> "According to the travel reports, the savages of North America...despised gold and silver. They lived freely and happily in the forests, dwelling harmoniously in small groups. They needed no laws and social constraints and ordered their lives by instinct rather than by reason; for example, they found their way home, instinctively the 18th Century thought, by following the direction of the sun at different times of day and not by reading the compass. As interpreted by Rousseau and others, the noble savage was the man who, with a wonderful set of emotions, and rejecting entirely his intellectual gifts, achieved for himself the simple and perfect life which civilization had since destroyed" (cit. Bronowski and Mazlish, 1960).

It was this construct of the "noble savage", although based on utterly incorrect assumptions and defective observations, which permitted a pre-Darwinian evolutionary benevolence. If "savages" could live without laws, i.e., only by their instincts, then those half-men half-beasts may deserve our attention as well. Daniel Beeckman is said to be one of the first to have visited orangutan territory. His account, *A Voyage to and from the Island of Borneo* (1714), suggests a certain ambivalence about these creatures:

> "The Monkeys, Apes, and Baboons are of many different Sorts and Shapes; but the most remarkable are those they call Oran-ootans, which in their Language signifies Men of the Woods: these grow up to be six Foot high; they walk upright, have longer Arms than Men, tolerable good Faces (handsomer I am sure than some *Hottentots* that I have seen), large Teeth, no Tails nor Hair, but on those Parts where it grows on humane Bodies... The Natives do really believe that these were formerly Men, but Meta-morphosed into Beasts for their Blasphemy." (cit. Harrisson, 1962).

Similarly and at about the same time, a Frenchman by the name of Benoit de Maillet wrote in *Telliamed* that [orangutans have] "the whole of the human form, and like us walked upon two legs." In captivity they are "very melancholy, gentle, and peaceable" (Schwartz, 1987).

One result of the New World discoveries and of the efforts during the 18th Century to classify and describe the world around them was to examine the origin of things and of all things human. As the natural world was scanned for information, so was the origin of language of great interest to 18th Century thinkers. In 1772, Johann Gottfried Herder presented his work, *Of the Origin of Language*, in which he argued that language was partly the result of a weakening of natural instincts and that for the sake of survival, other means of communication had to be found.

In 1774, Lord Monboddo published the first of six volumes called *Of the Origin and Progress of Language*. By that time, Rousseau's works had already been published and widely debated. Whether or not Monboddo knew of these works is almost secondary, although he might have, given Mary Woolstonecraft's thoroughgoing objection to *Emile*. It is important that Monboddo's sense of where to put the boundaries between humans and animals was blurred by his experiences with and study of orangutans. In addition, the idea of the noble savage, which came after all from England and from the age of Thomas More, inspired an openness toward other life forms, even if it was still tied to human self-interest.

Monboddo published anonymously, but his views were widely known when he wrote about the orangutan and its potential for language acquisition:

> "I still maintain, that this [the orangutan] being possessed of the capacity of acquiring it [language], by having both the human intelligence and the organs of pronunciation, joined to the dispositions and affections of his mind, mild, gentle, and humane, is sufficient to denominate him a man" (Monboddo, 1774).

Hence, in the spirit of the enlightenment, even the ape was accorded a status of humaneness and an anthropomorphic interpretation, a tradition that might well be evoked under the new banner of the *Great Ape Project* (Cavalieri and Singer, 1993). It is a tradition that accords equality to those who lack it and human status to those who are denied it (such as Aborigines who, in Australia, were classified under "flora and fauna" in British colonial administration and only received citizenship in 1967).

THE MODERN WESTERN WORLD AND ITS VIEWS OF NATURE

There is a vast literature on the early scientific history of anthropoid study, but one must see this in light of social theory, philosophy and economics. The latter, in particular, is of substantial interest. In 1776, on the very day of America's Declaration of Independence, Adam Smith published his most noted work, *An Enquiry into the Nature and Causes of the Wealth of Nations*. It was a blueprint for the Industrial Revolution to come, and it, as well as the profound experience of the Industrial Revolution, fundamentally changed our view of nature.

According to the present schemata, four types of attitude toward nature are distinguished. The purpose in describing them is to give an impression of the fundamental shifts in attitude toward the great apes. Such shifts are by no means along a simple linear development, as from negative to positive, but rather like a swinging pendulum, slowly evolving into a dialectic position to which we have perhaps, managed to progress today.

1. The competitive model: Nature is something to be feared and that fear is managed by attempting to assume control. It must be confronted and, as in the first words of the Bible, it is right and permissible to subjugate and control it. Humankind is the highest form of creation and hence, master of the earth.
2. The pragmatic, economic model: Nature, in a utilitarian sense, is simply a resource. We can take from it anything that we need. It offers us our subsistence and with some labor, it can be made to yield its riches. These riches are boundless and exist for our sake. Indeed, so natural is this provisioning that it requires no thought or ethical consideration in proceeding with the agenda of exploitation. Or indeed, there is an ethics, in the sense of allowing any action that contributes to the self-preservation of an individual, group or the human species.
3. The romantic model: Nature is benign, noble, romantic; a symbol for paradise. Nothing in it must be disturbed. With a nostalgic view, one can glimpse what humankind has lost and may never regain or may only recapture in part. Harmony reigns in nature and human society can only strive to imitate nature. Nature is perfect and needs no improvement or change.
4. The dialectic model: Nature and our relationship to it are complex (in a sense, this includes all three points above) and human society develops out of the natural world. It is indeed true that human society has the capacity to use and change nature, but not in the sense of domination and rapacious destruction. By recognizing that nature has its own laws and rules for survival, it is possible to live with nature and use it, but also put something back and allow it to be maintained.

Clearly, the views above are simplified and are unlikely to occur in pure form anywhere. They are not sequential historically, but may exist side by side. Except for the fourth position, however, all the views of nature described above relate to human society, to the way we regard nature and what it might be in relation to us. It is something "outside" us, an "other", but it exists, as it were, *only* in relation to us and what we may get out of it.

An example of the third model was already described in the works of Rousseau and his contemporaries. Model 2 (nature as resource), to some extent, holds for all human society, although degree, modes and purpose of use vary vastly according to the material basis and economic system of a given society. We shall come back to this.

Model 1 (nature as something negative and feared) has probably been the most prevalent one, at least since the onset of Christianity. We have to dwell on this, for this view of nature reemerged in a new garb and received a boost with the onset of the Industrial Revolution, which was to change our world irrevocably. In the literature, there is a notable change in attitude toward apes, including orangutans, from the second half of the 18th Century to the time of the Industrial Revolution, which occurred in England in the early 1800s and in the 1830s in France. Substantial and sudden technological advances were accompanied by a new recklessness and confidence. In this chilling wind of competition, the principles of capitalist accumulation of wealth began to set the only standard of conduct, a trend which has not halted (although lately modified) to the present.

The key word is *competition*, expounded just as eloquently in the evolutionary theories of Darwin and Wallace as by the social and economic theorists of the 19th Century. It spawned the theories of Joseph Arthur, Comte de Gobineau (1855) and Houston Stewart Chamberlain's widely influential book, *The Foundations of the Nineteenth Century* (1899), intent on declaring the supremacy of certain human "races" over others, of some peoples over others and of man over woman. It fostered and raised to a new height a concept of work and industry. It was at once divided into a coercive exploitative mode of cheap labor, on the one hand, and a further exultation of a "protestant work ethic", on the other. Nineteenth Century "laissez faire" attitudes toward commerce were based on rather undiluted views of natural law inherited from Thomas Aquinas and John Locke. Darwin's theory of evolution was not accidentally coeval with laissez faire capitalism. Competition was for growth, survival, profit, progress and productivity, understood in purely economic terms. These became the hallmarks of social thought and daily life.

Hence anything, and certainly anything in nature that was alien to or stood in the way of these endeavors and thoughts, could be despised, devalued, discarded, destroyed or ignored. Thus, Rennie (1838) condemned orangutans as slovenly and useless creatures: "Their deportment is grave and melancholy, their disposition apathetic, their motions slow and heavy, and their habits so sluggish and lazy, that it is only the cravings of appetite, or the approach of imminent danger, that can rouse them from their habitual lethargy, or force them to active exertion." (cit. Yerkes and Yerkes, 1929).

A way of justifying one's own competitive interests was also to resurrect the view that nature was dangerous and had to be fought. Fear of nature is as old as humanity itself. Nature harbored not only catastrophes, which still controlled human populations (see Malthus' theories), but it also housed beasts which made human entry into natural environments dangerous. Such danger was reason enough to destroy.

The travel literature of the 19th Century was dominated by "science" and by adventurers (often combined in one). It was the era of the big game hunters, of men whose one aim was to shoot the danger away and bring it back home as a trophy of their own greatness and power that the firearm had bestowed upon humans. It was in the interest of such adventurers to proclaim whatever they found as truly awesome and dangerous. Hence, Schlegel and Müller, in an expedition to Borneo in 1839-44, described the orang-utan thus:

> " His penetrating sharp glance and wild features, indicated him to be a more than untractable animal. Moreover, the long rough hair of his head and the heavy red beard under the chin gave him a wild appearance. Add to all this his terrible strength, of which he sometimes made use in a violent way, whenever one would torment him with a stick."(cit. Yerkes and Yerkes, 1929).

Indeed, this manner of describing wild animals has been exploited in ever more elaborate fashion until today, be this in horror movies or in literature. Here one finds an unbroken chain of films, literature, posters and drawings of the 19th and especially the 20th Century, portraying beasts arising in swamps, horrible prehistoric birds attacking and carrying away humans, and rats and mice infesting human quarters and eating human flesh. As another variant, insects (ants, bees) that attack in the thousands and devour human skin, evoke fear and nausea. In the ape world, one of the first epithets of the horrible monster ape image was created by Edgar Rice Burroughs in 1912 with his first publication of *Tarzan of the Apes*. Scattered through the pages of this work is an imagery that was later used in *King Kong*:

> "At last he saw it, the thing the little monkeys so feared—the man-brute of which the Claytons had caught occasional fleeting glimpses" (p.21). "The ape was a great bull, weighing probably three hundred pounds. His nasty, close-set eyes gleamed hatred from beneath his shaggy brows, while his great canine fangs were bared in a horrible snarl as he paused a moment before his prey" (p.22). " The man swung his axe with all his mighty strength, but the powerful brute seized it in those terrible hands, and tearing it from Clayton's grasp hurled it far to one side" (p.23).
>
> "The other males scattered in all directions, but not before the infuriated brute had felt the vertebra of one snap between his great, foaming jaws... With a wild scream he was upon her, tearing a great piece from her side with his mighty teeth, and striking her viciously upon her head and shoulders with a broken tree limb until her skull was crushed to a jelly" (p.27). " Standing erect he threw his head far back and looking fully into the eye of the rising moon he beat upon his breast with his great hairy paws and emitted his fearful roaring shriek. Once-twice-thrice that terrifying cry rang out across the teeming solitude of that unspeakably quick, yet unthinkably dead, world." (p.53)

The orangutan has never quite had the reception accorded the gorilla or been cast into similar lead roles of terror, but both in literature and drawings, some attempts were made to develop a similar story line for the orangutan. It is likely that Schlegel and Müller purposely mistreated the male orangutan with a stick in order to sample some of that "terrible strength". Stories were told of male orangutans who abducted fragile English ladies and raped them in trees. Likewise, drawings of awesome and/or ugly faces were no doubt intended to generate dislike and fear. Moreover, the orangutan exemplified what Hobbes described as the ultimate outcast image of "man" devoid of ordered social existence, namely a *solitary* life, *short, nasty and brutal* (emphasis authors). The orangutan was a mirror to humans of what they might, but ought not, be and it therefore, should be smashed. Reading the travel accounts of men such as Beccari, Shelford (see Yerkes and Yerkes, 1929), Hornaday or even Wallace, it is clear that there is not the slightest indication of pity expressed when they aimed at and shot the orangutans, that looked directly at them from the trees.

By the close of the 19th Century, the view of utopias, i.e., of beautiful human societies, and the idea of a benign Nature had moved far from view. Freud's writings were no comfort to the soul, for he reconfirmed our fears of nature and even worse, saw the darkest uncontrollable, possibly even evil forces, within ourselves. Rousseau's recommendations for an introspective psychology led only to the discovery that we harbor in ourselves eros and thanatos, competing powerful drives from time immemorial; that we have a subconscious that is not in our control and what we thought could be shot and eradicated in the natural world, was now also in us.

Rousseau's beautiful nature, inside or outside the self, as H.G. Wells described in his novel, *The Time Machine* (1894), did not lead to the utopian and benign anarchy that John Locke proposed. The harmony was false and the idea of progress towards a more humane and perfect world was bitterly disappointed. Instead of innocence, the time traveller found only ignorance. Instead of peace, he found that humanity in the distant future has collapsed back on itself into prehistoric origins with two species. One species was the Eloi, a beautiful peaceful people, ignorant and disinterested victims, only apparently, living in paradise and feeding on the fruits of Nature. The others were the Morlocks (We note here that Linnaeus conceived of two human species, *Homo sapiens* and *Homo nocturnus*). It was atavistic of H.G.Wells, that these "Morlocks" were fashioned in detail on Linnaeus' classification which is found in the tenth edition of his *System of Nature* (1758). In it, Linnaeus described a humanoid beside *Homo sapiens* which he called *Homo nocturnus* or *Homo sylvestris orang-outang*:

> "Body white, walks erect, less than half our size. Hair white, frizzled. Eyes orbicular; iris and pupils golden. Vision lateral, nocturnal. Liffe span twenty-five years. By day hides; by night it sees, goes out, forages. Speaks in a hiss. Thinks, believes that the earth was made for it, and that sometime it will be master again, if we may believe the traveller."

The time-traveller in H.G. Wells noted: They are small, about half size. In the dying moonlight at night he saw "white figures", a "solitary white, apelike creature" (p.56), a "second species of man" (p.60). "These whitened Lemurs" (p.64), were "whispering odd sound to each other" (p.70), and "those pale, chinless faces and great, lidless, pinkish-grey eyes" (p.71) frightened him in the dark, for the creatures were nocturnal and occupied an underworld. They had turned carnivorous and were feeding on the beautiful people living above them.

The imagery of an underworld, a devilish creature that was once human, is raised in reverse in William Golding's novel, *The Inheritors*. At the end of the novel, the more advanced hairless groups of *Homo sapiens* won, but amongst them was one survivor of the other, a little red thing which they were unable to kill, a thing they feared but were forced to carry with them - because they loved *and* feared it. Hence, the description used was like an externalized way of demonstrating the cargo of fear that Golding felt we have inherited from time immemorial. Interestingly, the features of the creature were conceptualized in terms of an ape that might well be an orangutan.

In all, there was a tradition that described nature as trauma, which in turn led to views and actions of plundering and fighting it. That battle, if anything, has intensified in the 20th Century. However, now the battle is no longer just vis-a-vis the "other," but within and amongst the human species, be this expressed in psychoanalytic theory from Freud to Melanie Klein or, alternatively, in the ultimate and perverted practice of biologism that informed the racial beliefs of the eugenics movement and of the Nazi regime. In the former, it first took the form of locating the fears of the external, of the nasty, ugly, the dangerous in ourselves. The savage deformed slave, as Shakespeare knew intuitively when he wrote *The Tempest,* is not Caliban, some poor wretch, but a mirror of ourselves or the hidden "id." It is something not quite human, something that tells us that the thing we feared most about nature was in fact within us. In the latter, the Nazi ideology, the fear of contamination and extinction, antedated by Ernst Haeckel, resulted in the most inhumane interpretation of "natural law". Ultimately this led to the Holocaust and the destruction of life that was not considered worth living, or considered not quite human. During the Nazi period and even before, at the height of the eugenics movement in Germany, the ape, the black and the Jew were often seen as belonging to the same sub-human stratum (cf. Kaplan, 1994).

We have now lived and seen and experimented with the two most extreme solutions of what can be done with our fear of Nature and Nature in us. Psychologically and socially there are a number of known strategies of how to deal with our own discoveries of self. Until very recently, probably the most common form of dealing with that fear was by denial. Denial of fear can take several forms, such as aggression, hatred, revenge or exploitation. There are examples of all forms of such action in human conduct. The most prevalent since the Industrial Revolution is exploitation. Exploitation has been most widely practiced and accepted by inadvertently or openly declaring that all of Nature is simply a supermarket for human needs. We pluck and pick what we like and as much as we like. Indeed, much of modern technology is aimed at making the exploitation faster, more efficient and on an ever increasing scale.

The problem is how we respond when some "article" in this natural supermarket does not strike our fancy. It is clear from the many accounts of the 19th Century that no particular value was ascribed to an orangutan. The orangutan was good for nothing — it grew too large and strong to be a pet, was only of limited use as a domestic helper (in 1892, Garner recorded that an orangutan worked as his domestic, performing chores like other "domestics" [cf. Yerkes and Yerkes, 1929]), did not have the same entertainment value as a chimpanzee and did not have any immediate medical or research value. It was difficult to observe, allegedly boring to watch because of its "sluggish" behavior and was difficult to keep confined. They were, if anything, superfluous within the scheme of human development. Since the Industrial Revolution, this attitude runs parallel to the confronting posture. We may gain a view of this attitude in Mrs. Pryer's Diary of 1894, where she recorded an incident of an orangutan encounter:

Diary Entry 15 August 1894

> "Later Mr. H. went, and after 2 1/2 hours absence arrived saying the Sooloos had got the poor beast but it was not dead so they brought it bound hand and foot and placed it before the house. I felt so sorry for it; for it had been shot and had also had a fall from a tree, but it was very brave and patient not making a sound altho' it must have been suffering severely. W. said it was a female of the large species. As it was tremendously powerful and too large to keep in captivity and was wounded and also they thought had an arm broken it was the most merciful thing to kill it, but it would not die, they had to shoot it three times before they succeeded in dispatching it. A horrid affair but orang utans can't be allowed to roam about at will devouring the sugar cane" (Tarling, 1989, p. 51).

The death of this female orangutan was pragmatically dismissed as a necessity and as the removal of an interference in human affairs. There was literally no place for it and upon that realization, Mrs. Pryer closed the matter, even though she felt sorry for the "poor beast." The view that orangutans were really not very interesting because one could allegedly "do" so little with them was echoed in 1924 in a scientific paper by Sonntag, who then proclaimed that

> "The Orang is the least interesting of the Apes. It lacks the grace and agility of the Gibbon, the intelligence of the Chimpanzee and the brutality of the Gorilla" (cit. Yerkes and Yerkes, 1929).

That prejudice was still present in the 1960s when Reynolds claimed of the orangutan that there was "nothing very spectacular about them" (Reynolds, 1967). It was thought to be less suitable for experiments (Drescher and Trendelenbury, 1927; Yerkes and Yerkes, 1929), less capable of problem-solving (Köhler, 1921) and less skilled in manipulating capacity and ability than the other apes, especially the chimpanzee.

In a sense, that lack of attraction to the orangutan has saved it from some severe exploitation on an organized level such as the rhesus monkey, chimpanzee and gorilla have experienced. On the other hand, a lack of interest can be disastrous for any species. The survival of any species now depends on human action. A lack of interest means also a lack of compassion and a lack of understanding that there is a Nature "out there" which is different from us, but which sustains us. If we destroy it, we destroy ourselves. This, our century, is a wholesale slaughterhouse of Nature, where species after species driven into extinction. We, as a collective species, have allowed the orangutan to approach the very edge of extinction through sheer neglect.

There is only one level at which orangutans have been "marketed" positively and that is in its demystification as a child's toy. Recent articles, such as "Born to be mild"

(Williams, 1991) and reports of the cruel transportation of orangutan babies have roused the sympathy of many. Williams described the orangutan as being "trusting and inquisitive", "endearing, loyal pets" and "one of the world's most delightful and rare animals". At one level at least, the mechanism is the same whether the imagery is terrifying or captivating, whether grossly exaggerated or belittling. The imagery has nothing to do with the actual being, but a good deal to do with our commercial world of marketing. Both forms of symbolism reveal an attitude toward Nature at least as alien, if not as hostile.

Not until the 1960s and 1970s were more positive and sober voices heard, such as that of Barbara Harrisson, the American primatologist, Duane Rumbaugh (Rumbaugh and McCormack, 1967; Rumbaugh and Gill, 1971) and in the work of the German primatologist, Jürgen Lethmate (1977); the latter's behavioral studies from 1974 to 1977 conclusively proved Rumbaugh's point that the chimpanzee's intellectual superiority over the orangutan was a myth.

There are signs of change. The environmental movement from the 1970s onwards has created a forum for examining our views of Nature on a wide scale. Words like "sustainable growth" have appeared in our vocabulary and our economic rationalism. Non-renewable resources are classified as such and the battle for rainforests has begun in earnest, but perhaps too late.

Amidst these events there has been a tradition which commenced in response to the Industrial Revolution (first noted by Friedrich Engels) and which needs only little modification now. This tradition is the dialectic view of Nature which we list as number 4) of our schemata. While the human species has a distinct identity from Nature and a clear separateness, it need not be hostile (The Frankfurt School). Beyond The Frankfurt School, Melanie Klein and C. Fred Alford have spoken of a reconciliation in which it is possible to entertain a passionate relationship in which each allows the other to live. Alford likens Nature to our relationship with our mother, as the source of nourishment and as an agent of control. He writes:

> "Mother Nature should be generally resented as the bad mother, the frustrating and ungiving mother, as in fact she frequently is. At the same time, many peoples — even, and perhaps especially, in those parts of the world in which nature exacts a harsh toll — feel a deep love and reverence for nature, as well as a desire to make reparation to her. But why? Klein answers that the issue is not so much the generosity of nature for which we feel gratitude, but rather the fact that we love nature in order to preserve a relationship with her while continuing to aggressively extract her resources... The love of nature allows us to remain close to, and part of, what we must also separate from: Mother Nature" (Alford, 1989, p. 158).

What C. Fred Alford and Melanie Klein were attempting to do was to propose an approach by which we abandon our fear and destruction of Nature without falling into the romanticized model of denying our aggression towards it (as the Frankfurt School proposed). The point is a dialectic process of acknowledgement, of giving and of taking in an interactive, interdependent way. As a species, particularly since the Industrial Revolution, we have been exceptionally skilled in "taking." We now have to do some "giving," and in so doing, as Klein suggested, we will be enriched as well by once again being able to believe in the "existence of something good that lies beyond ourselves" (Alford, 1989, p. 159).

REFERENCES

Alford, C.F., 1989, *Melanie Klein and Critical Social Theory. An Account of Politics, Art, and Reason Based on Her Psychoanalytic Theory*, New Haven and London: Yale University Press.

Alford, C.F., 1993, Reconciliation with nature, *Theory, Culture and Society*, vol.10: 207-227.

Beccari, O., 1904, *Wanderings in the Great Forests of Borneo. Travels and researches of a Naturalist in Sarawak*. London: Archibald Constable & Co.

Beeckman, D., 1714, *A Voyage to and from the Island of Borneo*, republished 1973, Dawsons of Pall Mall, Folkestone and London.

Bronowski, J. and Mazlish, B., 1960, *The Western Intellectual Tradition. From Leonardo to Hegel*, New York, San Francisco, London: Harper and Row.

Burroughs, E.R., 1912, *Tarzan of the Apes*, New York: Ballantine Books (1939 ed.)

Cavalieri, P. and Singer, P., eds. 1993, *The Great Ape Project. Equality beyond humanity*, London: Fourth Estate.

Drescher, K. and Trendelenburg, W., 1927, Weiterer Beitrag zur Intelligenzprüfung von Affen (einschlieβlich Anthropoiden), *Zeit. vergl. Physiol.* 5: 613-642.

Gobineau, J.A., 1855, *Essai sur l'inegalité des races humaines*. Trans. ed. by A. Collins, 1915, *The Inequality of Human Races*, New York: Putnam's.

Harrisson, B., 1962, *Orang-utan* . London: Collins.

Hendel, C.W., 1937, *Citizen of Geneva: Selection from the Letters of Jean-Jacques Rousseau*, New York and London: Oxford University Press.

Hobbes, T., 1651, *Leviathan*, (republished in many different editions post 1945), here: 1969, New York, Washington Square Press.

Hornaday, W.T., 1879, On the Species of the Bornean Orangs, with Notes on their Habits. *Proc. Am. Assoc. Adv. Sci.*, Salem, 20th meeting, pp.438-455.

Kaplan, G., 1994, Irreducible "human nature": Nazi views on Jews and women, in *Challenging Racism and Sexism: Alternatives to Genetic Determinism, Genes and Gender* , series, no.VII, eds, E. Tobach and E. Rosoff, New York: Feminist Press at the City University of New York: 188-210.

Köhler, W., 1921, *Intelligenzprüfung an Menschenaffen*, Berlin, Bern, Stuttgart: Springer Verlag.

Lethmate, J., 1977, *Problemlöseverhalten von Orang-Utans (Pongo Pygmaeus)*, supplement vol. 19 to *Fortschritte der Verhaltensforschung (Advances in Ethology)*, 69 p.

Monboddo, J.B., Lord, 1774, *Of the Origin and Progress of Language*, vol.1, repr. New York: AMS Press: 1973.

Reynolds, V., 1967, *The Apes. The Gorilla, Chimpanzee, Orangutan and Gibbon-Their History and their World*, New York: E.P. Dutton and Co.

Rumbaugh, D. M. and Gill, T. V., 1971, The learning skills of pongo, *Proc. 3rd Inter. Cong. Primatol.*, Zurich 1970, vol.3: 158-163.

Rumbaugh, D. M. and McCormack, C., 1967, The learning skills of primates: a comparative study of apes and monkeys, *Kongr. Inter. Primatol. Gesellschaft*: 269-306.

Russel, B., 1945, *A History of Western Philosophy*, New York: Simon and Schuster.

Schwartz, J. H., 1987, *The Red Ape. Orang-utans and Human Origin*, Boston: Houghton Mifflin Company.

Tarling, N., ed., 1989, *Mrs. Pryer in Sabah: diaries and papers from the late 19th century*, Introd. Nicholas Tarling, Centre for Asian Studies, University of Auckland.

Tawney, R. H., 1969, *Religion and the Rise of Capitalism*, Harmondsworth: Penguin.

Wallace, A.R., 1890, *The Malay Archipelago*. 10th ed., London: MacMillan.

Wells, H.G., 1971, *The Time Machine. An Invention*, (first published 1894), Cambridge, MA.: Robert Bentley Inc.

Williams, L., 1991, Born to be mild, *The Sydney Morning Herald Magazine*, 2 March: 34-35.

Yerkes, R. M., and Yerkes, A. W., 1929, *The Great Apes. A Study of Anthropoid Life*, New Haven: Yale University Press and London, Humphrey Milford: Oxford Univ. Press.

2

THE NEGLECTED APE?

NATO and the Imminent Extinction of Our Close Relative

H. D. Rijksen

Dienst Landbouwkundig Onderzoek
Institut voor Bos-en Natuuronderzoek (IBN-DLO)
Postbus 23, 6700 Wageningen, The Netherlands

ABSTRACT

The orangutan (*Simia satyrus* L. 1766) has certainly not been neglected by science. Moreover, it is a major attraction in zoos, and indeed, none of the other apes with the exception of our own species has so much appeal for the general public. For more than two decades the orangutan has been internationally acknowledged as an endangered species.

Since the late 1980s, the orangutan has come under a mounting threat of extinction; thousands of wild Bornean orangutans were displaced and killed as a consequence of expansive developments in Kalimantan. The Indonesian government is taking measures to halt further devastation of these apes and has invited international support. It is in assistance for the survival of the orangutan that neglect is prominent. Up until the present time, the international conservation community has largely ignored this call for support. The arguments to justify such neglect reveal a deplorable development in the large international conservation corporations. These arguments demonstrate a change in ideology and policy which is apparently unsuited for responding to emergencies, and which leads to exhaustive discussions which drain resources and prevent action. The scientific community is invited to persuade donors to act in the field, rather than to seek exhaustive bureaucratic proliferation.

INTRODUCTION

The "man of the forest", the translation of its Malay name *orang utan*, is perhaps the strangest and most appealing reflection of our evolutionary past (Rijksen, 1978; Schwartz, 1987). Since I have been so privileged to see the "red ape" alive in its forest, I feel that the great spirit of our kinship has captivated my mind. I attended the 1994 International Conference on Orangutans as an advocate, captivated by that spirit, to relate something of the recent sad history of this ape.

The Neglected Ape, Edited by Ronald D. Nadler et al.
Plenum Press, New York, 1995

IN WHAT SENSE IS THE ORANGUTAN NEGLECTED?

One may wonder why the title of this Conference was "The Neglected Ape". Can one say that science has neglected this ape? Do we know less of the natural history of the orangutan than of other apes or other mammals? I doubt that we do. Or, is it perhaps that people neglect what is known about this ape? I gather that the organizers of the Conference had especially the latter interpretation in mind.

I went to the Conference to tell those in attendance something about the problems of conservation with regard to the orangutan; something about error, negligence and euphemistic semantics. Error and deceitful semantics are, of course, of little importance as long as it concerns the virtual reality of science alone, but when negligence and deceit apply to the imminent extinction of our arboreal relative in the real world, it is reason for serious concern.

I doubt whether science is less informed about the red ape than it is of the other apes. As a matter of fact, I am convinced that the orangutan was the first of the apes known to science. The first accurate descriptions of the orangutan or *Indian Satyr* issued from specimens which were kept in the menageries of the Princes of Orange - the leading land-holders and royalty of the Netherlands since the early seventeenth century. However, for one reason or another the earliest Dutch references were misinterpreted and subsequently neglected.

At the turn of the century the ape was sadly dragged into the limelight of what people today consider to be "real science". Adventurers, now hailed as great scientists, came to the Indomalay archipelago to collect orangutans, shooting every specimen on sight (cf. Beccari, 1856; Wallace, 1865; Hornaday, 1885). When the public finally became interested in this "monster", zoos quickly developed a voracious appetite for having whole groups of this rather unsociable ape in their collections. During the first decades of the twentieth century, such vast numbers of orangutans were caught in Sumatra and Borneo to be transported to Western zoos (Dammerman, 1927; Brandes, 1939) that even the colonial authorities of the then Netherlands East Indies expressed concern over the survival of the ape in the wild. In 1925, the orangutan became the first formally protected mammal in Southeast Asia.

In spite of its formal protection, the orangutan has been subject to waves of persecution interspersed with only sufficient neglect to allow it to survive in the wild (Rijksen, 1982). When it was not persecuted, the reason was usually that people had other concerns. In the early 1970s, however, the Indonesian government for the first time invited international support to halt to such a wave. The strategy was to make law enforcement possible through the major instrument of rehabilitation (Rijksen and Rijksen-Graatsma, 1975). It worked for a while, at least in North Sumatra.

However, after a few years, the international community lost interest because it believed that the problem was solved. It left behind stations for a kind of rehabilitation which subsequently became a goal in itself. The stations, forced by a misplaced desire to become somehow self-sustaining, soon turned rehabilitation into a tourist attraction, contrary to the intent of the protection law. The idea that in such a way the public could get first hand information on the plight of the ape was little more than a convenient euphemism for the cash-generating circuses into which the stations readily developed (Fig. 1).

Now, twenty years later, we find ourselves in another, perhaps the last, tidal wave of orangutan extermination. The current state of affairs in conservation should make us wonder what went wrong. Reality shows me so far that the chances of survival for the red ape appear to be inversely related to the growing attention by science. It is awful to see that since the orangutan and its rainforest habitat became a hot topic in science, ever expanding tracts of its habitat were severely degraded and "converted" without the slightest concern for the ape. Indeed, in a world abuzz with ecological awareness campaigns, the red ape presently even

Figure 1. The orangutan circus at Bukit Lawang, North Sumatra, 1993.

came to suffer outright persecution from a booming human population which encroaches ever deeper into orangutan habitat in search of land and natural resources for exploitation.

Many a scientist will of course argue that such a correlation is nonsense. At the same time, no one would argue that the rainforest and its orangutan inhabitants would have suffered less if science had neglected them. A majority of people will undoubtedly argue that more information, rather than neglect, is needed before one should expect any positive effect on conservation. I shall not refute that argument, but I would like to emphasize that for the red ape, facing displacement and extinction, science has so far rallied very little effective support. The exchange of information has had little effect in the real world, and more discussion will not save the red ape from imminent extinction. Indeed, more than enough basic knowledge is available to warrant immediate action.

In 1988, it became evident that the 1986 policy of the Ministry of Forestry in Kalimantan to facilitate conversion of forest land for plantations caused the destruction of vast tracts of orangutan habitat. A few years earlier, the call for development in Indonesia required an administrative procedure to allocate different forms of land-use on the maps of

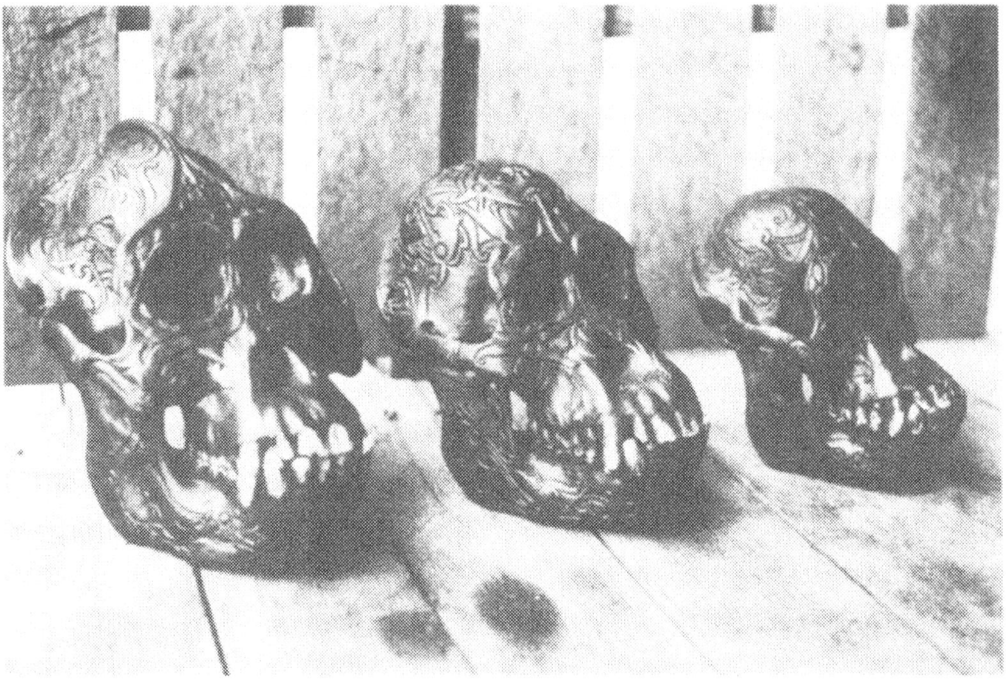

Figure 2. Carved orangutan skulls for sale at shops in Kalimantan, 1992.

state forest land for industrial production of tree crops. If the procedure took account of the existing boundaries of conservation areas, or even if it allowed for well-argued proposals to establish new conservation areas, it failed to consider the fact that many areas of forest which had been degraded by poorly controlled logging operations, still contained scattered groups of orangutans and other endangered species.

Soon the problem of displaced apes was highlighted because in at least one new plantation in East Kalimantan, the manager issued premiums for the eradication of orangutans in order to protect his young oil palms. At about the same time, the first illegal export of apes was detected in major East Asian ports. Tourists also reported that ape skulls, supposedly supplied by "traditional head hunters", were on sale in curio shops in most of the larger cities in Kalimantan, while young apes in significant numbers were offered for sale in Jakarta (Fig. 2).

The information took the Ministry of Forestry by surprise. At about the same time, the Survival Service Commission (SSC) of the International Union for the Conservation of Nature (IUCN) and the World Wide Fund for Nature (WWF) were informed. The Ministry's Directorate General for Forest Protection and Nature Conservation (PHPA) then sought discrete assistance in finding a solution to what soon became an embarrassing and wide ranging complex of problems. After due deliberation in Indonesian circles, it was decided to try first to contain the illegal trade through the well proven tactic of stepping up law enforcement, backed up by rehabilitation. However, in order to avoid the problems which had emerged during the 1970s campaign to enforce the protection law, a new type of rehabilitation, called reintroduction, was designed. I was asked to help seek international support.

Since the Netherlands branch of WWF had just run a massive media appeal which featured a "displaced orangutan seeking a new home", I expected that the organization would

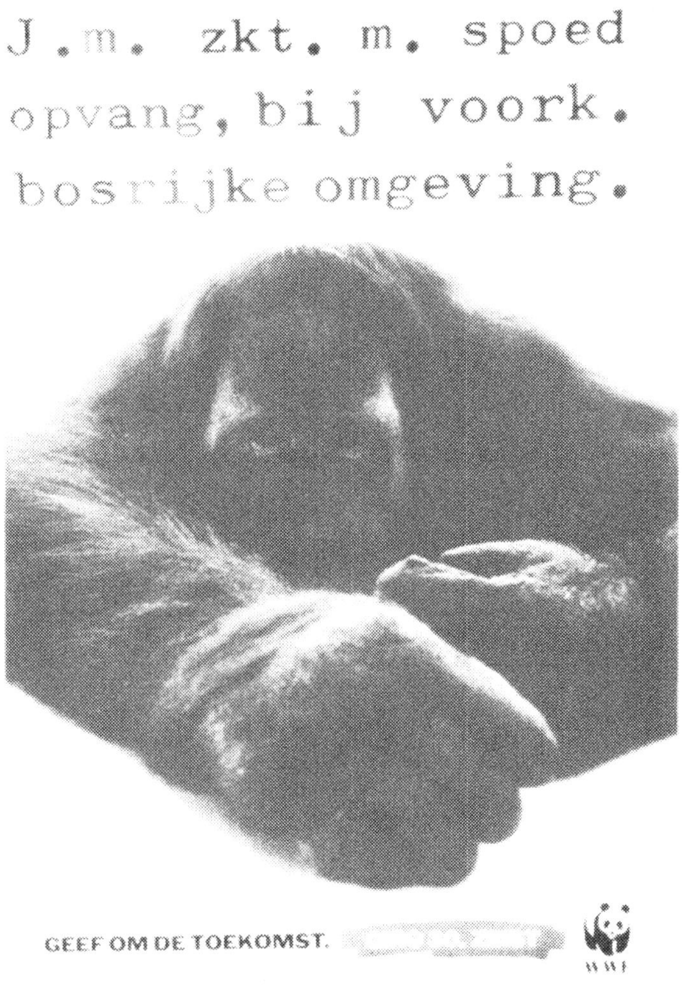

Figure 3. Advertisement to appeal for money to support WWF; the translation reads "Young mother (6 years), introverted character, is displaced due to house-demolition. Seeks immediate accommodation; preferably in forest-environment." (The donations were to support the organization, not the survival of the orangutan.)

not hesitate to support the rescue of the orangutan in Kalimantan (Fig. 3). That expectation was an error; the experts of the WWF Indonesia Program advised against involvement.

Thus, the organization which had been established especially to deal with emergencies of endangered species refused to participate in the problem of the imminent extinction of the red ape. Was this due, perhaps, to the embarrassment of admitting that the orangutan was suddenly being pushed to the brink of extinction, while their rescue program had been present in the country since the early 1970s? Or, was the alarm of one or two informed field workers not enough to encourage them to seek immediate verification? If that were the case, however, why did they launch the fund-raising campaign featuring the displaced orangutan? Certainly, it was argued that the organization did not want to be associated again with rehabilitation, which, in its current form, had been criticized by others and by me as dangerously outdated (Rijksen, 1982). The proposed rescue program, however, offered a new, more appropriate plan for rehabilitation, while the call for help was essentially unspecified.

I have tried to comprehend why there was such unwillingness to support a rescue operation. From the arguments which were used to turn down repeated requests for help, I gather that the refusals were due to a complex of at least two serious issues. First, poor corporate ideology, and second, the lack of consensus among those informants who should have known, but who, in fact, failed to be informed of the situation of the red ape. It is in the latter issue that scientific expertise plays a major role. Unfortunately, scientific opinion appears to be heavily influenced by territoriality and competition among the "experts", especially in the case of the red ape.

The main argument for negligence of the problem centered on the organization's change of policy from what was called "old fashioned species protection" to what they believed to be the "modern ecosystem conservation approach". Ironically, the fact that the orangutan had been neglected in the planning for development which caused such massive habitat destruction, is to some considerable extent due to the same cold analytical approach and administrative spirit that engendered the so-called "ecosystem approach". Since the early 1980s, it was widely held that "wise use of natural resources" is the essence of nature conservation (IUCN/WCS, 1982). Hence, protection was almost abolished as being supposedly ineffective and unpopular, while our endangered relatives were readily sacrificed to human development as long as the total population was not reduced to what some analytical biologists consider to be "critical" numbers. It is a frightening prospect if the biological accountants of the modern conservation corporations cannot be spurred into action before the population has dwindled to less than 5000 - or even 500 - specimens - while the mass of our own species rises beyond pest proportions.

From the devastation I have seen in the field under the aegis of the new policy, it is a dangerous illusion to believe that an ever-growing swell of people, with ever-expanding desires for material wealth, can live in harmony with nature or for that matter, use natural resources wisely. A policy of sustainable land use must certainly be applied to all areas designated for production, but when it is applied to conservation areas, it is no less than an euphemism for legalizing encroachment (Rijksen and Persoon, in press). Any wisdom pertaining to utilization is contained in law enforcement for protection and for constraints on exploitation of natural resources.

One may say that it is a miracle that the red ape has survived as long as it has. It is owing to the coincidence that only after the 1970s were much of the rainforests of Borneo and Sumatra subjected to the greed for land and natural resources, while the tribal societies had mainly persecuted each other until quite recently. The red ape, which is adapted to be secretive, could thus survive in some refuges. It was only recently that the rapidly growing human population, supplemented with transmigrants and aided by timber-exploitation, encroached into these refuges and ran into conflict with the last surviving red apes. It is evident that if we are seriously committed to keeping the wild orangutan on the face of the earth, we must remember that wherever human groups entered the great forests of Borneo and Sumatra, the red ape was hunted to extinction within a century. Whether it became extinct under traditional rituals for a good meal or for commercial profit makes little difference. The human species and its nearest relative cannot live together in the same forest habitat. Acknowledgement of that fact calls for effective, governmentally controlled segregation, rather than for euphemisms.

Conservation areas and endangered species must be protected, not exploited in any sense. It should not be difficult to understand that every organism of the incredible range of biological diversity on earth needs some space and its own conditions for survival. Such demands are invariably at odds with the interests and aspirations of an ever growing human population. It is certainly not difficult to understand why international socioeconomic corporations are content with opening up the last "reserves" for exploitation. Yet, one wonders why world conservation organizations gradually lose such simple ecological understanding and exchange it for ideological illusion.

One answer undoubtedly is that it is unwise for modern conservation corporations with voracious financial appetites to take sides openly against the mighty political interests of a growing mass of supposedly "poor" people and lucrative "development." As a consequence, many such organizations appear to be mainly concerned with seeking ways to appease a contingency in the exploitation of nature, if only it can be covered with the euphemism of "sustainability." Hence, much of the funds donated for conservation are applied to boost rural development in the unrealistic hope that people will show self-restraint, rather than applied to facilitate law enforcement and unpopular protection.

Finally, after embarrassing discussions over the last three years, the WWF recently offered some financial support for one of the first logical steps of a survival program, namely, a survey of the current distribution of orangutans. Although one must be grateful for any such support to implement a major strategic aspect of a survival program, it does not remove the uncomfortable feeling that something is fundamentally wrong in the operation of such a "conservation" corporation. Rather than eagerly seeking cooperation in a concerted effort to save the red ape by whatever means available, a well-calculated sum of money was provided to what apparently was considered to be someone's private project. Such an approach makes one anxious that a finding of more than 5000 orangutans may become the justification for further negligence.

The essential neglect of orangutan survival makes one question the value of at least three international symposia in which the plight of the red ape was raised. Why did the scientific gatherings so far fail to initiate the immediate, unconditional involvement of the international conservation community in this emergency? Why is it that interested people have such tremendous difficulty in cooperating in a survival operation that requires an international, cooperative approach? What makes conservation organizations and other potential financial donors demand such excruciating sophistication and detailed project proposals instead of seeking simple practical solutions to emergencies?

This brings me to the subtitle of my paper. What does NATO have to do with orangutans? Let me explain. Indonesian culture has developed great skill in creating appropriate acronyms. More than 45 years of independence have produced a word that expresses the frustration of endless meetings and discussions. The acronym NATO certainly bears a wicked secondary reference to the allied defense force of the Western world, but in Indonesia, it is understood first and foremost to stand for the saying, "No Action, Talk Only." And it is in NATO that a fatal negligence of the red ape comes to light most prominently. What has made our culture so muddled that it drains resources in endless discussions among a proliferating force of bureaucrats and takes no action whatsoever?

Since 1989, when the first alarm was raised, every year has seen another major international workshop or conference in which orangutans are prominently featured. The same problem was raised every time. The first meetings resulted in heated discussions, especially regarding the proposed rescue strategies. The Indonesian policy makers nonetheless readily accepted the newly designed reintroduction approach to boost protective law enforcement. Yet my pleading for support of this essential Indonesian policy from the international conservation community was sheer frustration. It was like standing before a burning house and being forced to discuss with bystanders whether or not the fire brigade should be called to extinguish the fire for fear of causing water damage and the loss of precious water. Be that as it may, the major effect of the discussions was that the bystanders decided that since there seemed to be a problem, they would first call for further discussion on the ways and means of its verification. No significant help or support has been forthcoming as yet.

Why do the international conservation corporations neglect serious information or use it exclusively for self-perpetuated discussions? Is it not peculiar that what the Indonesian authorities acknowledged in spite of obvious embarrassment, namely, that the red ape is in imminent danger of extinction, must first be verified in a sequence of meetings composed essentially of uninformed "specialists"? Are the masses of illegal captive apes, the sale of skulls and the haphazard information from reconnaissance trips into or over the area of former habitat not enough to warrant great alarm? Or is the fact that the protection of our closest relative against the swell of human development cannot be reconciled with "wise use" perhaps too embarrassing?

What happens with the information presented to the international conservation organizations? Why was the Newsletter of the Species Survival Commission silent on the plight of the Bornean orangutan over the last four years, in spite of the alarming information available since 1990? If I had not seen hundreds of illegally captured orangutan youngsters in the field and dozens of freshly prepared skulls in the curio shops of Kalimantan, but had been sitting at my desk reading the piles of glossy junk-mail on conservation, I would perhaps also be inclined to believe that there was no problem at all.

In any event, that frustration notwithstanding, quite a bit has been accomplished recently for the survival of the red ape. First, the Ministry of Forestry, under the guidance of a new and energetic minister, is actively seeking ways and means to call a halt to the ongoing devastation of orangutans and their habitat. My colleague Willie Smits and I are supporting the Indonesian Directorate General PHPA with organizing the very first steps in what we hope will expand into an orangutan survival program (see Smits, et al., this volume). Fortunately, our employer, the Institute for Forestry and Nature Research, allows us to do this as part of our regular activities.

At present, the Directorate General PHPA is attempting to increase law enforcement and stop the current surge of illegal capture, killing, and trading of orangutans, by means of a reintroduction project. Shortly, a representative of the Ministry of Forestry, Ir Widodo S. Ramono, will outline a policy framework of an Orangutan Survival Program to be implemented as soon as possible. An important aspect of this program is to seek control of further habitat destruction.

Elsewhere in this volume, Smits and his colleagues show that the reintroduction procedure is a great success, even on a shoestring budget of less than $95,000. It is unfortunately inappropriate to call it a complete success. In the short time it has been in operation, well over 120 orphaned apes have been gathered, most of them from East Kalimantan, and there is as still little indication that the tidal wave of illegal capture and habitat destruction is declining. The situation in Central and West Kalimantan is equally dismal; hundreds of displaced apes desperately roam Kalimantan in search of an acceptable place to survive.

Be that as it may, the creation of the reintroduction project by the Balikpapan Orangutan Society in East Kalimantan deserves special mention. Willie Smits and the headmaster of the Balikpapan International School, Joe Cuthbertson, needed little persuasion to become active. Step by step, a model quarantine station was established from a succession of donations. In view of the negligence of the international conservation community, it is sobering to note that the support raised thus far largely came from the industries that were at the root of the problem, the timber, coal and oil industries. At the present time, the Ministry of Forestry has provided an additional budget to support the quarantine and reintroduction project. However, the financial means are still too limited to run the full program and tackle the massive problems in the vast area of Kalimantan.

In view of the need for international support, I feel obliged to stress that what is being done for the survival of the orangutan should not be seen as one person's project. If any credit must be given, it should ultimately be given to the Indonesian PHPA, for it is essentially their task to conserve Indonesia's natural heritage. If the red ape were allowed to slip into oblivion it would be an appalling disgrace to our human existence, while helping the ape to survive, is a common moral obligation which should not be attributed to any individual or organization. I must also emphasize that it will not help to go on pointing fingers and laying blame on some government, agency or even people for what has happened. Further discussion will not help to save the red ape; what really counts is what happens in the future, with or without the assistance of the international community.

CONCLUSIONS

After more than 20 years of following the fate of the orangutan, I have come to the conclusion that what the red ape most seriously needs is to be neglected by the majority of mankind. After all, most of the contacts with humans during its evolution have meant untimely death. One must remember that until recently, the threat to survival of the red ape even included scientists who reaped renown for their destructive collecting zeal. As a solicitor on behalf of the red ape, I also feel obliged to state that the growing force of interested NATO bystanders may well be included among the threats to its survival, whether they call themselves scientists, conservationists, tourists or whatever other kind of consumers of its misery. I sincerely hope, therefore, that the International Conference on Orangutans is the last meeting to discuss the plight of the red ape and that instead, it will muster massive active support for the orangutan's survival. Support of all kinds is necessary, especially to spur the mighty international conservation corporations, such as the WWF, to start spending their funds in the field for the survival of the orangutan. With concerted international support for the Indonesian PHPA, the red ape could soon be neglected again in order to insure its survival in the wild, free from the persecution of its fatally assertive terrestrial relative.

REFERENCES

Beccari, O., 1904, *Wanderings in the Great Forests of Borneo*, London.

Brandes, G., 1938, *Buschi, vom Orang-Saugling zum Backenwulster*, Leipzig.

Dammerman, K.W.,1937, De Orang-Oetan. *Album voor Natuurmonumenten in Ned.* Indie, Batavia.

Hornaday, W.T., 1885, *Two Years in the Jungle*, London.

IUCN, 1982, World Conservation Strategy, Gland.

Rijksen, H.D. and Rijksen-Graatsma, A.G., 1975, Orang utan rescue work in North Sumatra, *Oryx 13*.

Rijksen, H.D., 1978, *A Field Study On Sumatran Orang Utans* (*Pongo pygmaeus abelii* Lesson 1827), *Ecology, Behaviour and Conservation*, 78-2. Wageningen, H. Veenman and Zonen B.V.

Rijksen, H.D., 1982, How to Save the mysterious "man of the rainforest" in *The Orang utan, Its Biology and Conservation* (Ed. L.E.M. de Boer), The Hague.

Rijksen, H.D. and Persoon, G., in press, Conservation beyond Natural religion, *Proc. Conf. Role Indig. People Conserv.*

Schwartz, J.H., 1987, *The Red Ape, Orang utans and Human Origins*, Boston.

Wallace, A.R., 1869, *The Malay Archipelago*, London.

3

THE IMPACT OF SOCIOECONOMIC DECISIONS ON THE STATUS OF THE ORANGUTAN AND OTHER EAST ASIAN FAUNA

A. A. Eudey

Asian Section
IUCN/SSC Primate Specialist Group
164 Dayton Street, Upland, California 91786-3120

ABSTRACT

The recent decline of the orangutan (*Pongo pygmaeus*) in Borneo, to an estimated range of 12,300 to 20,571 in 1993, may be attributed in part to the 1986 decision of the Indonesian government to open up Kalimantan to economic development. Habitat loss and degradation increased the vulnerability of orangutans to poaching and capture and made possible a surfeit of animals to fuel the pet trade in Taiwan, one of the major economic investors in Indonesia. The emergence of east Asia as the world's fastest growing economic sector has led to increased trade and communication among the countries comprising the region and new opportunities for illegal traffic in wildlife, including orangutans. Extraordinary measures will have to be initiated on both national and international levels to combat this threat to the region's fauna.

INTRODUCTION

In the final stages of preparing the Action Plan for Asian Primate Conservation: 1987-91 (Eudey, 1987), one problem that remained was assigning a priority rating for conservation action to the orangutan. Controversy existed over the degree of threat to the species. Differences stemmed primarily from the perceived status of the orangutan in Sarawak and Sabah (Malaysia) as contrasted with Kalimantan (Indonesia) and generalizations from these. Eventually, the orangutan was rated as "highly vulnerable," 4 on a scale of 1-6, "Surviving populations small or fragmented, and threatened by habitat destruction and/or hunting. Likely to move to category 5 (Endangered) by the year 2000 if no new conservation action is taken." No one appeared to comprehend that a political decision already had been made that would fragment the distribution of the orangutan and reduce its

The Neglected Ape, Edited by Ronald D. Nadler et al.
Plenum Press, New York, 1995

numbers to perhaps less than 25,000, thereby catapulting the species into the "endangered" category.

The threat of commercial logging and agricultural activity within the range of the orangutan in northern Sumatra was documented by Rijksen (1978) in the 1970's. In 1987, the Asian Action Plan recognized that increasing forest clearance and human settlement in the Alas Valley, which bisects Gunung Leuser National Park, threatened the integrity of the protected area. Although MacKinnon (1987) estimated the total orangutan population, based on suitable remaining habitat, to be 179,000 and that in protected areas to be at least 20,000, she noted that only 2.1% of what she considered to be original orangutan habitat was protected.

The 1986 decision by the Indonesian government to open up Kalimantan to timber concessions and other economic ventures increased the vulnerability of the orangutan to capture. Within Kalimantan, infant orangutans were reported to be purchased as "prestigious pets" by wealthy Indonesians, while the skulls of adult orangutans were sold to foreign tourists (Wildlife Conservation, 1990), even though the orangutan was given legal protection in Indonesia as early as 1925. Within a brief span of time, the number of Bornean orangutans may have declined to no more than 12,300 to 20,571, as determined at the 1993 Orangutan Population and Habitat Viability Analysis (PHVA) Workshop in Medan, Indonesia (Tilson, et al., 1993; Anon,. 1993a; also this volume). Prior to the workshop, the number of Bornean orangutans was estimated to be between 19,000 and 30,000 (Tilson, et al., 1993).

EXPLOITATION OF ORANGUTANS

During approximately two to three years bracketed by the period 1988-89 to 1990, as many as 1,000 to 3,000 young orangutans may have been smuggled by boat from Indonesia, especially Kalimantan, to Taiwan, where they were sold as pets and for publicity purposes to affluent buyers for the equivalent of between US$6,000 and US$15,000 (K. Nowell, personal communication, 1990). A popular television program featuring an orangutan was a major impetus to the trade. Publicity attendant upon the return of orangutans known as the "Taiwan Ten" to Indonesia in 1990 and a general anti-smuggling campaign appeared to result in a dramatic decline in trade in this species and other primates in Taiwan throughout 1991 (Phipps, 1992). According to the January 10, 1992 issue of *China Post* (Anon., 1992a), 283 orangutans were registered with Taiwan authorities, while at the same time the number of orangutans in Taiwan unofficially was estimated by the Orangutan Foundation Taiwan to be between 400 and 800 (Phipps, 1992).

The possibility of resumption of large-scale smuggling of orangutans into Taiwan probably is minimal (Phipps, 1992). A one-year-old male orangutan was confiscated in Hong Kong harbor, for example, on a ship of Thai registry that originated in Kalimantan and was en route to Taiwan (Anon., 1993c). The possibility of orangutans being smuggled out of Taiwan to other countries, however, is very real (Phipps, 1992),. In September 1993, two orangutans, along with four gibbons, were confiscated in a raid on a drug dealer in Hualin, Taiwan. Available information did not permit me to determine if these orangutans had been smuggled into Taiwan recently or if they were being held in anticipation of smuggling them out of the country.

It would be misleading to assume that all recent illegal trade in orangutans has been confined to Asia. In February 1990, six infant orangutans, which had been shipped as personal effects in bird cages and consigned to the Belgrade Zoo, were seized at Bangkok International Airport in transit between Singapore and Yugoslavia en route to the Soviet Union. In April 1993, Matthew Block, owner of Florida-based Worldwide Primates Inc., was sentenced on a felony count of conspiring to violate the U.S. Endangered Species Act and the Convention on International Trade in Endangered Species of Wild Fauna and Flora (CITES), having pleaded guilty to arranging the shipment (see Cater and McGreal, 1991; Anon., 1993e, f).

Taiwan has been the target of international efforts to get the country to enforce its wildlife laws, however. A petition filed with the U.S. Secretary of the Interior under the Pelly Amendment by the Washington, D.C.-based Environmental Investigation Agency (EIA) in November 1993 has led to formal certification that both Taiwan and China are in violation of U.S. law that could lead to broad trade sanctions against them for "engaging in trade and taking which diminishes the effectiveness of (an) international program for endangered or threatened species." The CITES standing committee also has recommended halting trade in animal and plant products with the two countries. On 17 January 1994, the San Francisco, California-based Earth Island Institute ran a full-page advertisement in the *Los Angeles Times* (and other papers) under the heading, "A bowl of tiger penis soup sells for $320 in Taiwan. Only while supplies last." The advertisement calls on U.S. President Bill Clinton to impose trade sanctions against Taiwan for failure to protect endangered species and for individuals to boycott goods labeled "Made in Taiwan" until Taiwan complies with "global conservation agreements." On 26 February 1994, the Republic of China on Taiwan (ROC) countered with a full-page advertisement in the *Los Angeles Times*, detailing the progress that has been made by the ROC against "abuses of animals on the endangered species list." In addition to enumerating prosecutions and convictions, the advertisement states that the ROC has won the support of the Chinese medicine community on Taiwan, "which has agreed not to serve elixirs made with by-products of endangered species".

In August 1993, at least 90 pygmy slow loris (*Nycticebus pygmaeus*), the survivors of 102 smuggled into Taiwan, were shipped without notification to Ho Chih Minh City, Vietnam, where all but four died (Anon., 1993b; Phan viet Lam, pers. commun., 1993). Although China Airlines, rather than the appropriate ROC governmental agency, the Council of Agriculture, appears to have been responsible for this decision, it is apparent that the ROC continues to experience problems in curbing wildlife smuggling. Private Taiwanese companies are reported to have invested so much money in Vietnam that Taiwan has surpassed industrial powers, such as France and Japan, to become Vietnam's largest single source of capital investment (Schoenberger, 1994). Phipps (1992) suggested that Vietnam might replace Indonesia as the main source of primates smuggled into Taiwan.

FUTURE PROBLEMS

The economic strength of southeast Asia may be measured by annual growth figures of 8% or 9% (Wallace, 1994). All indicators demonstrate the tremendous economic growth in east and southeast Asia. Although rank speculation contributed to the growth, the southeast Asian stock markets, for example, were among the "hottest" in the world in 1993. The Philippine market index rose 154%, Hong Kong 116%, Malaysia 98%, Thailand 88% (27% in the last two months of the year), and Singapore 59% (Wallace, 1994). Such economic growth constitutes an omnipresent threat to the environment. It also has led to increased trade and communication among the countries comprising east Asia and new opportunities for illegal traffic in wildlife, including the orangutan.

The region comprising southwest China (Yunnan province), northern Thailand, Laos and eastern Burma (Myanmar), the so-called "Golden Rectangle", for example, is considered to be one of Asia's last frontiers (Anon., 1993g). Already official plans are underway to promote trade and tourism in the region by developing a cross-border transportation infrastructure. The first major bridge across the Mekong River, which winds through the heart of the mountainous region, opened in April 1994, will link the northeastern Thai town of Nong Khai with Vientiane, the Lao capitol. Serious thought also is being given to the construction of a second, more southerly bridge across the Mekong River. With appropriate highway connections, this bridge would link Thailand and landlocked Laos with a Pacific Ocean port in Vietnam, conceivably in Da Nang, thereby bringing Vietnam into this informal economic partnership. Within the "Golden Rectangle" itself, illegal trade in primates and other wildlife appears to be well established and the potential for it to expand into adjacent regions is great. By November 1993, the Shan State (northern Burma) was opened to foreign tourists at the border crossing between Mai Sai, Chiang Rai province, Thailand and Tachilek, where an open wildlife market exists (Royal Forest Department, personal communication, 1993; personal observation, 1993).

Extraordinary measures will have to be initiated on both national and international levels to combat the threat to southeast Asia's fauna. The economic alliances among east Asian countries that may be placing more burdens on wildlife and their habitat paradoxically may offer a mechanism for curtailing illegal trade. In 1992, at the eighth meeting of the Conference of the Parties to CITES, Thailand was elected to represent the Asia Region on the CITES Standing Committee. For this reason, and, it might be assumed, because of the central position the country occupies both geographically and economically in continental southeast Asia, Thailand is taking a leading role in encouraging neighboring countries to become Parties to CITES. This is in spite of the fact that the CITES Secretariat had found it necessary to prohibit trade with Thailand in CITES appendices species from April 1991 to April 1992 because of numerous violations (Anon., 1991; Anon., 1992b). In February 1994, Vietnam acceded to CITES; the action became effective April 20, 1994.

In January 1994, at the General Assembly of IUCN - The World Conservation Union - in Buenos Aires, Argentina, an informal meeting occurred between delegates and observers from Vietnam and China. The recent affluence of China appears to have been accompanied by renewed exploitation of Vietnamese (and other) wildlife, including primates, especially for traditional medicine (W. Bleisch, pers. commun., 1993). For a brief period of time, even a Tonkin snub-nosed monkey (*Rhinopithecus avunculus*), perhaps the rarest and most endangered primate species in Asia, was on exhibit in the Kunming Zoo, in China's Yunnan province (Anon. commun., 1993). It is not unthinkable that orangutans from Taiwan also might make their way into China's myriad of zoos. The opportunity exists for this to occur since the People's Republic of China was the largest recipient of Taiwan overseas investment in 1993, where government-approved investment more than quadrupled to US$1.14 billion from US$246.9 million in 1992 (Anon., 1994). China acceded to CITES in 1981 but has been reluctant to recognize the wildlife laws of neighboring countries which are not Parties to the Convention, such as Vietnam, Laos and Burma. At the present time, Taiwan is politically constrained from becoming a Party to CITES in that the China seat is occupied by the People's Republic. Although the border between Vietnam and China does not lend itself to easy policing, China can exert some control over wildlife smuggling by regulating the factories that produce traditional medicines (J. MacKinnon, personal communication, 1994). As indicated above, Taiwan already appears to have taken a step in this direction.

The July 1993 meeting of the foreign ministers of the Association of Southeast Asian Nations (ASEAN), comprising Brunei Darussalam, Indonesia, Malaysia, the Philippines, Singapore and Thailand, approved the creation of an ASEAN Regional Forum, a multilateral institution to discuss political and security issues in the whole of Asia. Among other countries, the Forum would include China, Laos and Vietnam (see Anon., 1993d). The mandate of the Forum could well be expanded to address wildlife trafficking and environmental issues common to the region.

On the national level, serious consideration must be given to weighing the benefits and costs of nature conservation and protection as contrasted with sustainable development. In the case of the orangutan, enforcement of laws pertaining to hunting and trade and protection of conservation areas are critical for the survival of the species. Local people, local governments and national and international commercial interests must be required to take an active role in these actions, with the latter assuming a significant share of the financial burden. Let us hope that the 20 January 1994 decision of the Indonesian Minister of Forestry, Mr. Djamaludin Suryohadikusumo, to ban the export of wild-caught long-tailed macaques (*Macaca fascicularis*) and pigtail macaques (*M. nemestrina*), noting that these "wild animals which have been exploited by human beings for a long time without having been protected by law, are showing signs of decline in the wild" signifies a new commitment to primate (and other wildlife) conservation in Indonesia.

REFERENCES

Anonymous, 1990, Primate watch: In Indonesian Borneo, orangutans are for sale, *Wildlife Conservation* 93(6):26.

Anonymous, 1991, IUCN Resolution on orang utans, *Asian Primates* 1(1):1-3.

Anonymous, 1992a, Orangutan repatriation put on hold, *The China Post*, 10 January 1992, National News.

Anonymous, 1992b, CITES proposal withdrawn, *Asian Primates* 1 (4):1.

Anonymous, 1993a, Orang-utan population and habitat viability analysis workshop, *Asian Primates* 2(3&4):3-4.

Anonymous, 1993b, Smuggling of pygmy slow lorises, *Asian Primates* 3(1&2):5.

Anonymous, 1993c, Orang-utan update, *Asian Primates* 3(1&2):10.

Anonymous, 1993d, Developments of interest, *Asian Primates* 3(1&2):13-14.

Anonymous, 1993e, "The Bangkok six" - 1990-present, *IPPL News* 20(3):21-22.

Anonymous, 1993f, Orangutan's destination-Soviet zoos, *IPPL News* 20(3):38.

Anonymous, 1993g, Taming Southeast Asia's "Golden Rectangle," *Los Angeles Times*, 22 November 1993, p. D3, from Associated Press.

Anonymous, 1994, Taiwan: Increase in overseas investment OK'd, Pacific Watch, *Los Angeles Times*, 17 January 1994, p. D3.

Cater, B., McGreal, S., 1991, The case of the battered babies, *BBC Wildlife* 9:254-260.

Eudey, A.A., compiler, 1987, Action Plan for Asian Primate Conservation: 1987-91. *Gland, IUCN - The World Conservation Union.*

MacKinnon, K., 1987, Conservation status of primates in Malesia, with special reference to Indonesia, *Primate Conservation* 8:175-183.

Phipps, M., 1992, Recent incidents involving illegal primate trade in Taiwan, *Asian Primates* 2(2):5-6.

Rijksen, H.D., 1978, *A Field Study on Sumatran Orang Utans* (*Pongo pygmaeus abelii* Lesson 1827), *Ecology, Behaviour and Conservation*, Wageningen, H. Veenman and Zonen B.V.

Schoenberger, K., 1994, Champing at the bit - U.S. firms are eager to get into Vietnam with lifting of trade embargo, *Los Angeles Times*, 4 February 1994, pp. D1, D12.

Tilson, R.L., Seal, U.S., Soemarna, K., Ramono, W., Sumardja, E., Poniran, S., van Schaik, C., Leighton, M., Rijksen, H., Eudey, A., eds., 1993, Orangutan population and habitat viability analysis report. Apple Valley, MN, *Minnesota Zoo Conservation Office and IUCN/SSC Captive Breeding Specialist Group.*

Wallace, C. P., 1994, Stocks sizzle in S.E. Asia, *Los Angeles Times*, 8 February 1994, pp. H1, H4.

4

ORANGUTANS, SCIENCE, AND COLLECTIVE REALITY

A. L. Rose

Social Change Systems
Post Office Box 488, Hermosa Beach, California 90254

ABSTRACT

The future of orangutan conservation and research is tied to public perceptions, and to the government and corporate powers that react to and shape the collective realities of their constituents. It is time for primatologists to organize to influence those collective realities. Crucial to this effort is the reconciliation of contrasting metaphors based on the objective, interpretive and interactive perceptions of the orangutan which are reported by experimental scientists, field naturalists and clinical practitioners. This treatise begins such a reconciliation by comparing reports of profound interspecies events (PIEs) — social phenomena identified and studied by the author, that relate directly to primatologists' different value-laden perceptions of nature. The relative effectiveness of scientific, naturalistic and humanistic scenarios in evoking human concern for wildlife must be considered when deciding which stories to tell and how to tell them. Pitfalls in reporting orangutan behavior in two perceptually "hot" areas, sexual behavior and male dominance, are discussed. A collaborative strategy is presented for producing and promoting collective realities that will promulgate the safety and well-being of orangutans in captivity, in sanctuary and in their native habitats.

INTRODUCTION

There are countless ways to perceive orangutan behavior. All of those ways are both biased and true at the same time. It's truly a matter of what the human observers see — inside their unique minds. Those minds are shaped to perceive behavior by a variety of genetic, developmental, psychosocial, and situational factors. If we are to study the orangutan, we must study ourselves at the same time. This requires levels of self-awareness and a willingness to disclose personal facts that cannot be expected of everyone. Many scientists entered primatology wanting to be detached observers, to avoid introspection, even to avoid people, politics, and conflict. All these are legitimate desires, honorable escapes, reasonable ways to stay "cool". Unfortunately, however, it's become very hot in primatology. Political

The Neglected Ape, Edited by Ronald D. Nadler et al.
Plenum Press, New York, 1995

and ethical contradictions are straining our commitment. We are forced to look at ourselves and ask "what can we do to resolve confusion and conflict; how can we collaborate to assure that primates and their habitats are understood, appreciated, and protected?"

If the orangutan is less appreciated than the other great apes, it is not only due to its semi-solitary arboreal lifestyle in the rain forest. It also derives from contradictions that have arisen out of its scientific study. Altogether, primatologists are conflicted about the orangutan — its behavior, evolution, well-being. Ultimately, the crucial target for orangutan awareness is the general public and the government and corporate powers that react to and shape the collective realities of their constituents. What consumers, media moguls and governors think of orangutans is more important than what we primatologists debate about them. It is the former, after all, who will decide the fate of every endangered species on this planet. If we want to conserve the orangutan and its rain forest habitat, we must contribute our combined ideas and talents to influencing those who mold and manifest public opinion.

The Disney Corporation's recent rendition of Rudyard Kipling's *Jungle Book* is an enjoyable movie asserting the wonder and wisdom of nature and the greed of humanity. In it, a trained orangutan performs the role of Monkey King. Although the ape is cast in a positive supporting role, he is used primarily for comic relief, mimicking human acts and emotions in otherwise tense scenes. The impact of his performance on public perceptions of the species is likely mixed. At best, the audience comes away from the experience aware of the orangutan's human-like intelligence. At worst, they come away thinking, "Those big orange monkeys are goofy; why fuss over them?" And while primatologists rightfully grumble over public misinformation, animal rights activists consider picketing Disney and demanding that the poor "oh-rang-a-tang" be sent back to Borneo.

Few of these antagonists know that Mowgli's Monkey King friend is a captive bred orangutan actor named Lowell, who would be extremely distressed and depressed if he were taken from his human family. Fewer still know that Lowell would stand little chance of survival in the jungle, even if there were a place to send him and funds for his care. Knowing that, some still find it hard to acknowledge that unique individuals like Lowell can do more for their species working in Hollywood than they can starving in Sumatra. So can apes in zoos, in wild animal parks, in homes and in laboratories do more — for conservation — if all primatologists organize and act to help them.

The crucial issue here is organization. How can primatologists from field, zoo and laboratory collaborate when we come with such divided experience and contrary values? In their volume on orangutans in Borneo, Kaplan and Rogers (1994, p. 16) warn that "we are bound by views and attitudes to nature which in turn determine the way in which we think of orangutans or any other species." It is time for the unbinding of our views and attitudes. As "pure scientists," we may argue for the Occam's Razor edge of mechanical simplicity. But there is no longer a pure science of primatology. We are all conservationists now, and as such, we must strive for the greater utility. Whatever luxury of pride and prejudice with which we came here, we must now rise above it and change, for the welfare of our primate kin.

Richard Rorty (1982, p. 60) made the case with elegance, for our human potential to change, in his review of the philosophy of mind: "For beyond the vocabularies useful for prediction and control — the vocabulary of natural science — there are the vocabularies of our moral and our political life and of the arts, of all those human activities which are ... aimed at ... giving us self images that are worthy of our species. Such images are not true to the nature of species or false to it, for what is really distinctive about us is that we can rise above questions of truth or falsity. We are the poetic species, the one which can change itself by changing its behavior — and especially its linguistic behavior, the words it uses."

Rorty challenges humankind to redefine itself. In so doing we redefine the "other". As Edward O. Wilson (1984, p. 74) states, "the symbols of art, music, and language freight power well beyond their outward and literal meanings." Primatologists must take responsibility for the power of their words and use them with pointed intention to foster care and reverence for the other species we study and use.

THE ORANGUTAN: A METAPHOR FOR HUMANITY

It is often asked whether or not the orangutan is a biological model for humankind. More than a model, the orangutan is a metaphor. How we shape and fulfill that metaphor reflects our desire and our terror, our sense of origin and destiny, not only for humankind, but for all life on earth. Elizabeth Atwood Lawrence warns us that "Interpretation of an animal's behavior in metaphorical terms can result in the creature being classified as "good" or "evil" — with consequent effects on the preservation or destruction of the species. The symbolizing process can enhance positive affiliation, resulting in preservation, or it can cause alienation of that animal from the human sphere with consequent destruction." (Lawrence, 1993, p. 332).

Evidence is mounting that the greater part of human conflict emanates from our alienation from the rest of nature — our failure to honor our "biophilia ... the innately emotional affiliation of human beings to other living organisms" (Wilson, 1993, p. 31). Having lost our place in the natural order, humans heap chaos on the planet. Psychologists devote lifetimes to enhancing positive affiliation among varied factions in the human community — generations and genders, races and religions, occupations and cultures (Rogers, 1961; Rogers and Rose, 1971; Rose and Auw, 1974; Rose, 1983). We must expand the scope of these endeavors to include the affiliation between humanity and other life-forms (Rose, 1994c). This effort is best launched in primatology because this is the area in which humankind struggles to face the meaning and impact of its natural origins (Leakey and Lewin, 1977; Cavalieri and Singer, 1993). But within primatology, as in any professional pursuit, there are factions and conflicts.

At the XVth Congress of the International Primatological Society (IPS) held in Bali in 1994, well over 100 primatologists participated in a symposium on ethics (Rose, 1994b). Panel discussion and audience interaction during a 4-hour time span was intense and revealing. The bottom line was the discovery that harmonious collaboration among factions was entirely possible — for those who attended and stayed. This doesn't mean minds were changed. It means divergent viewpoints were expressed and accommodated. That is the first and essential step towards consensus. This chapter considers a further approach toward consensus — a symbolizing process that expresses and accommodates varied perspectives in writing and in action.

As Henri Walton asserted, "The unique image is not the point of departure. Perception begins by multiplying the points of view for the needs of practical action ..." (Clay, 1978). The need for practical action regarding orangutan conservation is eminently clear. Failure to act has come from the insistence of individual observers that their unique images of their orangutans must be the singular points of departure. Defining the metaphor of the orangutan to serve one's self rarely serves the orangutan. The optimal reality with regard to any species is the collection of convergent perceptions of all those who define, describe and relate to that species. To create such a reality requires movement through metaphor to metonymy, through relationship of similarity to relationship of contiguity (Jakobson, 1962). We must place the metaphors of disparate primatologists into contiguity, to form a collective reality that will stir the public to protect the orangutan and its rain forest habitat.

INTERSPECIES EVENTS AND PERCEPTIONS OF NATURE

Persons who have encountered orangutans know how evocative they can be. Still, in the scientific literature, it's quite rare to read of the impact of the red ape on humans.

Primatologists describe the animal in its environment, not their reactions to it. Even in biography one must cull hundreds of pages for occasional anecdotes about interspecies events, typically written for purposes other than the direct illumination of human-ape affiliation (Rose, 1994a). Galdikas' (1995) recent autobiography is an exception. It is a well crafted work aimed at symbolizing and enhancing affiliation between human and orangutan. Lucid description of the author's interaction with rehabilitant and wild apes evokes empathy for and understanding of primate and primatologist. Interpretation of natural events are cast in a consistent positive light.

At the end of the book, Galdikas (1995, p. 403) declares, "Looking into the calm, unblinking eyes of an orangutan we see, as through a series of mirrors, not only the image of our own creation but also a reflection of our own souls and an Eden that once was ours. And on occasion, fleetingly, just for a nanosecond, but with an intensity that is shocking in its profoundness, we recognize that there is no separation between ourselves and nature. We are allowed to see the eyes of God."

These are words that metaphorically reflect a distinct cultural and personal viewpoint. Persons with Christian background will recognize that what Galdikas describes is an epiphany — a manifestation of divinity, in religious terms. In the parlance of modern poets and philosophers, epiphany is any event that sheds fresh light on a profound mystery of life. It is the mission of natural scientists to enter those domains where life's mysteries are ripe for revelation, experience the situation and report our findings so others may benefit from our discoveries.

The personal recognition of unity between man and nature to which Galdikas alludes has been the subject of this author's intensive research. It was during the 1994 International Orangutan Conference that spawned this volume (Rosen, et al., 1995) that I introduced the concept of profound interspecies events (PIEs) and began direct study of its manifestation among primatologists. Results of the first phase of research on PIEs were presented at the XVth Congress of the IPS in Bali (Rose, 1994a). At the Congress and thereafter, interviews and workshops with scores of persons have documented epiphanies in primatology to be used as a data base for scientific scrutiny and as anecdotes for promotion of public empathy and political support for primate research, care and conservation (Rose, 1995).

Interspecies epiphanies, like any other natural event, are experienced and reported in different ways, according to who is involved. This chapter focuses on professional groups that manifest three distinct value structures identified by Stephen Kellert (1993) in his research into public attitudes toward nature: Experimental Scientists (scientific values), Field Researchers (naturalistic values), and Clinical Practitioners (humanistic values). While there are other value structures at play in the domain of interspecies experience (moralistic, ecological, utilitarian, aesthetic, etc.), they often appear to overlap and define subsets within the three professional categories on which we focus.

Clearly the Galdikas metaphor cited above is not scientific. Neither is it humanistic. In fact, it emanates from the naturalist's "sense of fascination, wonder, and awe derived from an intimate experience of nature's diversity" (Kellert, 1993, p. 45). The *naturalistic value structure* is a useful channel for attracting humans to affiliate with primates. The popularity of nature documentaries on public television and the rise in eco-tourism demonstrate the attractive power of ventures into wilderness. While there is concern regarding an adverse impact of these activities, the activities themselves do add value to endangered ecosystems. Among primatologists, *Field Researchers* are most affected by this attitude complex. Naturalistic PIEs, in which animals Show Extraordinary Elements of Nature to human observers (the SEEN scenario), provide rich *interpretive data* on the orangutan. Naturalists are typically satisfied with visual contact with wild animals. Others seem to need hands-on experience.

Kaplan and Rogers (1994, p. 94) reflect that "one may get to know orangutans to a point by merely observing them but we begin to relate to an 'other' only when we interact." They go on to "admit" that they had an encounter with a young female orangutan which "retrospectively proved of greater value and importance than all the textbooks we had read. ... This started a long love affair which is certainly one valid explanation why she (Kaplan) has felt compelled to return to the same spot of the Bornean forest again and again." This anecdote exemplifies the *humanistic value structure* which Kellert (1993, p. 52) says "reflects deep emotional attachment to individual elements of the natural environment ... usually directed at ... the larger vertebrates." Persons helping to return ex-captive orangutans back into the rain forests of Borneo are among the most deeply committed conservationists this author has interviewed. Humanistic PIEs in which animals Seek A Friendly Encounter with humans (the SAFE scenario) are reported most often by these rehabilitation workers. *Clinical Practitioners* of all ilk — veterinarians, keepers, etc. — contribute valuable *interactive data* to the understanding of the orangutan. Ironically, there is great controversy surrounding the efforts of these dedicated individuals. A confluence of attitudes, beliefs and affective factors makes it difficult for some primatologists to admit to or accept strong humanistic values. Conversely, humanists can be less than tolerant when dealing with persons and perceptions that contradict their more emotional connection with individual members of other species.

Preeminent among the adversaries of humanism is the *Experimental Scientist*, whose *scientific value structure* reflects "the urge for precise study and systematic inquiry of the natural world, and the related belief that nature can be understood (best) through empirical study" (Kellert, 1993, p. 47). This kind of primatologist is engrossed in exploring the details of the organism. The goal of the scientist is to produce *objective data* that Exhibit Natural Reactions which Illuminate Crucial Hypotheses (the ENRICH scenario). Success in this pursuit is profoundly exhilarating, and hard-sought. Nadler (1995) recalls the "great sense of satisfaction" felt when he finally realized why captive gorilla males court and aggressively initiate copulation with non-estrous females — a behavior rarely seen in the wild. At Karisoke, Nadler (1989) observed a male gorilla perform an increasingly complex display to a recalcitrant estrous female. "When I saw an example of the complete male chest-beat display (chest-beat, charge, stance) preceding copulation, the 'light bulb' really went off and I realized what had been going on in the laboratory tests. The behavior of captive males was an exaggeration of the normal wild male response to estrous females when the latter came closer than three meters. The males in the laboratory tests were responding to the close proximity of the non-estrous female as if the female was in estrus. It was an artifact of the captive environment." That "light bulb" experience is as profound to the scientist as the most awesome natural epiphany to the naturalist, or the deepest interspecies connection to the humanist.

Economics pushes many experimental scientists into using animals to solve human problems (Rose, 1968). This overlays a *utilitarian value structure* onto the scientific, stressing "the physical benefits derived from nature as a fundamental basis for human sustenance, protection, and security" (Kellert, 1993, p. 45). A utilitarian scientist can seem a cold character, often practicing detachment from the animals with which s/he works. At the same time, most are deeply devoted to the care of individual subjects and to conservation of the species. Scientists of this ilk need no personal reciprocation from primates in return for their devotion. Still, when they are personally touched by their animal subjects, it can be exceptionally profound (See the story of a laboratory research technician and a dying chimpanzee [Rollin, 1993, p. 213]).

This author's first profound interspecies event occurred in 1963 at the Brain Research Institute of the University of California at Los Angeles. A frightened janitor tracked me down to tell me that a pig-tailed macaque had escaped from its cage and was ransacking my laboratory. This had never happened to me before. Subduing a frightened monkey in a cage was one thing; catching an escapee was another. As I entered the room and peered through the haze and clutter, a familiar smacking sound drew my eyes to the far wall. Snicky, a three-year-old male, stared down at me from atop a bookcase, hair on end, eyes wide, teeth bared. I smacked back at him, our usual morning greeting. He shuddered and at once jumped down from the case, leapt into my arms and held on. In the distance he had seemed so huge, imposing, wild. Now in my arms he was small, vulnerable, dependent. I sat on the linoleum floor with this animal in my lap — cleaned the scab that edged his dental cement skullcap, checked his implanted electrodes to be sure they hadn't loosened and examined his dilated eyes. I remember thinking "after all I've done to him, he wants my friendship more than his freedom."

Uncontrolled personal encounters with primates are rare for most experimental scientists. We strive to avoid them, primarily because they complicate protocols and confound conclusions. Such events also induce human-animal bonds that have effects on the researcher: "If you bond with your animal you won't be able to dissect/shock/euthanize the animal. Save yourself the conflict." (Davis and Balfour, 1993, p. 2) To study cognition and language learning in apes, however, appears to require social bonds (Oden and Thompson, 1993). Those bonds help researchers like Miles (1993, p. 50, 54) make the case for the ethical treatment of apes: "We had a close relationship with Chantek. He became extremely attached to his caregivers, and began to show empathy and jealousy toward us. ... We have ... shared common experiences, as if he were a child ... have dreamed about him, had conversations in our imaginations with him and loved him. ... I have seen many people gasp with amazement as they conversed with Chantek, subjectively experiencing him as a person. If it were possible for all humans to have this experience, this book might be unnecessary."

The vast majority of humans require a bond with the other, before they will commit to protect the other. Interpersonal bonds, from infancy through old age, are grounded in a fundamental need to expand social contact outward, from self to other to community, from humankind to kindred species and beyond (Rose and Auw, 1974; Rose, 1988). Ego-, ethno- and homo-centric attitudes are prejudices produced for the most part by culturally induced fear, hatred and ignorance. Overcoming those prejudices regarding our primate kin requires a concerted public relations effort to report interspecies events that are profound and positive enough to transform fear into hope, ignorance into understanding, hatred into love.

THE SYNTHESIS OF PROFOUND EVENTS: REPORTING SAFE SCENARIOS

In the synthesis of profound events for the public domain, the *humanistic value structure* prevails. What people respond to most favorably is stories that reflect the SAFE scenario. My experience with the escaped macaque is an example of such a humanistic PIE in a scientific setting. When asked about profound interspecies events, Jane Goodall (1994) referred me to Marc Cusano, the animal caretaker who became a member of chimpanzee society at Lion Country Safari in West Palm Beach, Florida. Cusano has volumes of SAFE and not so safe scenarios to report — a gold mine of interactive data.

The event that Goodall (1990, p. 233-4) described, in which an alpha male chimp saves Cusano's life, contains all seven elements of the "ultimate PIE" (Rose, 1994a). Those elements are:

1. Initial insurmountable difficulty for the human to gain access to the animal.
2. Perseverance — patience and faith — by the human in pursuit of a connection.
3. Reversal of mistrust by the animal with regard to the pursuing human.

4. An arresting first contact, followed by successively closer and longer interaction.
5. Intervening forces that separate the pair, leaving one/both highly endangered.
6. Heroic acts by one/both members of the pair to reach/protect/save the other.
7. Profound shifts in perception of self/other/species by the member(s) of the pair.

A description of any one of these elements can evoke empathy for other species in a person who is reasonably open minded. A vivid report of all elements together can literally change a person's view of life.

There are scores of North Americans who have traveled to Indonesia to serve as foster parents to orphaned orangutans. These persons almost always go through the first four steps of the ultimate PIE. They cry when they must leave, and spend their separation fearing their "adopted children" will die in the rain forest (step 5). Some, unable to face the worst, never return. Others, often after working and sacrificing for years to scrape together funds, return and search the forest for their now grown, or possibly vanished wards (step 6). To be recognized and greeted by an orangutan friend after years apart is cement for lifelong interspecies kinship (step 7). Typically it is persons with these kinds of experiences who become the backbone of grass roots wildlife conservation organizations.

Some scientists and naturalists argue that these persons are not dealing with the full orangutan experience; that they transmit a falsely human picture of the great ape. It may backfire to hook people on the sociable young apes. As orangutans grow, the cuddly dependence vanishes. Of course these arguments are correct; the potential problems are real. But the solution is not to eliminate all human involvement with the orangutan, or even to replace warm volunteers with cool professionals. Learning to relate to adult orangutans can produce even more profound and useful interspecies events, for those who have the skill, patience and courage to meet the apes on their terms (Montgomery, 1991, p. 20-21).

There is controversy as to what those orangutan terms are and which human terms will best describe them. Much of that controversy is driven by gender. The vast majority of humanistic PIEs are reported by human females. This corresponds with studies in which female researchers appeared more "fused with the monkey's lives" and better able to "feel one" with the animals they were observing (Kawai and Asquith, 1981). Humans are prone toward dichotomous perceptions of animal behavior; women bond emotionally, men relate intellectually. Blending these perceptions is crucial when we tell the tale of the orangutan to a public deciding which species will survive and which will vanish from existence.

PERCEPTUAL HOT SPOTS: SEXUAL BEHAVIOR AND MALE DOMINANCE

Whether writing about PIEs or any other social phenomenon, we must be careful to use terms that will attract support from humans towards the species we are describing. Although the PIE elements are foundation for evoking empathy, they can be tainted by harsh labels and negative explanations. Careless descriptions of any primate social behavior can have destructive effects on public perceptions. Goodall's success demonstrates how tales of chimpanzee behavior can be written so sympathetically as to attract public interest. Still, she treads a fine line when reporting cases of infanticide and incidents of inter-group war (Goodall, 1990). Behaviors that arouse hostility in factions of human society must be approached cautiously. Two of these, sexual behavior and male dominance, require special artistry when writing about the orangutan.

MacKinnon's (1971, 1974) and Rijksen's (1978) classification of certain orangutan sexual behavior as "rape" is a case in point. Rijksen (1978, p. 266) capitulated that in those behavior chains called rape "females usually waited for the male to approach." He hypothesized that "... contact would certainly not have been achieved, if the female really had avoided or even fled". Nadler (1981) adds the words "forcible copulation" and eventually omits use of "rape", to avoid anthropomorphizing (Nadler, 1988). Kaplan and Rogers (1994, p. 82) resurrect the term, a risky semantic tactic in the political climate of the 1990's.

As Kaplan and Rogers (1995) report in this volume, culture and era influence the meaning of things. Intolerance of human male aggressiveness has been rising steadily over recent decades, sometimes blowing up into general intolerance for men. While this may be useful for the adjustment of power imbalances in our species, there is no evidence that such an adjustment is needed among wild orangutans. To say that male orangutans practice rape makes them less endearing and more endangered. The behaviorist's stricture against connotative labeling is a good first rule of public relations; it precludes careless judgment of both situation and subject by the greater audience.

Fortunately, rape doesn't fit the scientific evidence any more than it does the political climate. Galdikas (1979, 81, 84) observed and reported what MacKinnon and Rijksen had hypothesized — that orangutan sex in the wild is mostly civil; forcible copulation is a young adult affair. Maple (1980, p. 65) questions use of the rape label for captive animals. Nadler's (1982, also this volume) restricted access tests with captive apes confirmed field observation that when females are given control of the opportunity for contact, as is usually the case in the rain forest, the males act more like gentlemen, or "Romeos". The later term is used advisedly, since seduction is taboo in some political quarters. We could do worse however. Nadler (1981) reports that captive male orangutans use a penile display to attract estrous females, a wild behavior Schürmann (1981) calls "male presenting." If one wanted to cast negative aspersions, one might use the term "flashing" — a reprehensible behavior in most human societies. It seems foolish to project the human concept of flashing on male orangutan presenting. Such behavior in humans is a perversion, not a seduction.

But then, so is rape a human perversion. To consider forcible copulation a perversion in the wild orangutan may also be foolish. Perhaps it is a seduction. Or an exercise of youthful male libido that the females tolerate, bothersome but harmless, even necessary for otherwise isolated young males and females to learn the social and physical etiquette — a kind of practice courting and consortship. The female stops struggling. She doesn't cry or scream in pain; she rarely holds a grudge. Eventually those young males grow up and become the proper responsive adults that are chosen by these same females to father their offspring. Remember, we are talking about orangutans here — not humans.

Mental machinations over the meaning of forcible copulation among humans goes on in the minds of humans as they form their attitudes about the orangutan. Rapists are criminals in human society. Repeated rapists are jailed and even sterilized in parts of the world. Nature conservancies will have a hard time soliciting money to buy rain forest reserves for primate felons. It may serve certain human interests to declare that young orangutan males commit rape, but it doesn't serve the orangutan or science.

Related to the treatment of primate sexual behavior is the issue of political dominance. Adult male orangutans are bigger, stronger and more obviously aggressive. They seem dominant to most human observers. Is this because they dominate the scene when they come on stage, attracting human attention with their impressive hulk and bravado? Or do they truly dominate and control most aspects of the mostly solitary lives of all orangutans in "their territory"? Cusano (1995) reports that life inside chimpanzee society involves ever shifting relations among members, governed as much by situation as by strength, sex and seniority. If we could roam the canopies of Ketambe and Tanjung Puting for a few years, interacting with orangutans on their terms, we might tell a less simplistic tale.

The metaphors we use with orangutans will be confounded by human connotations. Feminists have argued that human male dominance achieves male success in areas where it is not warranted by need, intelligence, sensitivity, etc. Winning by force or intimidation is a poor tactic, one that doesn't lead to the best results for individuals, communities or humanity. Most people dislike socially dominant "bullies." To open the orangutan to this dislike should be done with great care. Wild orangutans spend over 95% of their waking hours in peaceful pursuits (Rodman, 1988). Behaviorally, competition is implied far more than observed, and conflict is rare. While certain males may be more inclined to coerce access to food and favors when they come across females, most live and interact in cooperative harmony. When two males meet, chances are they will pass quietly, not fight, not even exchange challenges. The fact that captive apes in cages and compounds become aggressive and sexually coercive, or morose and depressed, is more relevant to penology than to primatology.

Primatologists, like most humans, focus on the occasional acts of aggression because they attract attention, are exciting and memorable, and make a vivid story. Western society favors attacking destructive symptoms, rather than reinforcing constructive etiologies. Money and time spent on curing illness and punishing crime far outstrips preventive measures; aggression research is more easily financed than altruism research. Primate society is described in terms of macho and Machiavellian politics, not collaboration and friendship. It's okay to study these behaviors, but dangerous to escalate the importance of the results with dramatic metaphor or excessive generalization. This escalation is akin to the posture promoted by modern journalism — we learn a lot about dramatic and rare behavior on the six o'clock news. That's what attracts TV audiences. That's what sells books. That's what wins research grants. That's what builds fame. For natural scientists, this popular penchant has grave effects. It promotes false negative perceptions of the animals we study, it develops perceptual adaptation levels that make normal behavior seem vapid and uninteresting and it inclines us to see violence where it's not. It must be resisted.

Galdikas (1995, p. 356) likens the adult male orangutans to Sumo wrestlers preparing for "the few championship bouts that will determine whether their genetic identity will live or die." This metaphor is more sympathetic than sexual dominance. It puts a positive spin on the not-so-pretty selfish gene theory. But there's more we can do. Orangutan and human have evolved in natural ecosystems and survive in captivity as species and individuals that are mostly cooperative, sexually considerate and peace-seeking. We must promote these positive images, not only because they are closer to truth, but also because as metaphor they are more likely to engender positive affiliation of humankind with our primate kin.

A STRATEGY FOR CHANGING COLLECTIVE REALITY: ER/LC

Necessary to the promotion of public perceptions of the orangutan that will attract support for primate research and conservation is the development of practical strategies and tactics for changing collective reality. In the preceding section we considered problems created by presenting potentially negative realities to the public. Controversial issues like sexual behavior and political dominance require more than an "every primatologist for himself" approach. If we overlay our professional and personal biases on the data or study only the data that fit our biases, then we get what we deserve — personal satisfaction and professional conflict. That may be okay for thriving humans, but it's not okay for the threatened orangutan. We must overcome self-centered focus — sublimate self-image for the sake of orangutan-image, so the higher good can be achieved. Orangutan survival won't abide divisiveness and conflict among jealous and antagonistic power and fame seekers, any more than it does the costly corporate competition for market share among environmental charities. Rijksen's fatigue with No-Action-Talk-Only (NATO) calls for a practical acronym to correct the situation. It's time for the ER/LC solution — Everybody's Right, Let's Cooperate.

As stated at the outset of this treatise, our perceptions are truly what we see inside our unique minds. Everybody's right about their perceptions. And since we are talking here about managing human perceptions for the good of the orangutan, we can use all viewpoints. That's what a collective reality is — a collection of everybody's realities into an amalgam that works for the greater good. Let's take orangutan rehabilitation, for example, a hotbed of controversy. The clinical practitioners are right — "primate prisoners need socialization, orphans need parenting, refugees need economic aid and community building ... in all cultural contexts, for all social animals "(Bowman, 1994). The experimental scientists are right — we must minimize adverse human impact on the orangutans we rehabilitate and assure that they don't adversely effect their counterparts in the wild. The field naturalists are right — "the worst thing that can happen ... that will take millions of years to correct is the loss of genetic and species diversity by the destruction of natural habitats" (Wilson, 1994, p. 355). From the individual orangutan to the ecosystem, all must be accommodated, all protected, all revered. Rijksen (1995, this volume) is right — "survival of the orangutan should not be seen as someone's project ... it is a common moral obligation which should not be attributed to any individual or organization ... what really counts is what is going to happen."

This is not the time to neglect the orangutan, nor is it time to stop talking. Rather it is time to start talking collectively, to affirm one another's perceptions and take action to vastly expand public support for orangutan research, care and protection. The next International Orangutan Conference should be driven by the ER/LC principle. Interdisciplinary symposia should be organized to achieve consensus in crucial and controversial areas of science, clinical practice, and conservation.

Each consensus must have four elements: 1) an all inclusive mosaic of current "scientific truths", 2) optimal public perceptions to be induced by these truths, 3) cooperative methods whereby these truths and perceptions will be transformed into actions, and 4) mechanisms for assuring that actions have positive effects on our main targets — the public, the orangutans and the rain forests. It is important to note that steps 1 and 2 are necessary but not sufficient. In essence, they are all talk. But they are an essential prelude to successful lasting cooperative action — steps 3 and 4. This kind of interdisciplinary project is not new. The author has used it to launch long-term action-research programs in military, forestry and health care organizations (Rose, 1973, 1976, 1984; Stebbins et al, 1982). Basic to the use of such strategies among primatologists and others concerned about the future of the orangutan is our willingness to set aside prejudices and agree to cooperate with one another in the creation of collective realities to enhance the public's sense of kinship with and support for orangutans and their rain forest habitat.

CONCLUSIONS

As scientists, we can cite countless ways that our species is positively linked to the orangutan. As humanists, we can report interactions with orangutans that profoundly affect our kinship with them. As naturalists, we can portray the orangutan as perhaps the most sublime and sympathetic metaphor of our order. As primatologists, we can accept the fact that if we don't pull together to improve the lot of orangutans wherever they live, they will one day live no more.

REFERENCES

Bowman, K. W., 1994, Developing strategies for ethical pluralism. Abstract in *Proceedings of the XVth Congress, International Primatological Society*, Bali, Indonesia.

Cavalieri, P., Singer, P., 1993, *The Great Ape Project*, New York, St. Martin's Press.
Clay, J., 1978, *Modern Art*, Secaucus, New Jersey, Chartwell Books, p. 127.
Cusano, M., 1995, Personal communication, West Palm Beach, Florida, February 16th.
Davis, H., Balfour, A. D., 1993, *The Inevitable Bond: Examining Scientist-Animal Interactions*, New York, Cambridge University Press.
Galdikas, B.M.F., 1979, Orangutan adaptation at Tanjung Puting Reserve: mating and ecology. Pp. 195-233 in *The Great Apes: Perspectives On Human Evolution*, Hamburg, D. A., McCown, E. R., (eds.) Menlo Park, Benjamin/Cummings.
Galdikas, B.M.F., 1981, Orangutan sexuality in the wild. Pp. 281-300 in *Reproductive Biology of the Great Apes: Comparative and Biomedical Perspectives*, C. E. Graham, (ed.), New York, Academic Press.
Galdikas, B.M.F., 1984, Orangutan sociality at Tanjung Puting. *American Journal of Primatology*, 9:101-119.
Galdikas, B.M.F., 1995, *Reflections of Eden: My Years With the Orangutans of Borneo*, New York, Little-Brown.
Goodall, J., 1990, *Through A Window*, New York, Houghton Mifflin.
Goodall, J., 1994, Personal communication, Los Angeles, April 7.
Jakobson, R., 1962, *Selected Writings I*, The Hague, Mouton.
Kaplan, G., Rogers, L., 1994, *Orang-Utans In Borneo*, New South Wales, University of New England Press.
Kaplan, G., Rogers, L., 1995, Rich imagery — neglected practice: the plight of wild orang-utan in past and present. In: *The Neglected Ape*, R.D. Nadler, B.M.F. Galdikas, L. Sheeran, N. Rosen and (eds.). New York, Plenum Press.
Kawai, M., Asquith, P., 1981, *Life of the Japanese Monkeys*, publisher unknown.
Kellert, S. R., 1993, The biological basis for human values of nature. Pp. 42-72 in *The Biophilia Hypothesis*, S. R. Kellert and E. O Wilson, (eds.), Washington, D.C., Island Press.
Lawrence, E. A., 1993, The sacred bee, the filthy pig, and the bat out of hell: animal symbolism as cognitive biophilia, Pp. 301-344 in *The Biophilia Hypothesis*, S. R. Kellert and E. O Wilson, (eds.), Washington, D.C., Island Press.
Leakey, R. E., Lewin, R., 1977, *Origins*, New York, E.P. Dutton.
MacKinnon, J. R., 1971, The orang-utan in Sabah today, *Oryx*, 11:141-191.
MacKinnon, J. R., 1974, The behavior and ecology of wild orang-utans (*Pongo pygmaeus*), *Animal Behaviour*, 22:3-74.
Maple, T. L., 1980, *Orang-Utan Behavior*, New York, Van Nostrand Reinhold.
Miles, H.L.W., 1993, Language and the orang-utan: the old 'person' of the forest, Pp. 42-57 in *The Great Ape Project*, P. Cavalieri, P. Singer, (eds.), New York, St. Martin's Press.
Montgomery, S., 1991, *Walking With The Great Apes*, Boston, Houghton-Mifflin Co.
Nadler, R. D., 1981, Laboratory research on sexual behavior of the great apes. Pp. 191-238 in *Reproductive Biology of the Great Apes: Comparative and Biomedical Perspectives*, C. E. Graham, (ed.), New York, Academic Press.
Nadler, R. D., 1982, Reproductive behavior and endocrinology of orang utans. Pp. 231-248 in *The Orang Utan: Its Biology and Conservation*, L. E. M. de Boer, (ed.), The Hague, W. Junk.
Nadler, R. D., 1988, Sexual and reproductive behavior, Pp. 105-116 in *Orang-utan Biology*, J. H. Schwartz, (ed.), New York, Oxford University Press.
Nadler, R. D., 1989, Sexual initiation in wild mountain gorillas, *International Journal of Primatology*, 10:81-92.
Nadler, R. D., 1995, Personal communication, *E-mail*, May 5.
Oden, D.L., Thompson, R.K.R., 1993, The role of social bonds in motivating chimpanzee cognition, Pp. 218-231 in *The Inevitable Bond: Examining Scientist-Animal Interactions*, H. Davis and A. D. Balfour, (eds.) New York, Cambridge University Press.
Rijksen, H. D., (1978), *A Field Study On Sumatran Orang-Utans* (*Pongo pygmaeus abelli* Lesson 1827), *Ecology, Behaviour And Conservation*, 78-2. Wageningen, H. Veenman and Zonen B.V.
Rijksen, H. D., 1995, The neglected ape? In *The Neglected Ape*, R.D. Nadler, B.M.F. Galdikas, L. Sheeran, and N. Rosen (eds.) New York, Plenum Press.
Rodman, P. S., 1988, Diversity and consistency in ecology and behavior. Pp. 31-52 in *Orang-Utan Biology*, J. H. Schwartz, (ed.), New York, Oxford University Press.
Rogers, C. R., 1961, *On Becoming A Person*, Boston, Houghton-Mifflin.
Rogers, C. R., Rose, A. L., 1971, *Because That's My Way*, Pittsburgh, WQED-TV.
Rollin, B. E., 1993, The ascent of apes: broadening the moral community. Pp. 206-219 in *The Great Ape Project*, P. Cavalieri, P. Singer, (eds.) New York, St. Martin's Press.
Rorty, R., 1982, Mind as Ineffable, in *Mind In Nature*, R.Q. Elvee, (ed.), N. Y., Harper and Row.

Rose, A. L., 1968, Experimental alcoholism in monkeys and rats: meta-grumbles of a former rat-runner. Abstract in *Proceedings of the Annual Meeting of the Western Psychological Association*, San Diego.
Rose, A. L., 1973, A social systems approach to cross-cultural diplomacy, *Consultant's report to U.S. Navy Pers-P.C.*, Washington, D.C.
Rose, A. L., 1976, The emerging performance development process, *Consultant's report to U.S.D.A. Forest Service, Region 5*, San Francisco.
Rose, A. L., 1983, Improving the quality of patient-provider relations, *Consultant's report to Southern California Permanente Medical Group*, Los Angeles.
Rose, A. L., 1984, Building a service relationship program, *Consultant's report to Kaiser Foundation Hospitals and Pharmacies*, Los Angeles.
Rose, A. L., 1988, The quest for human integrity and cosmic Intent. *Social Change Systems Theory*, Hermosa Beach, CA.
Rose, A. L., 1994a, Description and analysis of profound interspecies events. *Proceedings of XVth Congress, International Primatological Society*, Bali, Indonesia.
Rose, A. L., 1994b, Ethical challenges to primate research and conservation, Abstract in *Proceedings of XVth Congress, International Primatological Society*, Bali, Indonesia.
Rose, A. L., 1994c, New paradigms for personhood in the age of atonement, *Proceedings of XVth Congress, International Primatological Society*, Bali, Indonesia.
Rose, A. L., 1995, Documenting the epiphanies of primatology, *Social Change Systems Research*, Hermosa Beach, CA.
Rose, A. L. and Auw, A., 1974, *Growing Up Human*, New York, Harper and Row.
Rosen, N., Sheeran, L, and Nevadomsky, J., 1995, The International Orang-utan Conference: The Neglected Ape, *Current Anthropology*, 37: in press.
Shürmann, C. L., 1981, Courtship and mating behavior of wild orang-utans in Sumatra. Pp. 130-135 in *Primate Behavior And Sociobiology*, Chiarelli A. B., Corruccini, R. S., (eds.), Berlin, Springer-Verlag.
Stebbins, M. W., Hawley, J. A., Rose, A. L., 1982, Long-term action research: the most effective way to improve complex health care organizations Pp. 105-136 in *Organizational Development In Health Care*, Margulies, N., Adams, J. D., (eds.), Reading, Addison-Wesley.
Wilson, E. O., 1984, *Biophilia*, Cambridge, Mass., Harvard University Press.
Wilson, E. O., 1993, Biophilia and the conservation ethic, Pp. 31-41 in *The Biophilia Hypothesis*, S. R. Kellert and E. O Wilson, (eds.), Wash., D.C., Island Press.
Wilson, E. O., 1994, *The Naturalist*, Washington, D.C., Island Press.

SECTION TWO – CONSERVATION, TRANSLOCATION, AND REHABILITATION

Introduction

Conservation is a multifaceted problem with political, economic, and social ramifications in addition to the more obvious biological ones. It is logical, therefore, to develop multiple approaches to the problem of species endangerment; if one approach fails, alternatives may still be feasible. Further, species do not recognize national and international boundaries, so governments must cooperate in the development of protection plans for most species. Such multilateral efforts may require a shift in focus from the immediate gain to the longer range profitability of habitat preservation. This is true not only for the country in which the species is indigenous, but also for those countries proposing that the species be protected. The 1992 United Nations Conference on Environment and Development (Earth Summit), for example, drew attention to the relationship between a country's environmental degradation and its poverty. Agenda 21 of the Earth Summit notes that the development of sound environmental policies in impoverished countries requires considerable investment of funds on the part of wealthy countries. The "G7" countries, while acknowledging this truism, could not agree upon monetary amounts.

A reorientation of national and international conservation goals may be facilitated by the use of "flagship species," organisms who come to embody in the public's mind the conservation crises faced by all members of their ecosystems. Dietz, Dietz, and Nagagata (1994; see also Stuart, 1991) note that the following characteristics are desirable in flagship species: high biological diversity of the habitat being represented, the keystone or central ecological role of the flagship species, and the public's likelihood of identifying with the flagship. They noted that large size, intelligence, close evolutionary relationship to humans, and complex social structure were all correlated with the development of a positive public view of the flagship. By these criteria, all primates are logical flagships; some primate species have been used, in fact, to spearhead rehabilitation and reintroduction programs. These are risky and costly exercises, however. Magin and co-authors (1994) note that in the past 400 years over 400 mollusk and vertebrate species have become extinct, but fewer than 2% of these have been successfully preserved and reintroduced into parts of their former range (see also Kleiman, 1989). The International Union for the Conservation of Nature developed a Reintroduction Advisory Group (RAG) in the late 1980s, but to date few species have been recommended for this expensive undertaking. The RAG drafted several guidelines for evaluating the need and probability of success for a proposed reintroduction project (Kleiman, Stanley Price, and Beck, 1994; Stanley Price, 1991; see also Beck, Rapaport, and Wilson, 1994). These include a consideration of several factors related to 1) the condition of the species (the need to augment size and genetic diversity, the existence of stock available for reintroduction, and the lack of endangerment to existing populations); 2) environmental conditions (eliminate or control original causes of decline, existence of suitable, protected habitat in original range, and low species densities where reintroduction is contemplated); 3) biopolitical conditions (no negative impact on the local human population, the existence of community support, support for the effort from government organizations and non-governmental organizations, and conformity of the reintroduction plan with all applicable laws); and 4) biological conditions (the availability of techniques of reintroduction for this or closely related species, knowledge of the species' biology so that success of the reintroduction can be evaluated, and the existence of resources to analyze the costs and benefits of the reintroduction). Since these criteria are met for very few species, RAG has rarely recommended reintroduction for any species, primates included. Chivers (1991) further notes that to assess the value of reintroduction realistically, such populations must be monitored for years. Experts agree, therefore, that the conservation of intact habitats is more feasible, ecologically sound, and less expensive than rehabilitation and reintroduction (Dietz, Dietz, and Nagagata, 1994).

The conservation of orangutans mirrors in miniature the broader issues cited above. Orangutans have been used as a flagship species and several attempts have been made to rehabilitate and reintroduce ex-captive animals. The continued survival of these animals, however, also depends upon the international community. One problem facing orangutan conservation is the lack of policing efforts in international wildlife trade, a lucrative business linked to the drug trade at the most recent Convention on International Trade in Endangered Species of Wild Fauna and Flora (see Eudey, this volume). Sugardjito opens this section with a strong statement which emphasizes the debilitating impact habitat degradation has had on orangutan densities. He notes that the most important conservation strategy is habitat protection. Hiong and her colleagues describe techniques for the safe translocation of wild orangutans from areas where they encounter humans to "safer" environments. The development of these skills are important since conflicts between humans and orangutans over habitat are likely to increase. The success of such a program, however, depends upon a suitable home for those individuals being translocated. Lardeux-Gilloux and Smits address the thorny issues of orangutan rehabilitation and reintroduction. Lardeux-Gilloux points out the difficulties inherent in long-term undertakings such as reintroduction programs, namely, the difficulty in securing adequate funding and staffing for indefinite periods of time. Smits and his colleagues describe one orangutan rehabilitation and reintroduction program in which an attempt has been made to meet the various criteria of the RAG have been met. This program contains innovations such as socializing the ex-captive animals prior to release into the forest and training them in the acquisition and processing of natural foods, activities relevant to their successful adaptation to a natural style of life.

REFERENCES

Beck, B.B., Rapaport, L.G. and Wilson, A.C., 1994, Reintroduction of captive-born animals. Pp. 265-286 in: *Creative Conservation: Interactive Management of Wild and Captive Animals*, (Eds. P.J.S. Olney, G.M. Mace, A.T.C. Feistner), New York, Chapman and Hall.

Chivers, D.J., 1991, Guidelines for reintroductions: Procedures and problems. Pp. 89-99 in: *Beyond Captive Breeding: Re-introducing Endangered Mammals to the Wild, Symposia of the Zoological Society of London 62*, (Ed. J.H.W. Gipps), Oxford, Clarendon Press.

Dietz, J.M., Dietz, L.A. and Nagagata, E.Y., 1994, The effective use of flagship species for conservation of biodiversity: The example of lion tamarins in Brazil. Pp. 32-49 in: *Creative Conservation: Interactive Management of Wild and Captive Animals*, (Eds. P.J.S. Olney, G.M. Mace, A.T.C. Feistner), New York, Chapman and Hall.

Kleiman, D.G., 1989, Reintroduction of captive mammals for conservation. *BioScience* 39:152-161.

Kleiman, D.G., Stanley Price, M.R. and Beck, B.B., 1994, Criteria for reintroductions. Pp. 287-303 in: *Creative Conservation: Interactive Management of Wild and Captive Animals*, (Eds. P.J.S. Olney, G.M. Mace, A.T.C. Feistner), New York, Chapman and Hall.

Magin, C.D., Johnson, T.H., Groombridge, B., Jenkins, M. and Smith, H., 1994, Species extinctions, endangerment and captive breeding. Pp. 3-31 in: *Creative Conservation: Interactive Management of Wild and Captive Animals*, (Eds. P.J.S. Olney, G.M. Mace, A.T.C. Feistner), New York, Chapman and Hall.

Stanley Price, M.R., 1991, A review of mammal re-introductions, and the role of the Re-introduction Specialist Group of IUCN/SSC. Pp. 9-25 in: *Beyond Captive Breeding: Re-introducing Endangered Mammals to the Wild.* Symposia of the Zoological Society of London 62. (Ed. J.H.W. Gipps), Oxford, Clarendon Press.

Stuart, S.N., 1991, Re-introductions: To what extent are they needed? Pp. 27-37 in: *Beyond Captive Breeding: Re-introducing Endangered Mammals to the Wild. Symposia of the Zoological Society of London 62.* (Ed. J.H.W. Gipps), Oxford, Clarendon Press.

5

CONSERVATION OF ORANGUTANS

Threats and Prospects

J. Sugardjito

The Indonesian Institute of Sciences
Research and Development Centre for Biology
Jl. Ir. H. Juanda 18, Bogor 16002, Indonesia

ABSTRACT

The estimated number of orangutans in the wild has been changing along with the development of methods used in censusing the species. The most appropriate method for estimating the population of orangutans in the wild is through a long-term study. Previous authors have provided several different estimates which range from 4,000 to as high as 145,000 individuals. Recent data, however, suggest that the number of orangutans in the wild is no more than 30,000 individuals. The varied estimates of the population suggest that a ground survey of orangutans in different habitats and categories of forest is of the utmost importance.

INTRODUCTION

The International Conference on Orangutans got off to an auspicious start by inviting individuals from orangutan habitat countries, a strategy for which the organizing committee deserves hearty congratulations. It is we in the habitat countries, after all, who are most directly involved in the conservation issues pertinent to the orangutan, "the neglected ape".

One reason we call the orangutan the neglected ape is because there are far fewer scientific publications on this species than there are on the other apes, gorilla and chimpanzee. There are, moreover, very few publications available on orangutan conservation. Bibliographic data on conservation of the great apes from 1980-1992 has been compiled by the Primate Information Center, University of Washington, Seattle. The figures show that among the great apes, there are fewer citations for the orangutan than for their African counterparts. From a total of 719 bibliographic citations available on conservation of the great apes, the orangutan has only 178, while the gorilla and chimpanzee, including the bonobo, have 278 and 263, respectively.

Despite the paucity of publications on the orangutan, conservation of this species has been recommended for over three decades. As a result of estimates by previous authors

The Neglected Ape, Edited by Ronald D. Nadler et al.
Plenum Press, New York, 1995

during the 1960s that only 4,000 orangutans were left in the wild (Schaller, 1961; Harrison, 1963; Milton, 1964; Davenport, 1967), this species was accorded the status of highly endangered. This immediately lead to an extensive campaign for its protection, including a global ban on trade. This effort, however, did not satisfy most conservation planners because of deficiencies in the methods that were used for estimating the number.

A prerequisite to conservation assessment for orangutans is some knowledge of the distribution, density, and size of its populations. Estimation of even simple parameters such as these, for orangutans, poses severe problems because visibility in the field is poor and the animals are unusually cryptic. To obtain an accurate estimate of the size of an animal's population, therefore, one must have accurate estimates of its density in each habitat type, and accurate estimates of the size of each habitat. Furthermore, a quick appraisal of larger populations requires sampling techniques based on sightings or nest counts over a wide area (see van Shaik, Priatna and Priatna, this volume). The best method of assessment under these circumstances is through intensive, long-term study at several sites. Although neither type of information is available with the precision that one would wish, there are enough data to permit some educated guesses.

THE POPULATION OF ORANGUTANS

The first reliable information on the orangutan's behavior and ecology was obtained only in the 1970s after scientists developed improved methodologies and facilities for intensive study of this cryptic ape (Horr, 1973; Rodman, 1973; MacKinnon, 1974; Rijksen, 1978; Galdikas, 1981). As a consequence of the improved knowledge of its habitat and density, the hidden ape became visible and the estimated size of the wild population increased. The total number of wild orangutans increased when scientists analyzed their average densities in Borneo and Sumatra in relation to its remaining habitat . At least 150,000 individual wild orangutans have been estimated to exist by extrapolating data on their average density and on the amount of forest cover suitable for orangutans (MacKinnon, 1985). This figure, however, did not consider the effect of hunting, human population pressure and differences of density in each habitat type. A more current finding suggested that the Bornean population consists of 19,000 to 30,000 individuals, while that for Sumatra consists of 5,000 to 7,400 individuals (Sugardjito and van Schaik, 1992). During the 1993 Workshop on Population and Habitat Viability Analysis of the orangutan conducted in Medan, North Sumatra (Section 3, this volume), it was determined that the size of the wild orangutan population is maximally 5,800 for Sumatra and 15,500 for Borneo.

THREATS TO ORANGUTANS

Whatever number is eventually estimated, one must still contend with rain forest degradation, the most significant threat confronting the conservation status of this species. Since the great majority of orangutan habitat is in lowland rain forest, the problems facing the conservation of orangutans are largely a function of the decline in this type of habitat. One of the differences between this Asian ape and the two African apes is that the orangutan is exclusively dependent upon trees for its existence. Trees are its home as well as its food source because it builds nests in trees to sleep and it moves arboreally while feeding predominantly on fruit. More than 50% of the orangutan's feeding time is spent on fruit, a type of food that is not continuously available in the forest.

Although most of the timber extracted from the forest have not been the food trees of this animal, they are the most popular stands for the orangutan's food supply, namely, the strangling fig. Removal of timber, therefore, ultimately destroys the strangling figs which are the most important food of this species. We can assess the importance of these trees as stands for the strangling fig by considering the data from the Ketambe study area in Sumatra. The stands apparently are determined by the height and bark of the trees. The higher the trees and the more cavities they have, the more suitable is the site for the growth of figs. Because the species favored as timber in Ketambe have these two characteristics, strangling figs are most frequently found in this family of trees. The density of Dipterocarpaceae trees in the study area was relatively low compared with Meliaceae trees that are also common there (Abdulhadi and Kartawinata, 1982). The trees that support the strangling fig, however, are primarily the Dipterocarp trees. More than 50% of the 75 tree stands examined were members of that family of timber trees.

A further problem facing conservation efforts is that the effects of habitat disturbance are often compounded by hunting. Humans rely on many species of wildlife as a source of meat. As with other primate species, the orangutan is a source of meat for human consumption in some areas; this constitutes a conservation problem especially in the interior of Borneo. The slow moving orangutan makes an easy target for an experienced hunter who can kill his prey from a distance. Rijksen (1978) suggested that the hunting of orangutans increased in historic times as it replaced the practice of human head-hunting and cannibalism. In Borneo, the sizable population of orangutans coincides with areas in which the indigenous people do not have a hunting tradition because they have been Muslim for centuries (coastal swampy areas). This pattern, however, may be confounded by the carrying capacity of the habitat. In some areas, the forest sustains so few orangutans that even slight hunting pressure is enough to decimate the local population. As a result of timber cutting activities, moreover, the area becomes accessible to humans. The more accessible the area, the more rapidly the population disappears. A recent survey in areas where selective logging had been conducted, indicated that the lowest density of orangutans was found in hill sites, the second highest was in lowland and the highest density was found in swamp forests (Yanuar et al., 1992). Although high densities were found in the lowland forests, this type of habitat is the most accessible after logging and, therefore, suffers the most from hunting. Furthermore, the orangutan prefers exactly those habitats which are preferentially converted to agricultural uses, namely, fertile lowland soils, usually close to the rivers. The data also indicated, however, that there remain extensive swamp forests in Borneo which could support a viable population of orangutans.

A less widespread but sometimes contentious form of direct orangutan exploitation is their trapping and trade. Although trapping for trade is generally a secondary threat, it merits careful attention because it is more amenable to control, both within source countries and through the Convention on International Trade in Endangered Species of Wild Fauna and Flora. There are presently 901 orangutans in captivity in 216 managed and unmanaged collections (380 Bornean, 309 Sumatran, 181 subspecific hybrids and 31 of unknown species). Of this number, only 632 individuals are included in a regional cooperative breeding and management program under the Captive Breeding Specialist Groups (Perkins, 1993).

PROSPECTS FOR CONSERVATION

Having considered the threats to conservation, we now consider the conservation remedies. First and foremost, it is important to recognize that conservation problems must be addressed from several directions at once. In order for there to be any long-term hope of success, strong collaboration must develop between donor countries and the orangutan habitat countries. Since habitat loss is the principal threat to the orangutan, it follows that habitat protection should be the highest priority. Hence, the most valuable direct means of assisting orangutan conservation is, in most cases, through the establishment and management of protected areas. The main aim of such a program is to protect representative examples of as many as possible of the habitat types. This principle is well illustrated by the protection system in Borneo, the most important biounit containing orangutans. There are some protected areas containing orangutans in Borneo, either gazetted or proposed, including the Danum Valley, Gunung Loton, Tabin and Kinabalu in Sabah, Lanjak-Entimau in Sarawak, and Kutai, Sangkulirang, Tanjung Puting, Bukit Baka-Bukit Raya, Gunung Palung, Kendawangan, Gunung Niut, Danau Sentarum, and Gunung Bentuang-Karimun in Kalimantan. In addition, the Gunung Leuser National Park has been designated for the Sumatran subspecies. Since forest cover decreases as a consequence of development and human population growth, these protected areas have become increasingly important.

In total, at least 27,500 km^2 of forest habitat are protected to support the orangutan's existence. There are only a few suitable orangutan habitats, however, either inside or outside the protected areas and the density of orangutans differs among habitats. Despite this, when the density of orangutans ranges from 0.4 to 3 individuals/km^2, it provides a corridor to link disparate units and establishes a good chance of survival for the orangutans. This is a vital element of the orangutan conservation strategy. Another important part of that strategy is to encourage spatial integration of land use, so that forms of land use that are compatible with biodiversity, such as sustainable timber extraction, or activities accompanied by low human population pressure, such as plantation forestry, become associated with the protected areas. In this way, larger areas can be created that have two major benefits. First, larger areas of integrated land use relieve pressure on the protected areas (buffer zone function). Second, they extend the living areas for a number of species living in the protected area, thus boosting the prospects for survival. In light of this potential benefit, it is all the more important that we determine which threats to the orangutan's survival derive from habitat exploitation. If human disturbance were the only problem, better concession management could greatly increase the amount of the habitat available to orangutans.

While many of the benefits of protected areas, and especially the reasons for conserving orangutan, are more apparent in a national or global perspective, most of the costs of refraining from exploiting the protected areas are borne locally. This asymmetry lies at the root of most conservation management problems and demands attention to the problems and welfare of ordinary people, whose lives are affected by the creation of protected areas. Increasing habitat encroachment and smuggling of animals outside the country is often related to this condition. Since all orangutan habitats are located in developing countries, the problem of their conservation must be viewed in the wider context of the economic and social development of the nation involved and the growth in the human population. This realization encourages a review of the role played by governments, private sectors, international aid agencies, and non-governmental organizations in the integration of conservation and development.

ACKNOWLEDGMENTS

I thank the Organizing Committee of the International Conference on Orangutans for the invitation to present this paper, Sutisna Wartaputera and Birute Galdikas for continuous support of work on orangutans and David Chivers for his assistance. The survey of wild orangutans described in this paper was supported by the Red Alert Program of the Flora and Fauna Preservation Society.

REFERENCES

Abdulhadi, R. and Karawinata, K., 1982, The pattern of forest vegetation at the Ketambe Research Station, Gunung Leuser National Park. *LBN-LIPI Rep.*, Bogor, Indonesia (in Indonesian).

Davenport, R.K., 1967, The orang-utan in Sabah. *Folia Primatol.*, 5:247-263.

Galdikas, B.M.F., 1981, Orangutan sexuality in the wild. Pp. 281-300 in *Reproductive Biology of the Great Apes: Comparative and Biomedical Perspectives*. C. E. Graham, ed. New York, Academic Press.

Harrison, B., 1963, Education to wild living of young orang-utans. *Sarawak Mus. J.* 11:220-258.

Horr, D.A., 1977, Orang-utan maturation: Growing up in a female world. Pp. 289-321 in *Primate Biosocial Development: Biological, Social and Ecological Determinants*. S.C. Skolnikoff and E.E. Porier, eds. New York, Garland Publishing, Inc.

MacKinnon, J.R., 1974, The behaviour and ecology of wild orang-utans (*Pongo pygmaeus*). *Anim. Behav.* 22:3-74.

MacKinnon, J.R., 1985, The conservation status of non-human primates in Indonesia. Pp. 99-136 in *Primates: The Road to Self-Sustaining Populations*. K. Benirschke, ed. New York, Springer-Verlag.

Milton, O., 1984, The orang utan and rhinoceros in North Sumatra. *Oryx* 7:177-184.

Perkins, L., 1993, Orangutan GASP Report, *Conserv. Breed. Spec. Grp. News* 4(3):16-17.

Rijksen, H.D., 1978, *A Field Study on Sumatran Orang Utans* (*Pongo pygmaeus abelii* Lesson 1827), *Ecology, Behaviour and Conservation*, 78-2. Wageningen, H. Veenman and Zonen B.V.

Rodman, P.S., 1973, Population composition and adaptive organization among orang-utans of the Kutai Reserve. Pp. 171-209 in *Comparative Ecology and Behaviour of Primates*. R.P. Michael, J.H. Crook, eds. London, Academic Press.

Schaller, G.B., 1961, The orang-utan in Sarawak. *Zoologica* 46:73-82.

Sugardjito, J., 1988, Some observations on the relationship between strangling figs (*Ficus spp.)* and the host trees in the tropical forest of the Ketambe area, Sumatra. *Ilmu dan Budaya* 10:932-937 (in Indonesian).

Sugardjito, J. and van Shaik, C., 1992, Orangutans: Current population status, threats, and conservation measures. *Proc. Conserv. Great Apes New Order Environ.* Pp. 142-151, Bohorok, Jakarta, Tanjung Puting. Ministry of Forestry and Ministry of Tourism and telecommunication of the Republic of Indonesia, Jakarta.

Yanuar, A., Chaerul, S., Wedana, I.M. and Sugardjito, J., 1992, Survey of endangered primates in Kalimantan with special emphasis on orang-utans. *Unpublished Report to the Red Alert Program of the Flora and Fauna Preservation Society*.

6

CAPTURE OF WILD ORANGUTANS BY DRUG IMMOBILIZATION

L. K. Hiong,[1] J. B. Sale,[2] and P. M. Andau[2]

[1] Sabah Wildlife Department
5th Floor, Block B, Wisma Muis, 88100, Kota Kinabalu, Sabah, Malaysia
[2] Sabah Wildlife Department
Sabah, Malaysia

ABSTRACT

Drug immobilization was used to address the increasing demand for a safe and reliable method of capturing wild orangutans that remain in remnant forest on agricultural plantations. This paper reports the capture of 70 animals by this means, mostly using a 5:1 ketamine-xylazine mixture at a mean dosage of 8.43 mg for the two drugs combined or 7.03 mg ketamine and 1.41 mg xylazine/kg body weight. An alternative drug used in a few cases was Zoletil (1:1 tiletamine-zolazepam) at a dosage of 6.90 mg/kg. The darting system of choice was the Telinject air-pressure powered vario 4V rifle, with reusable plastic darts activated by compressed air. A Dist-inject N 60 rifle, using either "Speedy" pre-loaded disposable plastic darts or aluminum darts assembled on demand, was used as a back-up. There was a wide range of tolerance to dosage levels, ranging from 2.60 to 22.64 mg/kg, indicating a wide safety margin for this mixture with orangutans. A high degree of individual variability in reaction to the drugs was also encountered and supplementary doses were required to complete immobilization in a large proportion of cases. A standard dosage for the 5:1 ketamine-xylazine mixture of 8 mg/kg estimated body weight is recommended for drug capture of wild orangutans.

INTRODUCTION

There are a number of circumstances in which it becomes necessary to capture wild orangutans (*Pongo pygmaeus*) in their natural habitat, the lowland rain forests of Borneo and Sumatra. Increasingly, their habitat is being cleared for plantation agriculture, chiefly oil palm, necessitating the capture and translocation of the affected apes to protected areas (Andau et al., 1994). In some cases, groups of orangutans which have become isolated in patches of forest begin feeding on young (up to 1-year-old) oil palm seedlings, most of which become damaged beyond recovery. Such "pest" animals must be captured and removed, or

The Neglected Ape, Edited by Ronald D. Nadler et al.
Plenum Press, New York, 1995

they are susceptible to persecution and injury by estate workers and their domestic dogs. From time to time, it is also necessary to capture individuals that have escaped into the nearby forest from a rehabilitation center, such as that at Sepilok in Sabah. In addition, the manager of a reserve containing orangutans occasionally sees an animal displaying signs of injury or ill health, indicating the need for capture, examination and possible treatment.

In all these scenarios, the capture methods normally available are unreliable, time-consuming and entail risk of injury to both the orangutans and the persons involved in their capture. Animals on the ground are typically chased until cornered and then restrained with nets or ropes. If the orangutan is in an isolated tree, this is cut down and the fleeing individual pursued as above. Payne (1988) described a method of greasing the tree trunk (to make it difficult to climb) and then luring the orangutan to the ground with bananas or other bait. Payne admitted that it might take several days to achieve success with this method. Forest workers sometimes beat an animal to weaken it prior to capturing it for use as a domestic pet. Accidents are commonplace with all the above manual methods of capture. There is, thus, an evident need for a quick and reliable method of capturing wild orangutans, which is efficient and safe for both the animal and its captors. Recognizing this need, the two senior authors conducted some trials in drug capture methodology in 1992 which resulted in a protocol for capture operations with orangutans (Sale, 1992). The logistics of drug immobilization of orangutans in their native forest required careful planning and the assistance of plantation managers who had problems with orangutans. A total of 95 animals were captured and the majority were translocated to the Tabin Wildlife Reserve (Andau et al., 1994). These operations provided an opportunity to develop the drug capture methodology; this paper examines the results obtained and discusses their implications.

METHODS

General

Animals were captured at four separate plantation locations between June and December, 1993. Each capture operation was carried out by a team that included a veterinarian as team leader (LKH), a ranger, a marksman to do the darting and six to eight laborers to help restrain and carry animals. The operations conformed to one of four basic patterns, briefly described below:

1. If the target animal was in a tree, it was darted up to 30 m above the ground. In most cases, the immobilized animal fell "gradually" through a tangle of climbers and made a "soft" landing in undergrowth at ground level. If no natural cushioning was present, a 2 X 2 m net was held out at the base of the tree by four or more workers to catch the falling animal. In a few cases, the immobilized orangutan remained in the tree, either on a nest or lodged in the branches, and had to be retrieved by a worker climbing up the tree.
2. Target animals hiding among vegetation or fallen trees at or near ground level were also immobilized by darting and then retrieved by two or more workers.
3. Some animals in ground-level situations were first partially restrained, either by a net or ropes, and then immobilized by manual injection of the drug. In several cases a pole-syringe or "jab stick" was used for this operation.
4. The capture of a number of immature or small adult animals, which did not pose a threat to their captors, was accomplished by manual restraint alone, without the use of any drug. Data derived from this form of capture are not considered in the

following account which is confined to the procedures involved in drug-assisted capture.

After capture, each animal was examined by the veterinarian for injury and then carried in a net or sack, slung beneath a pole borne on the shoulders of two workers, to a base camp. Here, the new captive was weighed, blood, hair and fecal samples were taken and a long-acting antibiotic (Betamox L.A., Norbrook) was administered intramuscularly. Animals perceived to be unduly stressed or dehydrated were, in addition, given an intravenous infusion containing dextrose and/or saline. In a few serious cases, a corticosteroid injection was also given. The animal was then placed in a strong metal cage, in deep shade, pending recovery, during which time body temperature and other vital signs were monitored. Signs of hypothermia were counteracted by showering with cold water and, in some cases, administration of an antidote to the immobilizing drug.

Darting Equipment

For the great majority of dartings, a Telinject vario 4V rifle was used, powered by air pressure from a foot-pump, together with 1 or 3 ml reusable plastic darts activated by compressed air (Telinject USA Inc., Saugus, CA.) (although this model of rifle is fitted to take a CO_2 cartridge). As a back-up system, a powder-charged Dist-inject N 60 rifle was used in a few cases, firing either 3.5 ml "Speedy" disposable plastic darts with built-in detonator caps or aluminum darts assembled on demand (Peter Ott AG, Basel, Switzerland). Various types and lengths of needle were used with both systems.

Areas of muscle suitable to receive a dart are limited to the shoulder and thigh of the orangutan. Before each operation, the team marksman was put through target practice with the rifle and the type of dart to be used on that occasion. Shooting practice is particularly important because of the relatively small target areas available on the orangutan.

The Type of Drugs Used

The immobilizing drug of choice was a 5:1 ketamine-xylazine mixture (Ketalar, Parke Davis and Rompun, Bayer). The addition of xylazine, as opposed to using ketamine alone, enables a reduction in the volume of drug required. This, in turn, allows the use of a small sized dart, which is much preferable for the limited target area and thin skin of the orangutan. It also produces a deeper state of sedation which is desirable when capturing an excited, wild animal and it maintains sedation for a longer period of time. This facilitates carrying the new captive to the base camp and its subsequent examination, before placing it in a cage for recovery and transportation to its ultimate destination.

To facilitate the use of a small dart (1 or 3 ml), it was desirable that the volume of the 5:1 mixture be kept to a minimum. Lyophilized ketamine was not available; in order to obtain the mixture, a 25 ml stock solution was prepared by the following two steps:

1. 5 ml of 10% (100 mg/ml) ketamine solution was added to 500 mg xylazine (Rompun) dry substance (in original 5 ml vial) and shaken until dissolved.
2. This 5 ml preparation was added to 20 ml 10% ketamine solution to give 25 ml of the 5:1 mixture, containing 100 mg ketamine and 20 mg xylazine per ml.

Initially, this mixture was given at a dosage of 6 mg (5 mg ketamine, 1 mg xylazine) per kg estimated body weight. Thus, an average-sized adult orangutan with an estimated weight of 30 kg would require 1.5 ml of the mixture.

Yohimbine hydrochloride (Reverzine, Parnell) was administered intravenously by hand in cases where quick recovery from ketamine-xylazine immobilization was required. Otherwise animals were left to recover naturally over a longer period.

An alternative immobilizing drug, used in a few cases where the field supply of the ketamine-xylazine mixture ran out, was Zoletil (Virbec), containing a 1:1 mixture of tiletamine and zolazepam. This was given at a dosage rate of 5 mg (2.5 mg of each component) per kg estimated body weight. This mixture does not have an antidote.

RESULTS

Of the 95 orangutans captured during the 1993 translocation operations, 70 involved drug immobilization. Of this number, 50 were darted either in trees or on the ground and 20 were injected manually after initial restraint. Animals estimated to be 7-years- old and older were designated "adults," of which 23 were males and 37 females; the remainder (10) were immatures up to 6-years-old.

In the early operations, no suitable weighing device was available and subsequently, the weighing of the occasional animal was omitted due to more pressing priorities at the time. Accurate body weights were obtained for 41 of the drug-captured animals, thus providing an objective basis for the calculation of actual dosages—as opposed to dosages based on estimated body weight prior to immobilization. The weights obtained are summarized in Table 1. As expected in a sexually dimorphic species, the mean weight of the males was significantly higher than that of the females. In fact, the largest male captured weighed exactly twice as much (72 kg) as the largest female (36 kg).

In the 50 captures involving darting, the mean estimated shooting distance was 13 m (range 3-30 m). Beyond 30 m, accurate shooting was not possible with such small target areas of muscle. For that reason, no darting was normally attempted above 30 m.

The great majority of the darts found their intended target in either the shoulder or thigh muscles. A few darts strayed from these optimal target areas, often as a result of the animal moving at the moment of firing. Stray areas included back musculature, ribs, forearm and abdomen; all minor injuries resulting from these dartings were treated successfully.

There were two fatalities. One resulted from an immobilized animal that remained recumbent in a high nest. It was not possible to remove it before excessive exposure to the sun had occurred. Post-mortem examination revealed symptoms of sun-stroke. The other fatality was a nursing infant which accidentally received the dart and drug dose intended for its mother (the mother had moved at the moment of firing) and thus, died of a large overdose.

Equipment Performance

The Telinject 4V rifle system was found to be more satisfactory for delivery of the drug mixture at darting distances up to 30 m. Its pressurized air propulsion is virtually silent and produces a gentle dart velocity, inflicting minimum trauma on impact. The adjustable

Table 1. A summary of the body weights (in kg) of 41 of the 70 orangutans captured by drug immobilization

Age/sex class	n	Mean	SD	Range
Adult Male	9	33.4	20.47	14.5 - 72.0
Adult Female	25	27.0	5.64	14.0 - 36.0
Immature	7	13.9	3.24	10.5 - 17.5

rear sight on the recent Telinject rifles enables more accurate marksmanship than the earlier type and the lightweight plastic darts maintained a satisfactory trajectory. (All these operations were conducted in the absence of wind, which might have caused significant dart deflections with such small target areas). The pressurized air activation of the plunger of the Telinject dart is also silent, minimizing disturbance to the animal at the time of impact. Of two needle lengths used with the Telinject system, a 30 mm one proved rather long for orangutans, but a 20 mm needle was optimal. Initially, smooth needles fell out before injection was complete, but needles with a retention collar avoided this problem and were therefore, used in all later dartings.

A disadvantage of the Telinject rifle, as used, is the clumsiness of the foot-pump pressurizing arrangement, particularly when working in thick forest and when a rapid shot is required before the target animal changes position. In addition, the robustness of the pump and its connections to the pressure chamber of the rifle proved inadequate for prolonged use under rigorous field conditions. Repeated repairs and replacement of components were necessary.

A constraint of the Telinject darts, also noted by other workers (Anon., 1991), proved to be considerable inconsistency in the time they were able to retain air pressure in the rear chamber (for activating the plunger). Initially, the practice was to load and pressurize about half a dozen darts before setting out for a capture operation. However, a number of injection failures demonstrated the necessity of pressurizing each dart only a short time (maximum 15 minutes) before use, in order to eliminate the risk of pressure loss before firing. This necessity reduces the marksman's ability to take a quick shot on first sighting a suitable target and considerably prolongs the average time taken to capture an animal.

The Dist-inject powder rifle has a very accurate sighting system and produces a flat trajectory which assists accurate placement of the dart, providing the correct strength of cartridge is selected for the dart weight and target distance. It is somewhat noisy on firing, however, which sometimes frightens the animal, causing it to move off after receiving the dart. The pre-charged "Speedy" dart, with fixed needle, facilitates rapid dart preparation but, with a volume of 3.5 ml, was larger than needed for many of these operations. The "clack" produced by the plunger-activating cap, moreover, can add to the disturbance of the target animal. Of the two needle lengths available with "Speedy" darts, the 20 mm one is ideal. The retention barb on these needles is satisfactory in preventing the dart from falling out of the animal prematurely. It can cause more tissue damage on removal, however, than a collar. Disposable darts such as the "Speedy" are somewhat expensive, especially when a large series of immobilizations is being undertaken.

The self-assembly aluminum darts of the Dist-inject system have the advantage over the "Speedy" of allowing selection of a syringe size to match precisely the desired volume of drug. A 19 mm collared needle is available, although of a larger diameter than is necessary for such a thin-skinned target as the orangutan. The added weight of these metal darts means that more momentum is generated than with the lightweight plastic darts. As a result, there is a greater possibility of tissue trauma on impact, particularly if an overly powerful charge is used for the weight of the animal and the distance traversed.

Drug Action

Of the 70 orangutans captured by drug immobilization, 65 received the 5:1 ketamine-xylazine mixture and 5 received Zoletil. In the first three capture locations, involving 57 animals, the initial dose of the mixture given was 6.0 mg/kg *estimated* body weight. As this proved too low in more than 30% of the cases, which had to be given supplemental doses, the initial dose was increased to 7.8 mg at the fourth location, in which the last eight animals reported here were captured.

Table 2. Actual dosages of the 5:1 ketamine-xylazine mixture (in mg/kg) administered to wild orangutans

Age/sex class	n	Mean	SD	Range
Adult male	9	9.35	4.19	3.89 - 18.95
Adult female	25	7.96	3.96	2.60 - 22.64
Immature	7	8.93	3.96	3.76 - 15.43
Combined	41	8.43	4.06	2.60 - 22.64

Table 2 summarizes data collected on the actual dosages of the 5:1 mixture, calculated after the animals concerned had been accurately weighed. In calculating the effective dosages shown, the following arbitrary rule was followed in cases where a second or third supplemental injection was needed to produce full immobilization: any doses of drug administered within 30 minutes of each other were added together to give the reported dose. If a supplemental injection followed an earlier one by more than 30 minutes, the earlier one was ignored in reporting the total dose given (and in calculating induction time - see below).

The mean dosage overall (n = 41) was 8.4 mg/kg body weight (equivalent to 7.0 mg ketamine and 1.4 mg xylazine). Only relatively small differences are shown between the means for the different sex and age classes and all have similar standard deviations. It is clear from the large range of effective dosages shown in Table 2 that there is great variability in the reactions of individual animals to this particular drug mixture. This is further borne out by an examination of the need for supplemental doses for different classes of initial dose, shown in Table 3. In all dose classes except the highest one (No. 5), more than 25% of the animals had to be given at least one supplemental dose in order to effect capture. In classes 1, 2 and 4 it was 50% and above. The mean actual dosage for the eight animals in the fourth capture location, only one of which required a supplement, was 8.25 mg/kg, but the standard deviation (1.5 mg/kg) is significantly lower than that of the overall mean for the series (4.1 mg/kg).

Induction and recovery times are shown in Table 4. Induction time was defined as the time between injection of the initial effective dose (see above) and manifestation of complete immobilization. Recovery time, based only on animals that did not receive any antidote, was the time between immobilization and the first signs of bodily movement, indicating the beginning of recovery. The completion of recovery was a much less clear-cut event and therefore, it was difficult to assign a precise time.

Mean induction time for all age and sex classes was 7.8 min (n = 53) but there was great variation about this mean (SD = 7.1 min) and the overall range was 1-35 minutes. Immature animals showed a lower mean induction time (5.3 min) than either adult males (7.4 min) or females (8.8 min), but all classes showed high variation. However, if induction

Table 3. Administration of supplemental doses of the 5:1 drug mixture for various classes of initial dose

Initial dose			Number of supplemental doses given			
Class	Dosage (mg/kg)	n	1	>1	Total	%n
1	3.1 - 5.0	10	3	2	5	50
2	5.1 - 7.0	12	4	2	6	50
3	7.1 - 9.0	11	1	2	3	27
4	9.1 - 11.0	6	3	1	4	67
5	11.1 - 13.0	2	0	0	0	0

Table 4. Induction and recovery times

	Induction times				Recovery times			
Age/sex class	n	Mean	SD	Range	n	Mean	SD	Range
Adult Male	17	7.41	5.35	3 - 20	12	48.33	26.89	20 - 117
Adult Female	27	8.78	8.39	2 - 35	20	44.84	20.10	19 - 85
Immature	9	5.33	4.99	1 - 19	4	40.20	12.62	24 - 56
Combined	53	7.76	7.13	1 - 35	36	45.36	21.98	19 - 117
1 dose only	35	4.37	1.76	1 - 12	26	48.19	22.48	19 - 117
> 1 dose	18	14.33	8.83	4 - 33	10	38.00	18.71	20 - 86

times of animals receiving only a single injection of drug are examined separately from those of animals which required one or more supplemental injections (bottom two lines of Table 4), the former showed much briefer induction, with a mean of only 4.4 minutes and relatively low variability (SD = 1.8 min; range 1 - 12 min). The animals that received supplemental injections showed a longer induction time (mean = 14.3 min) and greater variability (SD = 8.8; range: 4 - 33), which is at least partially explained by the time lag between the (inadequate) initial dose and supplemental one. The distribution of induction times between single and multiple dosed animals is shown in Fig. 1, which also shows that the most common induction time was 4 minutes.

Mean recovery time (Table 4) was 45.4 minutes, but, like induction time, there was considerable variation between individuals (SD = 22 min; range: 19 - 117 min). Differences between the means of the three sex/age classes and between single and multiple dosed (supplemented) animals were small. The correlation between overall dosage and recovery times was r = 0.26. The correlation is greater, however, when single-dosed animals only are considered (r = 0.57).

The reversing agent yohimbine was used on 10 occasions to speed recovery from the ketamine-xylazine mixture, where it was felt that more prolonged natural recovery presented a danger to the animal. After the injection of yohimbine the animal normally revived in less

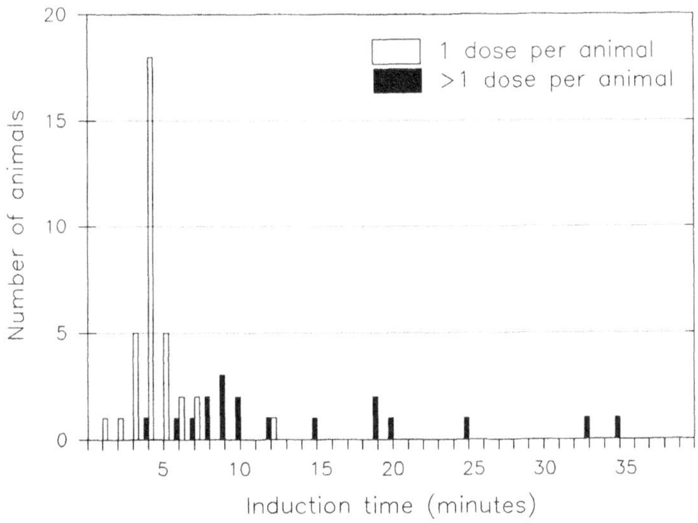

Figure 1. The frequency of induction times.

than 5 minutes. The mean dosage of Zoletil used (n = 5) was 6.9 mg/kg body weight and the range was 4.7 - 9.1 mg. Induction times were similar and recovery times were slightly longer than for the ketamine-xylazine mixture.

DISCUSSION AND CONCLUSIONS

The series of drug-assisted captures reported here, totaling 70 animals, demonstrates that chemical immobilization is an effective and reliable option for the large-scale capture of an ape such as the orangutan. The fact that such a large series of captures produced only two fatalities indicates that the procedures used were safe for orangutans in remnant rain forest. It should be noted, however, that such forest contains no especially tall trees compared to primary, unlogged rainforest and in these operations, the upper darting limit was 30 m above the ground. It would not be advisable to dart orangutans in primary forest trees that greatly exceeded this height.

The pressurized air or gas rifle used in these operations, while having some advantages over powder charged rifles, requires improved reliability for continuous long-term use. In particular, the connections for both the gas cartridge and foot-pump must be strengthened to withstand constant robust handling in the field.

Air-powered lightweight plastic darts proved ideal for this type of short-range darting, in the absence of deflecting wind currents. The efficiency of the capture operations would be considerably improved, however, by a dart with greater ability to retain rear chamber pressure prior to use and would save valuable time in darting an appropriately positioned animal. It would enable the preparation of a variety of darts and drug volumes prior to setting out on an operation. For larger orangutans, above 45 kg, use of a dart of greater volume than 3 ml for the initial shot would reduce the need for supplemental dartings and thus save time and money. For example, a 4 or 5 ml plastic dart, fired from a 13 mm barrel rifle, would be appropriate.

The 5:1 ketamine-xylazine mixture proved highly suitable for the capture of orangutan under the conditions described. The high individual variation in reaction to the drug mixture made precise determination of the optimal dosage difficult. However, the very large range of dosages tolerated by all age and sex classes is a clear indication of the margin of safety of the 5:1 mixture with this primate species.

A mean induction time of under 8 minutes for all dartings and under 5 minutes for single-dose dartings is satisfactory, particularly for animals in isolated trees where there is little tendency to re-locate between darting and induction. For animals in more mobile situations at ground level, clearly the shorter the induction time, the less likelihood of losing the animal during this critical period.

A desirable goal in improving the methodology is to reduce the present high incidence of supplemental injections. Apart from ensuring that the larger animals get an adequate initial dose by using a large volume dart, it seems justified to increase the dosage of 6 mg/kg estimated body weight which was used first. For example, it is significant that when the estimated dosage was increased to 7.8 mg/kg in the last eight captures, there was a much lower variance (SD = 1.5 mg/kg versus 4.1 mg/kg) and only one of the animals required a supplement, although the mean of the actual doses was similar to the overall mean (8.3 mg/kg versus 8.4 mg/kg). On the basis of rounding up the higher dosage figure used (7.8), for convenience of calculation, a dosage of 8 mg of 5:1 ketamine-xylazine mixture (6.67 mg ketamine and 1.33 mg xylazine)/kg estimated body weight is proposed for future immobilization of wild orangutans. Thus, an animal estimated to weigh 30 kg would require 240 mg of the mixture or 2 ml of the solution, requiring a 2 ml dart; a 45 kg animal would require 360 mg and a 3 ml dart, and so on.

ACKNOWLEDGMENTS

We thank our colleagues Jumrafiah Abdul Shukor, Dr. Rika Akamatsu and Christovol Jomitin and their respective staff for support and assistance during the translocation operations. The work was often extremely demanding on the skill and resources of the rangers and laborers of the Wildlife Department and we express our appreciation to all of those who were involved. The cooperation of the management staff of the respective plantations is also acknowledged. The Sabah Government also thanks the United Nations Development Programme for the services of J.B.S. as Chief Technical Adviser to the Wildlife Department, through its Project MAL/88/009.

REFERENCES

Andau, P. M., Lim Khun Hiong and Sale, J. B., 1994, Translocation of Pocketed Orang-utans in Sabah, *Oryx* (in press).

Anon, 1991, Wildlife Restraint Series, *Int. Wildl. Vet. Serv., Inc.*, Salinas, Calif.

Payne, J., 1988, Orang-utan Conservation in Sabah, WWF Malaysia Project No. 96/86 and WWF International Project No. 3759. *WWF Malaysia, Kuala Lumpur*, 137 pp.

Sale, J. B., 1992, Orang-utan Capture Operations, *Recommendations prepared under UNDP Project MAL/88/009*. Sabah Wildlife Department, Kota Kinabalu, 5 pp.

7

REHABILITATION CENTERS

Their Struggle, Their Future

I. Lardeux-Gilloux

50 Rue E. Dalanglade
13006 Marseille, France

ABSTRACT

There are no systematic analyses or studies dealing directly with the pros and cons of rehabilitation centers for orangutans (RCOs), simply because there are no hard data on the subject and no applied research has been carried out. The purpose of this paper does not reside in rating the performance of RCOs because this is neither a scientific report nor an evaluation of conservation as it relates to the interests of eco-tourism. Based on a review of the literature, personal behavioral observations and ethological studies, this paper was written in an effort to focus attention on the process of orangutan rehabilitation.

INTRODUCTION

In this paper, I briefly discuss the historical background to the creation of rehabilitation centers for orangutans (RCOs), as it is important to our understanding of the different approaches that have been taken to the concept of rehabilitation. I then discuss the importance of behavioral management in the rehabilitation process. Last, but not least, I conceptualize the role and the place of orangutan rehabilitation in the context of wildlife management and eco-tourism.

HISTORY OF THE REHABILITATION CENTERS

The relatively large number of orphaned orangutans entering the pet trade was the reason that rehabilitation centers were first created in the 1960s. Such a project was first established in 1961 in the Malaysian Bako National Park by the Sarawak Museum Staff and Barbara Harrisson. The objective at that time was to focus attention on the illegal trade in this species and to encourage international interest in the conservation of orangutans (Schaller, 1961; Harrisson, 1962). The Sepilok RCO was established in Sabah in 1964, while the Bako project failed and was replaced in 1977 by the Semenggok RCO in Sarawak. In

The Neglected Ape, Edited by Ronald D. Nadler et al.
Plenum Press, New York, 1995

1971, the Ketambe area of the Gunung Leuser Reserve was established as the first Sumatran RCO; in the same year, the Tanjung Puting Bornean RCO was set up in Kalimantan Tenggah. In 1973, a second Sumatran RCO was started near Bohorok in the Langkat part of the Gunung Leuser Reserve.

No other RCOs were established until five years ago. Then, in 1990, Tanjung Harapan RCO was founded at the Tanjung Puting Park boundary; in 1991, Semboja RCO was set up at the Wanariset Research Center in East Kalimantan (Wartaputra, 1991). The reason these more recent rehabilitation centers were created, 20 years after the original ones, was again the illegal traffic in orangutans. This time, however, the main aims had changed slightly. In the late 1960s, the objective of rehabilitation programs was to secure the enforcement of the Animals Protection Ordinance (of 1931 in Indonesia and 1963 in Malaysia) and to re-introduce ex-captive orangutans into the wild. At the present time, the potential of the orangutan to promote tourism has been realized and the activities of rehabilitation projects have shifted towards public education in conservation issues.

With recognition of the need to avoid 1) social stress on wild orangutan populations by introducing rehabilitated orangutans (Sugardjito and van Schaik, 1991) and b) the transmission of zoonotic and enzootic diseases from those animals to the natural population (Amsel, 1991; Sayuthi, et al., 1991), there was a reassessment of the role and functions of RCOs. The result was the closing of both Ketambe (1979) and Tanjung Puting (1990), where the priorities of management and conservation were to maintain untouched ecosystems. It is notable that in the past, attempts were made to separate rehabilitation centers from areas of field research and wildlife population management. The more recent policy is guided by the benefits that accessible rehabilitation centers give to eco-tourism.

It has long been recognized that RCOs situated in conservation areas or areas susceptible to the activities of an ever-expanding tourist industry cannot maintain the strict protection standards required for the survival of an endangered species (Frey, 1975, 1976; Rijksen, 1974, 1991; Rijksen and Rijksen-Graatsma, 1975; Aveling and Mitchell, 1982; Sugardjito and van Schaik, 1991; Bennett and Gombek, 1992). What has happened in the operations of the RCOs is that more than ever before, the differing demands of conservation, rehabilitation and tourism have come into conflict. Increasingly, research is needed to accommodate the disparate interests of different institutions. Rehabilitation programs have been of great value in repairing some of the damage of poaching and logging and in increasing conservation awareness in local and international communities. They have sometimes failed, however, to care humanely for the displaced, illegal orangutan populations and to demonstrate that the RCOs actually have a conservation value. The whole question of RCO management is fraught with scientific disputes and political complications. Rarely has any serious attempt been made to consider the problem from the ethological point of view taken here: What are the behavioral needs of the displaced orangutan population?

BEHAVIORAL MANAGEMENT

The rehabilitation process has been and remains largely a matter of educated guesswork. Far from being an established technique, as typically conducted, the rehabilitation process has failed to benefit from its ongoing experience; and it has not contributed unequivocally to the well-being of rehabilitant orangutans. Currently accepted methods of husbandry for this species are inherently unsuitable for planned captive management, long-term breeding, population stability and genetic diversity (Markham, 1991; but see Smits, et al., this volume). The basic husbandry problems faced by the rehabilitation centers are similar to those encountered with the captive population of orangutans in zoos (Maple,

1980). The former, however, are complicated by increased human intervention. In effect, the orangutan is being mismanaged because its behavioral needs are not being properly met. One reason for this deficiency is that its biology, psychology and sociology remain unclear and the stigma of being a "semi-solitary" species has resulted in the isolation of individuals. Various studies (Markowitz, 1982; Poole, 1988; Ghaffar, 1989; personal observations), however, and zookeepers' accounts have revealed the great social capabilities of orangutans in captivity and the benefit of group-living as an environmental enrichment tool (see also Zucker and Thibaut, this volume). The natural solitary behavior patterns of orangutans, their cognitive abilities and their social propensity make orangutans especially conducive to rehabilitation because, 1) the orangutan's development of maintenance activities and social behavior does not depend on learning by imitation or through observation and 2) its intelligence gives it an extreme behavioral plasticity and adaptability. Given this potential for rehabilitation, an appropriate program of species-specific husbandry can be defined as an active management plan designed to reintroduce the displaced population of ex-captive orangutans successfully into the natural habitat.

Special Conditions to Address

Considered here are the management problems facing a species that has entered the world of humans and lost contact with its natural world. The behavior patterns of these animals have been altered by environmental and emotional changes and will be further affected by the human intervention required for their rehabilitation.

Problems that Counter Rehabilitation

The reintroduction of displaced orangutans, as defined by Rijksen (1978), consists of two stages: 1) *ecological rehabilitation*, in which orangutans develop independent maintenance activities (foraging, nest-building, etc.) and are no longer dependent on the center for food and 2) *social rehabilitation*, in which orangutans develop positive social contacts with their peers and with wild orangutans. These two stages are greatly affected by an individual's past history and the specific process of rehabilitation. Several factors may compromise the success of the process.

1. The chances of successful rehabilitation are impaired by abnormal behavior which originates in captivity. Abnormal behavior prevents individuals from dealing effectively with conflicts and anxiety and must be addressed through careful behavioral engineering (Lardeux-Gilloux, 1988). It is a sad fact that stress in orangutans is frequently apparent only after the animal has died because orangutans exhibit very few behavioral stereotypies. These few stereotypies, however, can be used to assess each individual's progress. All displaced orangutans are orphans, suffering from an abused childhood, lack of confidence, absence of mother, rejection and isolation. The effects of these traumas show up in the form of bodily rocking, tantrums, regurgitation/reingestion, finger-sucking and heightened aggression (Forsythe, 1992).
2. The chances of successful rehabilitation are impaired by a lack of self-confidence which creates emotional instability (Bowden, 1980). From the time of arrival at the center to the time of reintroduction into the wild, the orangutans encounter successive changes in their physical environment, i.e., quarantine, nursery, platform and ultimately, the forest. Insecurity and fear are repeatedly exhibited at the onset of each stage.

3. The chances of successful rehabilitation are impaired by improper nutrition. Often overlooked, nutritional problems lead to behavioral problems and a failure to rehabilitate. Bloat is one of the most common problems of improper nutrition; it results from regurgitation/reingestion behavior. Overly frequent feeding is a cause of center dependence and fixed, short feeding times are a cause of increased competition and aggression between individuals.
4. The chances of successful rehabilitation are impaired by retarded climbing skills. The "flat foot syndrome" is the result of living a terrestrial life and lacking the opportunity to grasp the mother's hair early in infancy. This results in a weak grip, falling from trees and the inability to climb effectively.

Husbandry Techniques Required for Rehabilitation

The RCOs are a combination of captive and wild environments. They have all the artificiality of the captive environment and all the unpredictability of the wild environment, creating a complex, dynamic system to address. Human intervention is an integral part of the system, but interaction is not. Too often, one is mistaken for the other.

Human attention should be given to relieve anxiety bouts, to guide and to reassure, but only when necessary, as it is essential to avoid strong human attachment. The best condition for the animals is the group situation with a variety of age-sex classes. This situation provides a sense of security, a way to deal with the new, bewildering environment and the opportunity for learning and developing social relationships. The formation of these relationships cannot be forced upon individuals, but can be actively encouraged through careful monitoring of the social interactions occurring in the group. This is especially important for infants. It has the added advantage of eliminating humans as their major models during ontogenetic development.

Robinson's (1992) comparative study on the sociality of rehabilitants showed that adults were more solitary than immature animals and immature animals were more interactive than the adults. The incidence of play and other interactive behavior decreased with age; the amount of time spent gnaw-wrestling, for example, was significantly different in adults and young. Thus, adults became increasingly solitary, spending significantly more time sitting alone and grooming, whereas, young orangutans were more interactive and playful, just as in zoos (see Zucker and Thibaut, this volume).

Nutritional problems must be addressed in RCOs, similar to zoos (Table 1). Feeding should be selective, i.e., at any given time, some individuals should be fed while others are not. Feeding should be carried out randomly with respect to time and space to encourage foraging activities. Food supplementation should be stopped or reduced in conjunction with the fruiting season. Food should be given above ground (Bowden, 1980) and distributed over time to encourage food handling by the animals (Lardeux-Gilloux, 1988). Low interest food, such as sweet potatoes, carrots and star fruits should be provided regularly, but high interest food, like bread, milk and bananas should be avoided in group situations as they create fights, bloat and regurgitation (personal observation). Providing the animals with cold fruit is another effective way of increasing handling time.

Lewis' (1992) comparative study of the modes of locomotion used by rehabilitants showed that there were differences between age classes. Older animals performed significantly more arboreal quadrupedalism, cross suspensory locomotion and brachiation than younger animals, indicating that orangutans become increasingly adept at travelling arboreally as they age. The use of ropes as an environmental enrichment tool in Lewis' study brought about a significant increase in the rehabilitants' arboreal behavior. Locomotion was correlated with the time of the day, suggesting that the daily activity budget in captivity, as

Table 1. A list of indexed food plants known to be eaten by wild orangutans (from Ghaffar, 1989) compared with a list of non-indexed food plants eaten by rehabilitant orangutans at the Sepilok Rehabilitation Center, Sabah (compiled by Lardeux-Gilloux and bin Majanil, 1992)

Indexed Food Plants			Non-indexed Food Plants		
Family	Genus	Species	Family	Genus	Species
Anacardiaceae	Dracontomelom	dao			
Annonaceae	Polyalthia	sumatrana			
Araceae	Soindapsus				
Burseraceae	Canarium				denticulatum
	Canarium				dichotrinum
				Santiria	rostrata
			Connaraceae	Cnestis	palala
Dipterocarpa	Dryobalanops	lanceolata			
				Parashorea	malaanonan
	Shorea				gibbosa
	Shorea				leprosula
Ebenaceae	Diospyros				cf. maritima
	Diospyros				curraniopsis
	Diospyros				tuberculata
Euphorbiaceae	Aporusa				frustescens
				Ptychopixis	kingii
			Fabaceae	Saraca	declinata
				Cynometra	elmerii
Graminae	Dinochloa				trichogona
Lauraceae				Litsea sp.	
Lecythidaceae				Lecythidaceae	ashtonii Pa.
			Leeaceae	Leea	mali Mali
Leguminosae	Fordia			gibbsiae	Dun et. Bak F.
	Fordia			splendidissima	
				Caesalpinia	oppositifolia
				Cynometra	elmepi Mevr.
	Spatholobus				hirsutus
Meliaceae	Aglaia	odoratissima			
Moraceae	Artocarpus	dadah			
	Ficus	acompathophylla			
	Ficus	annulata BI			
Myristicaceae	Khema				woodii
				Horsefielda	brachiata
			Oleaceae	Jasminum	crassifolium
Palmae	Calanus				caesius BI
				Karthalsia	rigida blume
				Karthalsia	Macrocarpa B.
Pandaceae	Pandanus				
			Polygalaceae	Xanthophylum	adenotus Miq.
Rutaceae	Glycomis	rambutan-ake			macrantha
Sapindaceae	Nephelium				
			Sapotaceae	Madhuca	
			Verbenaceae	Teijsimannio-dendron	
					bogoriense
				Teijsimannio-dendron	globrum
			Zingeberaceae	Alpinia	

in the wild, is governed in part by food availability. This study also showed that the daily activity budget of the rehabilitants is significantly altered by the need to conform to the regimented provision of food. Locomotion was also correlated with the amount of time that elapsed since the animal was first released, which reflects the acclimatization phase of exploring the new environment.

I do not discuss here the medical attention required in the process of rehabilitation to avoid disease transmission, epidemics and repair of physiological damage, as others have drafted quarantine and veterinary procedures (Karesh, et al., 1991; Smits, et al., this volume). Medical treatment must not be forgotten, however, as it is as much an integral part of the rehabilitation process as behavioral engineering.

The main problem to overcome in the process of rehabilitation is the "humanization" of orangutans, which has a negative behavioral and psychological impact. Humanization is also the cause of another potentially dangerous situation. Semi-feral orangutans that interact with humans can be extremely dangerous (de Silva, 1971; Bowden, 1980). Not all orangutans become aggressive toward humans, but those that do may become fierce and uncontrollable as adults. Aggressive orangutans generally show a common personality profile: past experiences of abuse, strong attachment to the original caretaker and rejection of all others, jealousy, resentment toward humans, an understanding of their physical advantage over humans and cunningness (personal observation).

How much of an effect does rehabilitation have on a socialized, semi-solitary species? The diurnal activity patterns of six free-ranging adult rehabilitants, for example, did resemble those of wild orangutans and included an increase in social interactions, but this was not sufficient to form a cohesive group (bin Majanil, 1993). Thus, these orangutans were not adversely affected by the socialization process and were capable of being reintroduced as a semi-solitary species. There is a paucity of such data, however, which requires further research to document the long-term adaptation of rehabilitants to the natural habitat.

ROLE AND PLACE

The immediacy of the displaced orangutan population demands a rigorous analysis of rehabilitation methodology, results and strategy. One must be wary, however, of developing expensive, but ineffective rehabilitation programs that address only medical problems. The planner's responsibility in establishing a rehabilitation operation is to coordinate all resources in the pursuit of the overall goal of species conservation, including management of the RCO, conservation education, research, humane management and husbandry techniques and eco-tourism. The integrity of rehabilitation projects can be preserved in spite of the many factors working against them, but the establishment of a rehabilitation management policy is essential if we are to follow the philosophy and ethic behind the conservation of wild animals and biodiversity as a whole.

With respect to primates, we must properly reintroduce ex-captives according to the dictates of law, the principles of conservation and the demands of appropriate husbandry, if we are to be credible. This requires that rehabilitated orangutans be classified according to three categories of action (Fig. 1):

1. Release into the wild after quarantine using translocation techniques.
2. Release after social and ecological rehabilitation.
3. Do not release; either humanely cull or sacrifice for captive breeding and tourist activities. An effective management option for de-institutionalizing rehabilitants is to release them into areas where orangutans lived in the past. There may also

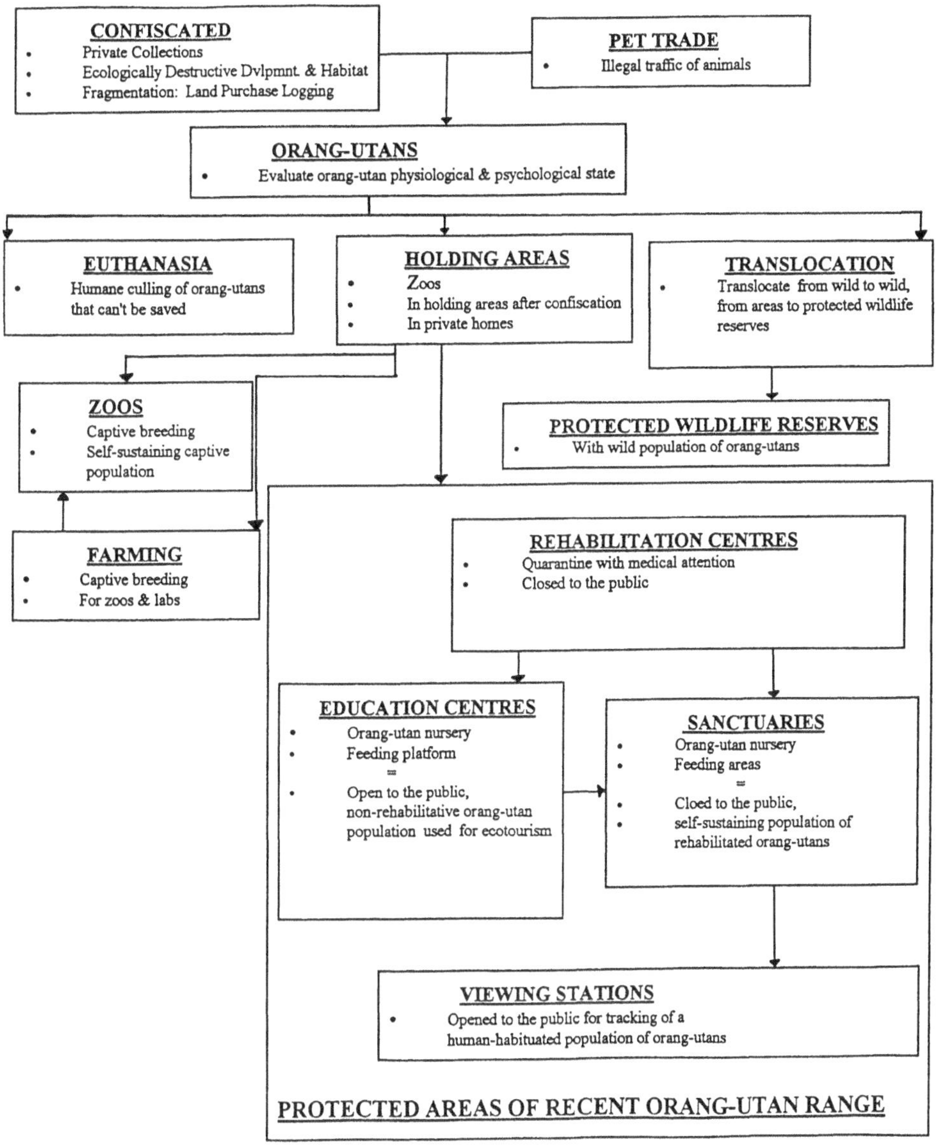

Figure 1. Integrated rehabilitation system.

be conservation value in increasing genetic diversity by establishing new populations.

Rehabilitation programs must accommodate the conflicting requirements of eco-tourism and conservation education. The current market-driven attitude of eco-tourism violates the accepted concepts of sustainable utilization and protection of the wildlife heritage. Currently, RCOs engender a false sense of security and an anthropocentric view that displaced orangutans are being “saved” (Bennett, 1991; see also Rijksen, this volume). With respect to the responsibilities of RCOs, the management of people must be viewed as an integral subsystem of the management of the primates.

REFERENCES

Amsel, S., 1991, Wildlife rehabilitation: the conservation tie that binds biologists and veterinarians; a veterinarian's perspective. *Proceedings on Conservation of the Great Apes in the New World Order of the Environment, Indonesia.*

Aveling, R. J. and Mitchell, A., 1982, Is rehabilitating orang-utans worth while? *Oryx* 16: pp. 263-271.

Bennett, J. and Gombek, F., 1992, Orang-utan ecotourism in Borneo. *Paper presented at BRC 2nd Biennal Int. Conf.*

Bowden, E., 1980, Evaluation of the orang-utan rehabilitation process at Sepilok Rehabilitation Centre in Sandakan, Sabah-Malaysia. *Unpublished BA thesis*, Colorado College, USA.

Frey, R., 1975/76, Orang-utan rehabilitation centre Bohorok, (Ed. P. Jackson) *WWF Annual Report*, pp. 12-123.

Frey, R., 1975, Management of orang-utans. *Wildlife Management in Southeast Asia*, pp. 199-215.

Forsythe, F., 1992, Expression of social behaviour in a group of ex-captive orang-utans at Sepilok Rehabilitation Centre in Sabah, Borneo. *Unpubl B. Sc. thesis*, Cambridge University.

Ghaffar, N., 1989, Social behaviour in a solitary species? A comparison of behavioural interactions in captive groups of orang-utans, chimpanzees and gorillas. *Unpublished B.A. Thesis*, University College, London.

Harrisson, B., 1962, The immediate problem of the orang-utan. *Malay. Nat. J.* 16:4-5.

Karesh, W., Sayuthi, D., McManahan, R, Martin, H., Amsel, S. and Kusba, G., 1991, Recommendations to the Department of Forestry of The Republic of Indonesia on the medical quarantine of orang-utan intended for reintroduction. *Proceedings on Conservation of the Great Apes in the New World Order of the Environment*, Indonesia.

Lardeux-Gilloux, I., 1988, Environment enrichment of captive great apes at London Zoo: chimpanzees, gorillas, orang-utans. *Unpublished B.Sc. thesis*, Queen Mary College, London, U.K.

Lewis, S., 1993, Locomotor behaviour of a group of orang-utans at Sepilok Rehabilitation Centre, Sabah. *Unpubl. B. Sc. thesis*, Cambridge University.

bin Majanil, A., 1993, Social grouping of six free ranging semi-wild orang-Utans in Sepilok, Sabah. *Unpubl. Veter. thesis*, Universiti Pertanian Malaysia.

MacKinnon, J. R., 1977, Rehabilitation and orang-utan conservation. *New Scientist* 74 (1057):697-699.

Maple, T. L., 1980, *Orang-Utan Behavior*. New York, Van Nostrand Reinhold Company.

Markham, R. J., 1991, The husbandry and management of captive orang-utans: problems and opportunities. *Proceedings on Conservation of the Great Apes in the New World Order of the Environment*, Indonesia.

Markowitz, H., 1978, Analysis and control of behaviour in the zoo. In: *Research in Zoos and Aquarium*. pp: 77-90. (Ed. G. Rabb) Washington, The National Academy of Science.

Payne, J., 1988, Orang-utan conservation in Sabah. *WWF Malaysia Project No 96/86.*

Poole, T., 1988, Social behaviour of a group of orang-Utan (Spp) on an artificial island in Singapore Zoological Gardens. *Zoo Biol.* 6:315-330.

Rijksen, H. D., 1974, Orang-utan conservation and rehabilitation in Sumatra. *Biol. Conserv.* 6:20-25.

Rijksen, H. D. and Rijksen-Graatsma, A. G., 1975, Orang-Utan rescue work in North Sumatra. *Oryx* 13(1):63-73.

Rijksen, H. D., 1991, Sustainable development and the extinction of a relative: Witnessing the eradication of the people of the forest. *Proceedings on Conservation of the Great Apes in the New World Order of the Environment*, Indonesia.

Robinson, K., 1993, The activity budgets of a group of orang-utans at Sepilok Rehabilitation Centre, Sabah, Borneo. *Unpubl B.Sc. thesis*, Cambridge University.

Sayuthi, D., Karesh, W., McManahan, R, Martin, H., Amsel, S. and Kusba, G., 1991, Medical aspects in orang-utan re-introduction. *Proceedings on Conservation of the Great Apes in the New World Order of the Environment*, Indonesia.

Schaller, G. B., 1961, The orang-utan in Sarawak. *Zoologica* 46:73-82.

de Silva, G. S., 1971, Notes on the orang-utan rehabilitation project in Sabah. *Malay. Nat. J.* 24:50-77.

Sugardjito, J. and van Schaik, C. P., 1991, Orang-utans: current population status, threats, and conservation measures. *Proceedings on Conservation of the Great Apes in the New World Order of the Environment*, Indonesia.

Wartaputra, Ir. Sutisna, 1991. *Proceedings on Conservation of the Great Apes in the New World Order of the Environment*, Indonesia.

8

A NEW METHOD FOR REHABILITATION OF ORANGUTANS IN INDONESIA

A First Overview

W. T. M. Smits,[1] Heriyanto,[2] and W. S. Ramono[3]

[1] International Ministry of Forest Tropenbos-Kalimantan Project
P.O. Box 319, Balikpapan 76103, Indonesia
[2] Quarantine Center, East Kalimantan /Orangutan Project Wanariset Samboja
Km 38 Balikpapan-Samarinda, Kalimantan Timur, Indonesia
[3] Subdirectorate for Species Conservation of Flora and Fauna
Directorate General for Forest Protection and Nature Conservation, Ministry of Forestry of Indonesia
Manggala Wnabakti Bldg. Bl, L8, Jl. Gatot Subroto, Jakarta 10270, Indonesia

ABSTRACT

In this paper, an overview is provided of a new project involving orangutan rehabilitation in Indonesia. The approach used differs from the traditional method in that groups of medically screened orangutans are socialized and trained to use natural food items before transfer to a forest release site devoid of natural populations of orangutans. Contact between orangutans and humans, including tourism, is minimized to prevent the potential spread of hazardous diseases. By early 1994, the project had processed 114 orangutans; 95% of the 40 animals already released adapted well to the new forest environment. The project is supported by an active confiscation program in Indonesia and efforts to end international trade in orangutans. The project is linked with international research groups and schools to promote conservation of orangutans and their habitat.

INTRODUCTION

In spite of more than two decades of experience with orangutan rehabilitation and a large number of visitors to the rehabilitation centers in Sumatra (Bohorok and Ketambe), Kalimantan (Tanjung Puting) and Sabah (Sepilok), with consequent world wide exposure for the plight of orangutans, there still remains much to be done. It has become clear over the past five years that the smuggling of infant orangutans still takes place. The widely

The Neglected Ape, Edited by Ronald D. Nadler et al.
Plenum Press, New York, 1995

publicized "Bangkok six" and other orangutans kept in Taiwan as pets give some insight into the seriousness of the problem. One must conclude that the rehabilitation projects for orangutans aimed at stopping poaching and smuggling and which are supported by considerable public exposure, have not been able to eliminate the aforementioned problems. One may assume, however, that if these projects had not been so active, the problems would have been even worse. Unfortunately, the publicity and popularity of orangutans, which were featured in films like "Every Which Way You Can" and in television serials about orangutans living with a family in Taiwan, have contributed to a worsening of the situation as well.

Even during the time Herman Rijksen was working with rehabilitant orangutans in Ketambe (before the stations at Bohorok and Tanjung Puting were established), he concluded that the approach he was using was not optimal and suggested a basic outline for improving the rehabilitation of captive orangutans (Rijksen, 1978). John MacKinnon of the World Wildlife Foundation also proposed an orangutan survival program based on the same principles as those proposed by Rijksen (MacKinnon, 1992). It is this approach, also termed, "orangutan reintroduction", that has now been applied for the first time in Indonesia in East Kalimantan at Wanariset Samboja.

At the same time that the new method was initiated, a number of other actions were taken to make the total program more effective for the protection of orangutans and the preservation of their habitat. These include the national registration of all protected fauna (Decree SK 301, 1992), concerted searching for poachers and people holding unregistered animals, contacts with other organizations like the Association of Indonesian Concession Holders and the establishment of high level international collaborations designed to reduce international trade in orangutans as well as other protected plant and animal species. In addition, national surveys are under way to assess the orangutan situation further throughout their entire present distribution in the wild.

The project at Wanariset Samboja discussed in this paper, therefore, is to be regarded as only one of the activities in a larger framework of programs. The project in East Kalimantan involves searching for and confiscating illegally held orangutans and transporting them to the quarantine facilities at Wanariset for medical screening, followed by socialization and introduction to various foods available in the wild. It also entails the selection of suitable release sites, security measures for the orangutans after release, research on a range of subjects related to orangutan rehabilitation, such as the influence of orangutans on the surrounding vegetation and conservation education.

The research results of this project were presented at the XVth Congress of the International Primatological Society which was held in Bali in 1994. A more thorough evaluation on how the released orangutans are interacting with each other, especially interactions between groups, will be conducted in 1995. The present document, therefore, serves to describe this new project and its basic organization.

BACKGROUND

The Wanariset forestry research station is located 38 km north of the oil city of Balikpapan along the main road leading to Samarinda. The station is part of the Forestry Research Institute Samarinda, which operates under the Agency for Forestry Research and Development of the Indonesian Ministry of Forestry (MOF). The station is the location for the "International Ministry of Forestry Tropenbos-Kalimantan Project". One of the research groups in this cooperative effort is the biodiversity/forest ecology group. This research group was established to put more emphasis on the entire ecosystem and not just on the tree-growing aspect which, until 1991, dominated the project's activities. Studies on orangutans are now part of the research conducted by this group on interactions between fauna and the surrounding vegetation. This group benefits from the extensive botanical knowledge available at the project, as well as the expertise of other disciplines, such as the use of remote sensing techniques for monitoring the forest condition and the propagation of local tree species.

The project is centrally located within the triangle east of the Barito River and south of the Mahakam River, an area believed to be free of natural populations of orangutans. Old orangutan skulls have been recovered from Dayak villages in this area and there are now reports of a small number of orangutans living at the point where the Mahakam and Barito tributaries approach each other. There was one report, unconfirmed by other observers, that three orangutans were seen at Prampus, which is located further into the triangle. Further away there have been no reports of orangutan sightings during the last 20 years.

The forest near the project site, some of it on alluvial soils, seems to be ideal orangutan habitat, based on the botanical work conducted here and the presence of Sepans (salt lick places) and extensive fresh water swamps. Since the area was virtually uninhabited by humans until some 70 years ago, it is possible that the orangutans disappeared because of some disease.

THE NEW APPROACH

The new approach to orangutan rehabilitation is based upon two important assumptions. First, establishing social groups of orangutans is important for their successful rehabilitation and well-being. Second, because many orangutans contract diseases in captivity, rehabilitants must not be released into an area with an existing population because of the risk of spreading potentially dangerous diseases. Releasing additional orangutans into an area with an existing population, moreover, may disturb the balance between the number of orangutans present and the supply of food, in terms of the carrying capacity of the forest, specifically as regards heath forests with extremely poor soil.

Table 1 provides an overview of some of the major differences between the "traditional" approach and the new approach now practiced in East Kalimantan. The risk of diseases brought in by ex-captive orangutans proved to be very high. This is based on the experience gained with 114 orangutans at the Wanariset station as well as checks on the health of the people who formerly held those animals. Seventy percent of the orangutans had health problems. The problems included diseases such as tuberculosis, hepatitis A, B, C and E, cholera, malaria and others. Parasites were common. Also, cross-reactions were found with HIV in some special tests, but not in the more common tests like the Western Blot. The HIV tests were conducted on 4 individuals, some of which were recently captured from the forest. We assume, therefore, that this virus is already present in the natural population. Undoubtedly more medical problems will be encountered when we further expand the medical screening. Based on these facts, it appears that any further release of ex-captive orangutans into forests with free-roaming conspecifics would threaten the survival of the whole population and, therefore, should be discontinued. This conforms, moreover, with the conclusions of the Medan Population and Habitat Viability Analysis Workshop conducted in 1993 (see Section 3, this volume).

Because contact with humans poses a risk of disease for orangutans, the handling of animals older than two years of age is limited as much as possible. The personnel at the project are screened for diseases at the start of their employment and they receive a wide range of vaccinations. The highest possible standards are maintained for animals in quarantine. An attempt is made, moreover, to overcome the problems of stress by establishing social relations among the orangutans. It is anticipated that this will also discourage the orangutans

Table 1. Some differences between the traditional approach of orangutan rehabilitation and the new scheme applied in Indonesia

Traditional approach	New approach
Poor quarantine and medical care	Intensive medical screening and after care, including all personnel involved
Release of animals one by one	Release of animals in socialized groups
Release into a forest with existing orangutan population	Release only in forest devoid of a natural population of orangutans and without natural connections to them
Limited preparation in terms of food training	Intensive food training during quarantine and socialization supported by botanical experts and tree climbers
Camps and human presence in the release forest	Quarantine completely separated from the release forest site
Much contact between handlers and orangutans	Minimal contact between handlers and orangutans
Tourism related, physical contact possible in the release forest	No tourism in the new population, no direct contact with orangutans after release
Overstocking happening	Carrying capacity carefully assessed before establishment of new groups
Incomplete record keeping	Complete record keeping and reporting for all individual orangutans

from returning to human settlements and thereby, avoid renewed risks of exposure to hunting and diseases.

All orangutans arriving at the station undergo a series of routine checks and treatments that include serological testing, fingerprinting and picture-taking for identification purposes, sampling of hair and nails and a physical medical examination which includes recording temperature, blood pressure, pulse, respiration, etc. This is followed by complete isolation in a quarantine building, where they can establish eye contact with other orangutans.

SELECTION OF RELEASE SITES

In order for an area to be selected as a suitable release site, two criteria must be met. First, the area should be sufficiently large and contain the types of vegetation and species composition which are capable of supporting a self-sustaining population of the size anticipated. This must also include the consideration of occasional climatic extremes. Second, the security of the area must be protected against further human encroachment and consideration must be given to possible protection measures.

At present, two sites have been selected for the release of orangutans, the first being the Sungai Wain forest. This forest provides most of the water for Balikpapan and the oil industry and as such, is of extreme importance to the city. The city has serious problems with fresh water because of the rapid expansion of the city and its industry. The problems are exacerbated by the reduced capacity of the other drinking water reservoir through shifting cultivation in its watershed and the consequent erosion of soil. Balikpapan is almost surrounded by sea. In parts of the city, intrusion of sea water is already taking place, compromising the use of deep wells. The former artesian fresh water wells are also overexploited, while many wells yield a mixture of water and oil and sometimes warm water.

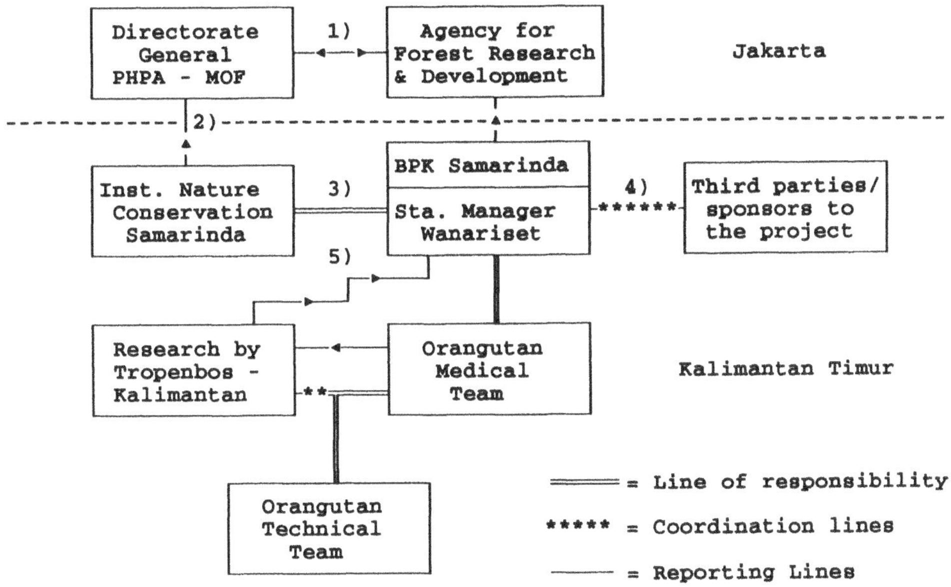

Figure 1. Present organization of the Warnariset project. MOF stands for Ministry of Forestry, BPK for Forest Research Institute Samarinda, AFRD for the Directorate General for Forestry Research and Development of the Ministry of Forestry, PHPA for the Directorate General for Forest Protection and Nature Conservation of the Ministry of Forestry. Tropenbos-Kalimantan is the field name of the cooperative effort on tropical rain forest research between the Indonesian Ministry of Forestry and the Dutch Foundation Tropenbos.

The Sungai Wain forest is located some 15 km north of Balikpapan. It comprises an area of 10,000 ha of lowland and mixed dipterocarp rain forest with extensive swamp areas which contain some sea crocodiles in the southern part. In the same vicinity, there is another 10,000 ha of logged forest and swamp. In the south, a canal and fencing provide physical isolation, while a series of seven security posts around the forest provide additional security. In the north, the area borders on the uninhabited Inhutani I concession. In the west, the forest borders on the bay of Balikpapan.

A total of 422 botanical plots were established, evenly distributed over the entire area. Observations were made on species composition of the trees and the wildlife present. Combined with checklists of the food items of tree species known to be used by orangutans elsewhere and the phenological data of the project at Wanariset, an estimate was made of the number of orangutans that could survive there.

The second release site, Meratus, is more remote and larger than the Sungai Wain forest. It is completely covered by virgin lowland to hill rain forest and also contains some swamps and many fig species. Also relevant, there is a complete absence of humans in the area for about 70 km. Initial surveys of trees covering some 200 ha have been completed and another survey covering the whole transect, from top of the mountain to the lowland, is planned.

The Meratus area consists of protected forest surrounded by a number of commercial timber concessions. There is no encroachment by shifting cultivators in this remote area. The selectively harvested forest around the protected forest is now abandoned and constitutes a large expanse of potential habitat for orangutans. The site can be reached from the Wanariset station to the east in about four hours over good concession roads. Figure 1 presents the present organization of the project at the Wanariset site.

ORGANIZATION OF THE PROJECT

1. The Agency for Forestry Research and Development reports its progress to the Directorate General for Forest Protection and Nature Conservation (PHPA) and coordinates the orangutan activities at the Wanariset station with this Directorate General.
2. The institute in Samarinda reports directly to its Directorate General on numbers of orangutans confiscated or surrendered, official documents accompanying the orangutans and spending of PHPA funds.
3. The station manager officially receives the animals from the Samarinda institute. In some cases where the owners bring their animal directly to the station, he signs for acquisition of the animal with copies to the Samarinda institute and the provincial forestry office.
4. The station manager coordinates relations with third parties who want to sponsor the project. One of these parties is the Balikpapan Orangutan Society, which, in consultation and coordination with the project, arranges fund-raising activities.
5. All data on medical, research and practical aspects are reported by the head of the fauna research group to the Tropenbos-Kalimantan project manager and team leader, who subsequently report the combined data to the station manager.

Sponsorship of the project has been provided almost entirely by the Balikpapan Orangutan Society and schools in Balikpapan, which have raised funds through a variety of activities. These include writing letters for support to companies in East Kalimantan, organizing special fund-raising days with many activities, producing an interactive database on orangutans distributed to schools around the world on CD-ROM and actively corresponding with other schools around the world through E-mail and other methods. The sponsors also support an "adoption" program whereby individuals contribute to the support of an orangutan undergoing rehabilitation and raise money through the sale of T-shirts.

FACILITIES

The Wanariset project now has a well-equipped clinic, laboratory and attached quarantine facilities, including a quarantine extension which can hold some 40 orangutans for extended periods. The station has four socialization cages and a fifth one is being constructed. Other facilities include food sheds, security posts and watch towers.

The project is staffed by a team of 20 people, including a full-time veterinarian who lives at the site, assisted by three more local veterinarians and two local doctors. There is one data analyst, two tree-climbers who collect food for the orangutans from the forest, technicians who work with the orangutans at the station and technicians who monitor the orangutans in the forest after their release. All of the facilities were financed by local sponsors, while some of the technicians are paid by the International Ministry of Forestry-Tropenbos-Kalimantan project. All operational funds come from the "adoption" program.

FOREST PROTECTION

A number of activities fall under this heading. They include regular patrolling of the total area by the project personnel assisted by officials from the Ministry of Forestry, thereby creating jobs for people living around the Sungai Wain forest. Other activities include replanting of fruit trees suited for orangutans in areas that experienced damage (e.g. a few helipads constructed for seismic exploration and some previously shifting cultivation fields), totalling so far some 60 ha with 58 different fruit tree species. Information is also provided to people living around the Sungai Wain forest and in Balikpapan, through regular meetings and other means, in cooperation with local village heads.

RESEARCH

Topics of research include the influence of orangutans on the germination of seeds which they consume, social interactions between individuals and the socialized groups before and after release, identification of orangutans by means of fingerprints, hair samples and morphology and the medical problems of orangutans. In the Sungai Wain forest, a botanical plot will be established in cooperation with the Smithsonian Institution, Washington, D.C., for long-term monitoring of forest succession and the influence of newly introduced orangutans on succession.

PROCESSING OF THE ORANGUTANS

The project is very active in locating and confiscating illegally held orangutans. In just over two years, the project has received 114 orangutans. Among these were 20 orangutans from Taiwan, the large majority confiscated relatively close to Balikpapan and Samarinda. Figure 2 shows the different ways orangutans arrive at the Wanariset station. It also shows the minimum number of orangutans expected to be processed by the project in the coming three to four years.

After arrival at the station the orangutans are processed as shown in Figure 3. The monitoring of the animals after their release sometimes leads to the return of one or more for treatment of wounds, such as those inflicted by clouded leopards and by falls from trees.

As of 1994, a total of 114 orangutans had arrived at the Wanariset station. Forty animals have been released, the first group of which has now been in the forest for almost two years. Two orangutans died in the forest; one small one was eaten by a clouded leopard

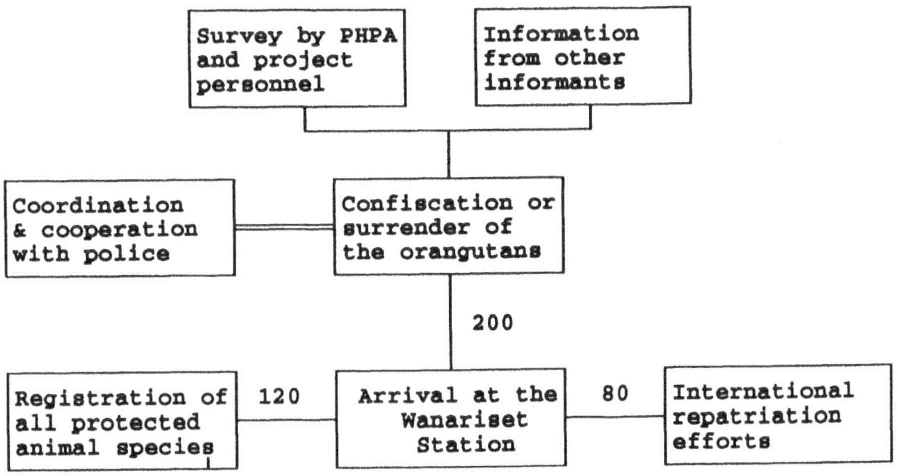

Figure 2. Schematic presentation of the processing of orangutans from different sources for reintroduction through the Wanariset project.

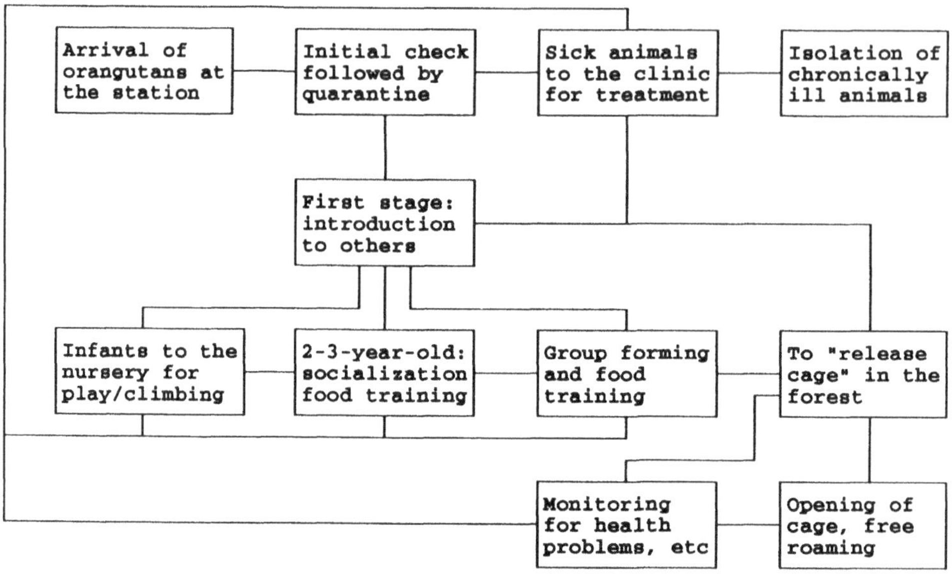

Figure 3. Schematic presentation of the rehabilitation process at Wanariset.

and another was killed by a wild boar when it fell from a tree as a result of a branch breaking. All the other orangutans have made nests and are independent of further food provisioning. In the fruiting season, released orangutans tend to remain in the forest after only two weeks, whereas, outside the fruiting season, this period may be closer to three months.

At the station, another 13 animals have died. Three of these were brought in with serious wounds, e.g., one had 60 cuts from a machete, another had part of its intestines protruding from a wound and the third one had broken arms with open wounds and a skull fracture. Seven others died within a few weeks of arrival. The autopsy reports suggested that some died of liver problems due to hepatitis, whereas, others likely died of tuberculosis. Two infants died from enteritis caused by massive worm infestations and probably salmonella. One infant died from a fall in its cage which caused a rupture of the lung.

Treatment of tuberculosis resulted in disappearance of the positive reaction and complete clinical recovery of the animals. Most of the orangutans developed antibodies for hepatitis after six months in quarantine. Anorexia sometimes occurred with isolated animals. Other problems encountered were blindness, paralysis and in one case, hyperactivity leading to the infliction of wounds. One case of filaria lead to swollen lymph nodes that persisted for over a year and a wide range of other parasites was common as well.

The number of locally confiscated orangutans (East Kalimantan) is slowly decreasing, as can be seen in Figure 4. This has occurred in spite of the continually increasing efforts to locate more illegally held orangutans. At the same time, the average age of newly confiscated orangutans is dropping significantly. One can draw a number of conclusions from the graph in Figure 4. Firstly, the project is having a significant impact on the number of people holding orangutans. Secondly, poaching of infant orangutans is still going on in East Kalimantan.

Most of the orangutans are caught under circumstances that are related to the loss of their habitat, such as a group becoming isolated in a patch of forest because of mining activities, orangutans living in areas that are clear-cut for timber estate establishment and orangutans that are captured by shifting cultivators. Conversely, the hunting of orangutans

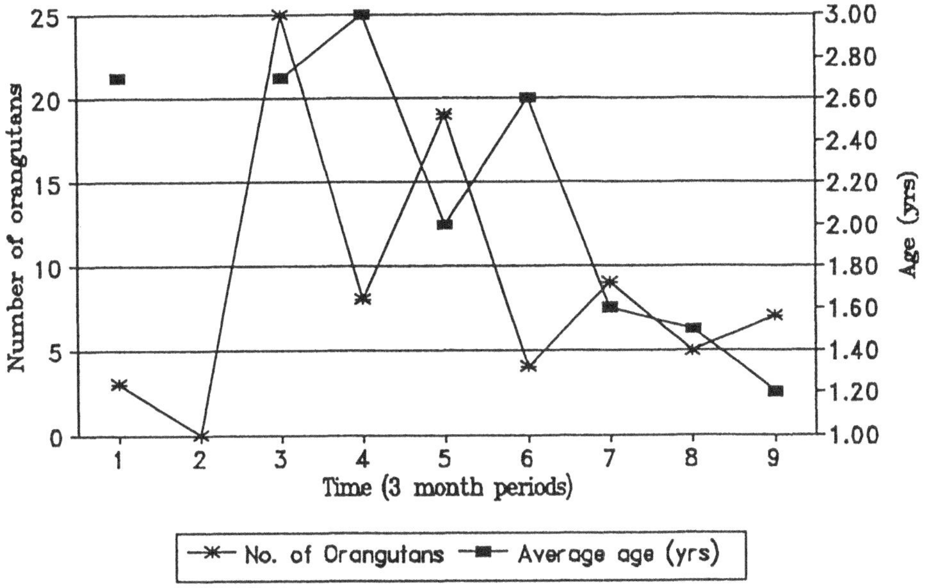

Figure 4. Fluctuations in the number and age of orangutans received at the Wanariset station, East Kalimantan, Indonesia, over a 27-month period.

by persons looking like Gaharu seems to have stopped. Deep in the interior, people sometimes still offer an orangutan for sale, but almost everywhere in East Kalimantan people realize the legal risk. At present, there are 10 people in jail in East Kalimantan for offenses against orangutans of the Wanariset project. The imprisonments were widely publicized locally. It is likely, therefore, that recently encountered young orangutans in Taiwan and Hong Kong originate in other parts of Borneo (see Eudey, this volume).

CONCLUSIONS

1. There are still a large number of captive orangutans that require rehabilitation and more action is needed to eradicate poaching.
2. The process of reintroducing orangutans into a habitat without a natural presence of orangutans seems to be successful after socialization and food training.
3. The large number of medical problems in captive orangutans necessitates high quality medical facilities and staffing.
4. In view of the frequent medical problems among the ex-captives, they should not be introduced into a forest with an existing population of orangutans.

REFERENCES

MacKinnon, J., 1992, Species survival plan for the orangutan. Pp. 209-219 in: (Eds. G. Ismail, M. Mohamed, and S. Omar), *Proc. Int. Conf. Forest Biol. Conserv. Borneo*, Kota Kinabalu: Yayasan Sabah.

Rijksen, H.D. (1978) *A Field Study on Sumatran Orang Utans* (*Pongo Pygmaeus Abelii* Lesson 1827), *Ecology, Behaviour and Conservation*, 78-2. Wageningen, H. Veenman and Zonen B.V.

SECTION THREE – ORANGUTAN DEMOGRAPHY AND POPULATION AND HABITAT VIABILITY ANALYSIS

Introduction

What is the probability that a species will survive given a certain degree of protection and a measurable set of environmental circumstances? How many individuals must be conserved, and how large and ecologically diverse an area must be committed? Answers to these complex questions can be obtained in part through the methods of population biology and the mathematics of Population and Habitat Viability Analysis (PHVA). It is now recognized that endangered species, whose numbers are increasing, must be studied as *metapopulations*, comprised of *core* and *satellite populations* (Primack, 1993). Core populations are crucial because they are the primary source of migrants for satellite populations, whose sizes may fluctuate and periodically crash. Destruction of core populations and/or the elimination of dispersal routes between the core and satellite populations thus have a profound impact on the satellites. Shaffer (1981) described a species' *minimum viable population size* as the number of individuals required to yield a 99% probability of survival for at least 1000 years. He noted, however, that this calculation is complicated by several factors. The likelihood of extinction in both the short and the long term is influenced by the complex interactions among a species' morphology, physiology, behavior, genetics, distribution, habitat, and demography. Conservation biologists further point out that there is a tendency for small populations to decline toward extinction, i.e., *extinction vortices* (Primack, 1993). Environmental and demographic variability coupled with a loss of genetic diversity is the cause of this downward spiral in small populations.

A strategy to oppose extinction vortices is the development of population and habitat management procedures that minimize the effects of small population size and habitat fragmentation, degradation, and isolation. These plans must meet the species' requirements for survival in both typical and exceptional years, i.e., when primary food resources are abundant and when they are unavailable. The Conservation Breeding Specialist Group (CBSG) of the International Union for the Conservation of Nature (IUCN) developed PHVAs to meet these diverse requirements, yielding separate conservation plans tailored for individual species and their ecological conditions (e.g., tigers, gibbons and orangutans). The IUCN cooperates with host countries in developing PHVA workshops. The involvement of government officials, indigenous scientists, and interested lay people insures that recommendations are meaningful and applicable to the host country. Section 3 contains the results of the first PHVA Workshop for Orangutans held in Medan, Sumatra in 1993.

Soemarna and his co-workers provide the background for the Workshop which was jointly organized by the CBSG of the IUCN and the Indonesian Directorate of Forest Protection and Nature Conservation. They succinctly describe the characteristics of orangutans that make them particularly susceptible to extinction (e.g., a relatively long interbirth interval), many of which are elaborated upon in subsequent papers in this section.

Faust and his colleagues emphasize that different habitats and microhabitats currently occupied by orangutans support different numbers of animals. Through the use of a Geographic Information System (GIS), they analyze the likely ramifications of implementing various policies, such as building a road that bisects a reserve, and thereby, threatens the dispersal routes of orangutans in the area. Other variables that are examined with GIS are the effects of deforestation and changes in the boundaries of the Gunung Leuser National Park, North Sumatra.

Leighton and his co-authors describe the influence that environmental variation (e.g., drought, loss of key food species) exerts on the future of orangutan populations, particularly noting the impact of losing animals of different age/sex classes. Through the use of VORTEX modelling, they estimate the changes in population size contingent upon differing rates of reproduction and mortality.

van Shaik and his collaborators propose that the orangutans inhabiting the Gunung Leuser National Park are the most important core population in Sumatra. This population is vital in terms of its overall size, likelihood of continued protection, and as a major source for dispersal of orangutans to other areas.

Rijksen and his colleagues describe the isolated orangutan populations found in the fragmented forests of Borneo. Using results from the PHVA workshop, they describe the impact of population isolation on the survival of Bornean orangutans and the alarming decline in this subspecies which has continued unabated in recent years.

The PHVA workshops culminate with the development of models and guidelines for the host countries. Soemarna and his co-authors summarize the results of the PHVA and present the conclusions that follow from their deliberations. The result is a list of concrete recommendations to government officials of both Sumatra and Borneo which illuminate the probable outcome of various economic and environmental decisions. One noteworthy recommendation is that the supplementation of natural populations will not enhance the population viability. The most critical problem which the governments must resolve is the loss and degradation of orangutan habitat.

A final paper in Section 3 which was not a part of the PHVA workshop is that of van Schaik and colleagues. These investigators used line transects of nests to estimate the geographic distribution and densities of orangutans in a wide range of habitats in Sumatra. They describe the relationship between habitat type and population density and then present preliminary evidence regarding the importance of floodplain forests for the conservation of orangutans more widely distributed.

REFERENCES

Primack, R. B., 1993, *Essentials of Conservation Biology*, Sunderland, MA: Sinauer Associates Inc.

Shaffer, M. L., 1981, Minimum population sizes for species conservation. *Bioscience* 31: 131-134.

9

INTRODUCTION TO THE ORANGUTAN POPULATION AND HABITAT VIABILITY ANALYSIS (PHVA) WORKSHOP

K. Soemarna,[1] W. Ramono,[1] and R. Tilson[2]

[1] Indonesian Directorate General of Forest Protection and Nature Conservation
[2] IUCN/SSC Conservation Breeding Specialist Group and The Minnesota Zoo

ABSTRACT

The first Population and Habitat Viability Analysis (PHVA) Workshop for Orangutans (*Pongo pygmaeus*) was held in Medan, North Sumatra, Indonesia on 18-20 January 1993. PHVA workshops use a computer model (VORTEX) to simulate the deterministic and stochastic, or random, processes that threaten small populations and to explore what effects various management options may produce on the survival of the population. The workshop was coordinated by the Indonesian Directorate General of Forest Protection and Nature Conservation (PHPA) and the Survival Service Commission (SSC) of the International Union for the Conservation of Nature (IUCN) Conservation Breeding Specialist Group (CBSG). Close to 40 people were in attendance, primarily Indonesians and a significant number of orangutan field workers.

INTRODUCTION

The Workshop focused on the status of wild populations of orangutans on Sumatra and Borneo, with major emphasis on the Sumatran population *(Pongo pygmaeus abelii)*. Three working groups were established: Orangutan Distribution and Status in Sumatra, Orangutan Distribution and Status in Borneo and Life History Characteristics and Vortex Modelling.

SUMATRA

Estimates of habitat and population numbers for orangutans were derived in the working groups through consensus of the field biologists; for Sumatra, the estimates were

The Neglected Ape, Edited by Ronald D. Nadler et al.
Plenum Press, New York, 1995

probably more reliable because of the larger database than they were for Bornean populations of orangutans. For Sumatra, the exact boundaries of orangutan distribution are not known, but there are several distinct populations. The Greater Leuser orangutan population, which extends beyond the national park boundaries, is thought to cover approximately 11,710 km^2 and number about 9,200 animals. Within the more restricted boundaries of Gunung Leuser National Park, the total number of orangutans is estimated to be about 5,800. This population was judged to be among the best in the world, in terms of numbers and the potential for protection, and Gunung Leuser National Park was considered to be vital to the long-term survival of the Sumatran orangutan.

BORNEO

For Borneo (Kalimantan, Sabah and Sarawak), the known distribution of orangutans comprises eight regions with currently isolated populations. No published or unpublished data are available on population numbers for any of these areas, with the exception of Gunung Palung, Kalimantan. The total area of orangutan habitat on Borneo was calculated at 22,360 km^2, and the estimate of total population numbers ranged from a minimum of 10,282 to a maximum of 15,546. These figures suggest a more serious decline has occurred in the Bornean population than was previously thought.

LIFE HISTORY AND EXTINCTION

The working group on Life History Characteristics and Vortex Modelling relied primarily on unpublished data collected at Ketambe, Gunung Leuser National Park, and Tanjung Puting National Park, Kalimantan. The orangutan appears to be the ultimate *K*-selected species, in that survivorship is high, interbirth interval is long (mean = 8 years) (Galdikas and Wood, 1990), and the female makes a high parental investment in her offspring. Vortex modelling indicated that adult females are the most valuable members of an orangutan population and that the death of an adult female has the greatest influence on increasing extinction rates of all life history variables.

RECOMMENDATIONS

A comprehensive set of recommendations for the conservation management of orangutans were developed:

- Management strategies for orangutans and protected areas occupied by orangutans in Sumatra and Borneo include stronger protection measures for orangutans and forests, prevention of fragmentation (both species and habitat) and restoration of degraded habitat.
- Reintroduction of captive orangutans into wild populations has no conservation value in terms of enhancing population viability and may even have negative effects through the introduction of diseases or inappropriate genetic subspecies.
- Questions concerning the role that captive populations of orangutans may have in reintroduction schemes were not specifically addressed; they should be examined at a workshop of the IUCN/SSC Reintroduction Specialist Group.

- Establishment of new viable populations of ex-captive orangutans in habitat formerly occupied by orangutans, but where they do not now occur, may contribute to the viability of metapopulations.
- Updated recommendations were presented on medical procedures used during quarantine of orangutans intended for reintroduction.
- Additional surveys and more comprehensive map-linked databases are needed for Borneo before an in-depth PHVA can be performed for Bornean orangutan populations.

The recommendations of the working groups formed the basis for developing the *Indonesian Orangutan Action Plan.* Information generated by each working group and their conclusions are presented in the following reports.

ACKNOWLEDGMENTS

The database, interpretations and recommendations contained in the papers in this section were derived at the Orangutan PHVA Workshop held in Medan, North Sumatra in January 1993, which was coordinated by the Indonesian Directorate General of Forest Protection and Nature Conservation (PHPA) and the Survival Service Commission (SSC) of the International Union for the Conservation of Nature (IUCN) Conservation Breeding Specialist Group (CBSG). The authors of this chapter and those of the other chapters in this section acknowledge all of the participants who contributed to this effort: Sutisna Wartaputra, Raja Inal Siregar, Adjisasmito, Muladi Wijaya, Suharto Djojosudharmo, Kunkun Gurmaya, Kuppin Simbolon, Tatang Mitra Setia, Oliver Nelson, R. Steenbeck, Antong Hartadi, S. Poniran, Dolly Priatna, Sukianto Lusli, Totak Sediyantoro, Ahpal Effendi, Suci Utami, Thomas Faust, Lori Perkins, Ardith Eudey, Jito Sugardjito, Ian Singleton, Ling-Ling Lee, Norm Rosen, Lizzie Green, Gary Shapiro, Marcus Phipps, Li-hung Lin, Carel van Shaik, Herman Rijksen, Mark Leighton, Dondin Sajuthi, Agus Lelana, William Karesh, Mike Griffiths, Ulysses Seal, Kathy Traylor-Holzer and Richard Tenaza. We also thank the sponsors of this workshop: the IUCN/SSC CBSG, Indonesian PHPA, Zoo Atlanta, the Orangutan AAZPA Species Survival Plan, Taronga Zoo and the Orangutan ASMP (Australia) and Jersey Wildlife Preservation Trust, U.K.

REFERENCES

Galdikas, B.M.F. and Wood, J., 1990, Great ape and human birth intervals. *Am. J. Phys. Anthropol.* 83: 185-192.

10

USING GIS TO EVALUATE HABITAT RISK TO WILD POPULATIONS OF SUMATRAN ORANGUTANS

T. Faust,[1] R. Tilson,[2] and U. S. Seal[3]

[1] Conservation Biology Program, University of Minnesota
St. Paul, Minnesota 55108
[2] Minnesota Zoo
Apple Valley, Minnesota 55124
[3] IUCN/SSC Captive Breeding Specialist Group
Apple Valley, Minnesota 55124

ABSTRACT

A geographical information system (GIS) was used to map the current distribution of the Sumatran orangutan (*Pongo pygmaeus abelii*) population in Gunung Leuser National Park (GLNP), North Sumatra. From this, vegetation cover of the orangutan's distribution was calculated. Vegetation cover and population density estimates within different vegetation types were used to estimate overall population numbers. In addition, the effects of potential land-use changes on the orangutan's distribution were analyzed using GIS. The land-use scenarios modelled with GIS were: (1) forest loss in and surrounding GLNP; (2) road construction through GLNP; (3) the loss of all low-hill forest in GLNP; and (4) changes in the legal status of park boundaries. This geographical analysis helped guide recommendations for the conservation management of wild orangutans in GLNP.

INTRODUCTION

The development of effective conservation management plans for wild populations of orangutans is predicated on the ability to model their population dynamics and habitat. Leighton et al. (1993) used "VORTEX," a computer simulation program, to model the population dynamics of wild orangutans to help guide the development and direction of management recommendations. In this paper, we document the use of geographical information systems (GIS) to model changes in orangutan habitat for the purpose of guiding conservation management recommendations for the Sumatran orangutan (*Pongo pygmaeus abelii*) population in Gunung Leuser National Park (GLNP).

The Neglected Ape, Edited by Ronald D. Nadler et al.
Plenum Press, New York, 1995

GIS are computer programs that allow for the encoding, analysis, and display of spatial data (Smith et al., 1989). GIS programs are currently being employed in many areas of conservation; in the U.S. and Australia, GIS is used to determine gaps in protected networks where species are not being protected (Davis et al., 1990; Cocks and Baird, 1991; Scott et al., 1993). Other researchers are combining GIS technology with radio-telemetry to understand how species utilize their environment (Haslett, 1990). In addition, GIS is being used to map the distribution of many species (Margules and Stein, 1989; Yonzon and Hunter, 1991).

Through the use of GIS, the effects of predicted changes in orangutan habitat can be analyzed to estimate their effect on the orangutan's distribution. Current threats to the wild Sumatran orangutan population are logging, shifting agriculture, and poaching (Rijksen, 1978; Aveling, 1982; Sugardjito and van Schaik, 1993; Tilson et al., 1993). Through the use of GIS, logging and shifting agriculture in and around GLNP is evaluated as to its effect upon orangutan distribution and thus, the eventual population numbers of the orangutans. These results, in or outside the current park boundaries, are evaluated as to their acceptability. Recommendations can then be made to alleviate their effect on the Sumatran orangutan population.

METHODS

Atlas-GIS (Strategic Mapping Inc., Santa Clara, CA) was used to map GLNP. Geological and Land-use/Forest Status maps from the Ministry of Forestry were used to create the spatial database. The thematic layers incorporated into the database were protected area boundaries, vegetation cover, roads, rivers and settlements.

Indonesian Land-use and Forest Status maps (series RePPProT 1988; scale 1:250,000) were used for protected area boundaries and vegetation cover. Only vegetation cover within GLNP was digitized from these maps. The forest types included in the database were: (1) lowland forest (below 1000 meters); (2) sub-montane forest (between 1000-2000 meters); (3) montane forest (above 2000 meters); and (4) inland and mangrove swamp. In addition, other vegetation types such as bush and agriculture (along with minor forest types such as calcareous forest) were included in the database. Roads, towns and rivers were digitized from geological maps (Geological maps 1988; scale 1:250,000).

To estimate vegetation cover outside of GLNP, the World Conservation Monitoring Center (WCMC) provided a digitized coverage of vegetation on Sumatra (series RePPProT 1990; scale 1:2.5 million). WCMC's database only distinguishes between lowland forest (below 1000m), montane forest (above 1000m), inland and mangrove swamp, and non-forest. Thus, all areas without forest, such as bush and agriculture, are treated as a non-forest category (Cox and Collins, 1991).

Orangutan distribution was obtained at the Orangutan Population and Habitat Viability Analysis (PHVA) Workshop held in Medan, Sumatra, January 18-20 (Tilson et al., 1993). The Sumatran Orangutan Working Group report provided the best known and most current information available regarding orangutan distribution data in and surrounding GLNP. These data were based on surveys by Rijksen (1978) and other field research by members of the working group. The estimated distribution database was then entered into Atlas-GIS.

Van Schaik (unpublished manuscript) found 300m to be a critical determinant of orangutan densities: below 300m, densities are estimated to be 2.5 orangutans/km^2; above 300m, densities are estimated to be 1.8 orangutans/km^2. The GIS database for GLNP had elevation only in relation to forest type. Lowland forest is from 0-1000m, submontane from 1000-2000m and montane is greater than 2000m. In order to estimate the location of the 300m contour line, a 500m contour line was entered into the GIS database from a working

Table 1. Estimated orangutan densities at different altitudes and vegetation types

Vegetation altitude	Orangutans/km^2
0 - 500m	2.5
500 - 1000m	1.8
1000 - 2000m	0.4
Secondary forest	0.5
Logged	1

map used at the Orangutan PHVA Workshop (500m was used, since this working map did not have a 300m contour line).

The number of vegetation types within the orangutan's distribution was approximated using the split command in Atlas-GIS. Also using Atlas-GIS, several possible scenarios which would alter orangutan habitat were projected to determine their effects on both the distribution and total number of orangutans.

RESULTS AND ANALYSIS

Vegetation cover in the orangutan's distribution in and surrounding GLNP was multiplied by density estimates for each vegetation type to obtain population numbers for the greater Gunung Leuser area (Table 1). The estimated population numbers were then corrected because: (1) remote sensing has not yet found a method to differentiate between disturbed and undisturbed habitat, (2) many of the areas may not have orangutans because of hunting and other human-influenced factors and finally, (3) while there has been and will continue to be severe deforestation pressures around GLNP, the maps represent only past vegetation and forest cover. Thus, the original uncorrected number was multiplied by 0.75 to obtain the estimated corrected number (van Schaik, personal communication).

Current Population

Table 2 contains the results of the GIS vegetation analysis of the orangutan's distribution.

Table 2. Vegetation analysis for the greater Gunung Leuser population

	Western population		Eastern population	
Vegetation type	Area (km^2)	Number	Area (km^2)	Number
Agriculture	12	6	11	5
0 - 500 m	832	2079	237	592
500 - 1000 m	2312	4161	1831	3015
1000 - 2000 m	2846	1138	2021	809
Above 2000 m	1299	0	161	0
Non-forest	60	0	74	37
Swamp	51	255	0	—
Logged	0	—	176	176
Bush	0	—	4	0
Total orangutans		7639		4636

This analysis covers the greater Gunung Leuser orangutan population (inside and outside GLNP) and does not include other possible areas where orangutans may occur in Sumatra. The Working Group decided that the greater Gunung Leuser population was the only viable population in the near future, as the other confirmed sites are all too small.

The Kotacane-Blangkejeren road bisects GLNP, dividing the resident orangutan population into two sub-populations (Fig. 1). The western population is estimated at 5700 individuals (7639 uncorrected) and the eastern population at 3500 (4636 uncorrected), for a total population of 9200 individuals.

The current orangutan distribution was used to analyze the effects of the following scenarios on the distribution and number of orangutans: (1) the loss of all production and protection forest surrounding the park, (2) the construction of roads through GLNP, (3) the deforestation of all remaining low-hill forest within park boundaries and (4) encroachment and the loss of protection status of GLNP.

The Loss of All Production and Protection Forest

Shifting agriculture and logging occur in most areas surrounding GLNP. Therefore, the Sumatran Orangutan Working Group decided to model the effects of losing all forest surrounding GLNP on the orangutan distribution and population numbers. Projecting this scenario into the GIS database, in which all production and protection forest surrounding the park is lost, revealed that both populations lose a significant amount of habitat (Fig. 2). This loss of habitat would result in a significant population reduction.

The western population would decline to approximately 3500 individuals and the eastern population would decline to approximately 2400 individuals. The total population would be reduced to an estimated 5900 individuals, an overall reduction of 35% (Table 3).

It was concluded that the GLNP provides the only future secure habitat for orangutans in Sumatra at the present time (van Schaik, personal communication). Thus, the orangutan distribution obtained from this scenario is used as the base distribution for the following three scenarios: (1) road construction, (2) the loss of all low-hill forest in Gunung Leuser National Park and (3) changes in the legal status of park boundaries.

Road Construction

The Provincial government of Aceh has proposed building a road through GLNP from Bukit Lawang (in the eastern side of the park) to Kutacane (in the area between the eastern and western park sections), in the interest of developing ecotourism. Tourism currently is concentrated around the orangutan rehabilitation center at Bohorok, North Sumatra (Poniran, personal communication). The government of Aceh has proposed that a new road be built to develop their part of GLNP for easier and quicker access by tourists and visitors from the city of Medan, located east of GLNP (Fig. 3).

If the road is built, it would divide the eastern population of orangutans in GLNP into two sub-populations. The original eastern population would be divided into a northeastern population of 1825 orangutans and a southeastern population of 575 orangutans. Both populations would be isolated without the ability to exchange genetic material. Further, the southeastern population of 575 individuals would be at greater risk due to intrinsic factors because of its small size. Given these potential risks on the eastern orangutan population, it was recommended in the *Sumatran Orangutan Action Plan* that the road should not be built (Tilson et al., 1993).

Another road is proposed that would connect Lubuk Keranji to Pucuk Lembang and isolate the Kluet swamp in the western section of the park (Griffiths, pers. comm.; Fig. 3). This area contains the only swamp forest in the park, and of all vegetation types, swamp

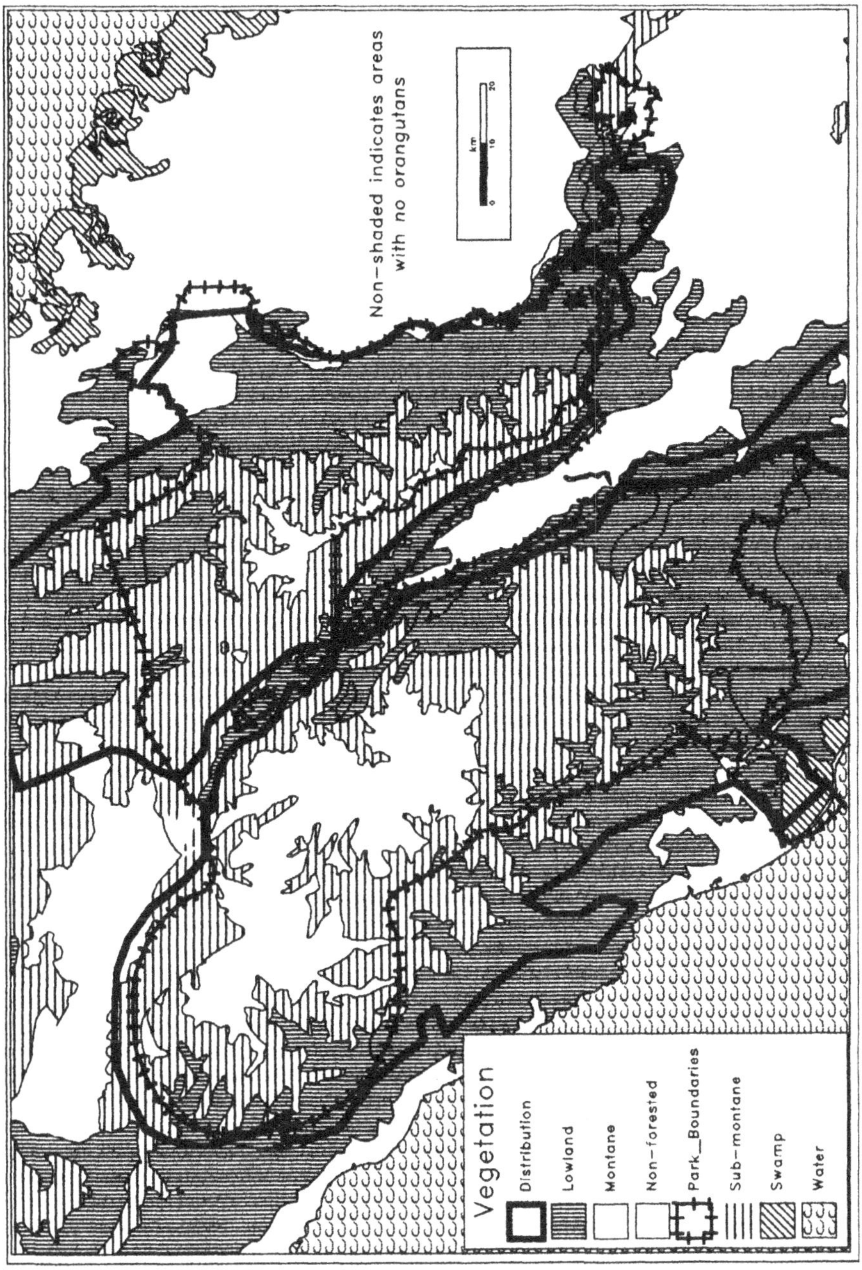

Figure 1. Current orangutan distribution around Gunung Leuser National Park. Vegetation cover with distribution.

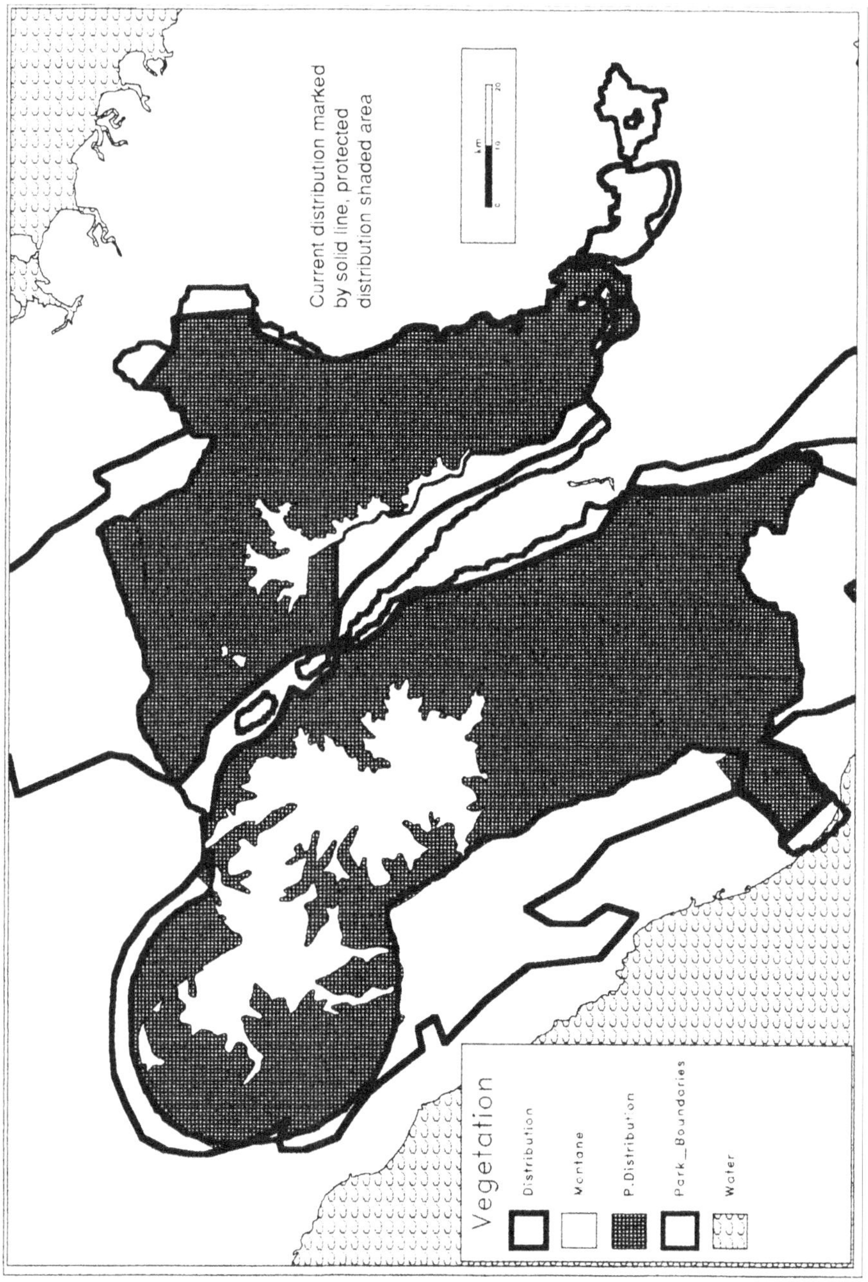

Figure 2. Current protected orangutan populations in Sumatra. Present and protected orangutan distribution.

Table 3. Vegetation analysis for the Gunung Leuser National Park population

	Western population		Eastern population	
Vegetation type	Area (km^2)	Number	Area (km^2)	Number
Agriculture	12	0	11	0
0 - 500m	216	541	237	592
500 - 1000m	1576	2838	1038	1868
1000 - 2000m	2427	971	1344	538
Above 2000m	1299	0	161	0
Non-forest	4.6	0	37	
Swamp	51	255	0	
Logged	0		176	176
Bush	0		4	0
Total orangutans		4605		2400

forest harbors the highest density of orangutans (van Schaik, pers. comm.). The isolation of the Kluet swamp would create a sub-population of only 350-450 orangutans. The rest of the western orangutan population would then have only slightly more than 4000 individuals. Again, due to intrinsic factors that threaten small isolated populations, the participants of the PHVA workshop recommended that the Lubuk Keranji to Pucuk Lembang road not be built. Likewise, they recommended that no new roads be built through National Park land (Tilson et al., 1993).

The Deforestation of All Remaining Low-Hill Forest within Park Boundaries

Within the boundaries of GLNP, deforestation from logging and shifting agriculture is presently underway. Rijksen (pers. comm.) reported that an area within park boundaries became a logging concession and the Ministry of Forestry was powerless to cancel the contract. Along the road from Kotacane to Blangkejeren, shifting agriculture has deforested both sides of the road for two kilometers at many sites, which is an impassible barrier for many species, particularly orangutans. Given logging and agricultural pressures which occur mainly in the low-hill areas, the Sumatran Orangutan Working Group decided to model the amount of available habitat in the event that all forest below 500 m were lost (van Schaik, personal communication; Fig. 4).

Due to loss of high quality habitat, the eastern population of orangutans would decline from 2400 to 1800 individuals and the western population would decline from 3500 to 2850 individuals. Given this reduction in orangutan numbers, workshop participants recommended in the *Sumatran Orangutan Action Plan* that deforestation within park boundaries should be stopped (Tilson et al., 1993).

The Loss of Forest Due to Encroachment and Changes in the Legal Status of Gunung Leuser National Park

In addition, to deforestation within GLNP boundaries, there has also been the loss of forest due to changes in the legal status of its boundaries. The boundary of GLNP is in a constant state of change and the total size of GLNP has decreased through the years. For example, the Sebolangit region was originally sited within the Park boundaries, but was reclassified to protection forest and is now subject to logging. The Sikundur reserve in GLNP was selectively logged in the 1970's. Other areas, have changed from HSA to HL or HP and been selectively logged. Due to these pressures, the Sumatran Orangutan Working Group

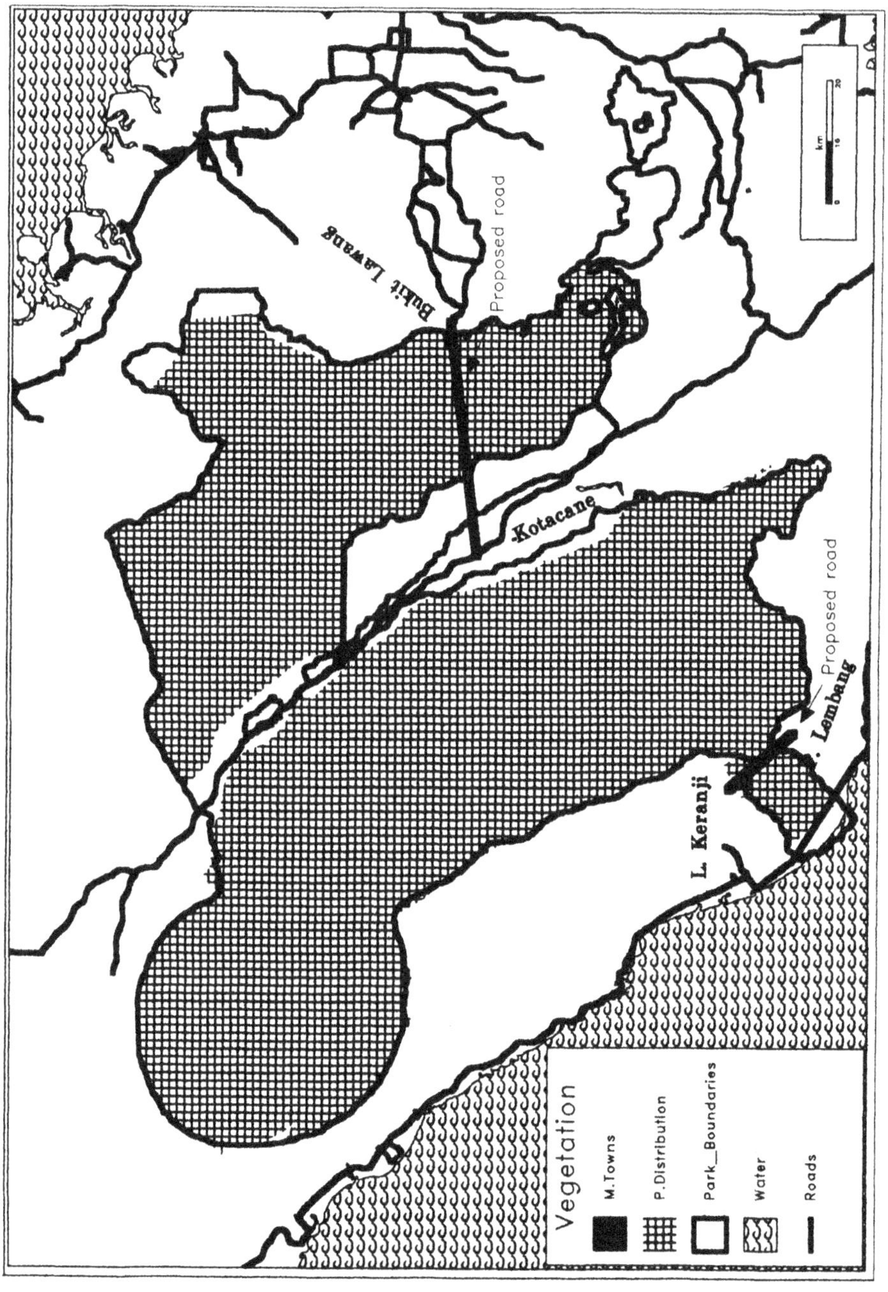

Figure 3. Current protected orangutan populations in Sumatra. Possible fragmentation of orangutan distribution due to proposed roads.

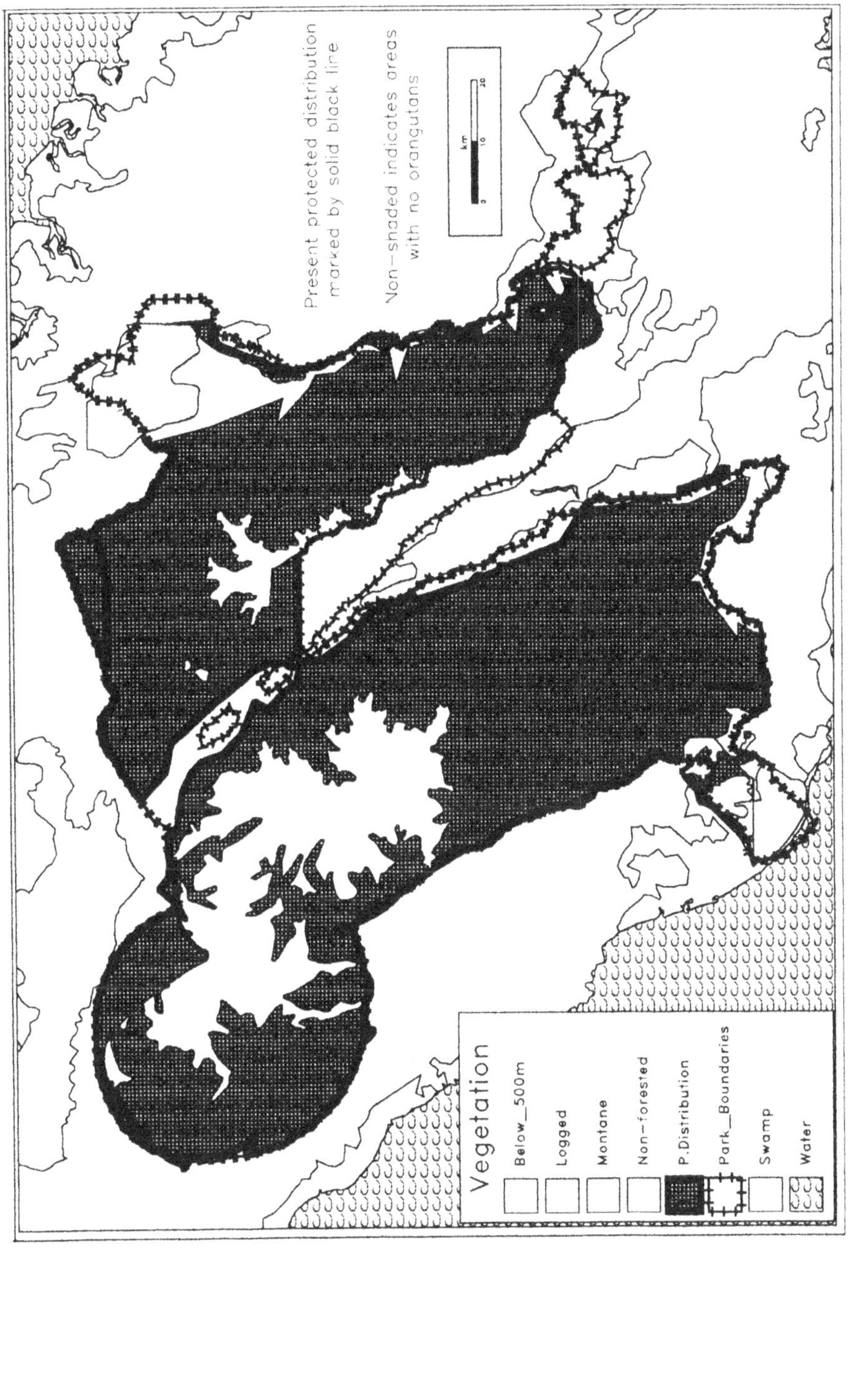

Figure 4. Current protected orangutan populations with loss of low-hill forest. Orangutan distribution with the loss of all forest below 500m.

Table 4. Encroachment and changes in legal status vegetation analysis for Gunung Leuser National Park population

	Western population		Eastern population	
Vegetation type	Area (km^2)	Number	Area (km^2)	Number
Agriculture	12	0	11	0
0 - 500m	13	24	54	94
500 - 1000m	948	1280	783	1057
1000 - 2000m	2034	610	831	250
Above 2000m	1299	0	161	0
Non-forest	4.6	0	37	0
Swamp	0	0	0	0
Logged	0	0	0	0
Bush	0	0	4	0
Total orangutans		1925		1400

decided to model a scenario that would examine the continuing loss of protection status for GLNP (van Schaik, pers. comm.).

The loss of forest to encroachment and loss of protection status would have a major impact on the currently estimated orangutan populations. The eastern population of orangutans would be reduced to about 1,400 individuals and the western population would be reduced to about 1,925 individuals. In addition, the western population would be fragmented into two separate populations; a northwestern population of 325 individuals and a southern population of 1600 individuals. There would be little chance of genetic exchange between the populations (Table 4). Given this reduction in the number of individuals (over 60%), the workshop participants recommended in the *Sumatran Orangutan Action Plan* that an absolute commitment be made to maintain the protection of the park (Tilson et al., 1993).

DISCUSSION

The conservation of wild orangutans will require an even better understanding than we now have of its distribution and preference of habitat types within that distribution. From the information provided in this paper, population size and fragmentation of the orangutan population in GLNP can be estimated. This estimate can serve as a starting point for examining how humans impact the orangutan's habitat and its numbers. GIS programs can play a large role in these types of analysis. In this paper, we document the use of GIS in analyzing the effects of changes in the orangutan's habitat on future numbers and distribution.

At the Orangutan PHVA Workshop, the use of GIS permitted estimates to be made of habitat types within the orangutan's present distribution, as well as an analysis of threats to its habitat. From these analyses, it was shown that if current trends in the land-use practices of the Indonesian Ministry of Forestry continue, there will be significant changes in the overall number and distribution of the wild Sumatran orangutan populations. For example, if all orangutan habitat outside of GLNP is lost, the total population will be reduced by over 30%. In addition, the analysis revealed that if current land-use trends continue inside GLNP, the population will decline further (Table 5).

From this geographical analysis, the Sumatran Orangutan Working Group at the Orangutan PHVA Workshop was able to make several recommendations, which gained the consensus of all workshop participants, to improve the long-term viability of wild Sumatran orangutans. The four main recommendations that came from the spatial analysis were: (1) restore the connection between the eastern and western populations of orangutans in GLNP;

Table 5. Orangutan population estimates for possible land-use scenarios associated with Gunung Leuser National Park (GNLP)

Possible land-use patterns	Population numbers
Current	9200
Loss of all forest surrounding GLNP	5900
Loss of low-hill forest in GLNP	4650
Changes in legal status	3325

(2) do not permit any new roads to be built through GLNP; (3) reduce future forest loss in GLNP; and (4) maintain a commitment to the protection of GLNP.

The database on the distribution of wild orangutans in GLNP must be expanded to develop fully the management strategies that will ensure the long-term viability of this species. Important areas for future geographical assessments are: (1) the updating of the vegetation cover in the database with current remote-sensed data, (2) an accurate description of how orangutans use their landscape (for example, research needs to be conducted on orangutan survivorship without access to the high productivity lowland forests) and (3) the identification of geographic areas that are under the most threat of alteration by humans. If these and other areas are incorporated into the geographical database, conservation strategies for the Sumatran orangutan will be enhanced.

CONCLUSION

The results of the geographic analysis of the wild Sumatran orangutan population in GLNP indicate that the present population is fragmented into two subpopulations. These two subpopulations are considered safe as long as no further development occurs in the GLNP. To safeguard these populations for the future, the Directorate General for Forest Protection and Nature Conservation (PHPA) must guard against threats that further subdivide or decrease these populations. Without such protective measures, the current population would be further reduced to a size that would severely threaten its long-term viability. As long as the present situation is maintained, the wild northern Sumatran populations of orangutans in GLNP appear to be viable and self-sustaining.

ACKNOWLEDGMENTS

We are grateful for the generous contribution of data from the PHVA Workshop participants upon which this manuscript is based, and for the insights provided by the field biologists and PHPA staff working in Gunung Leuser National Park (see Soemarna, Ramono and Tilson, this volume, for common acknowledgements). This manuscript is part of a M.S. thesis of the first author at the University of Minnesota, Faculty of Conservation Biology, and was supported by grants from the Minnesota Zoo and the Survival Service Commission of the International Union for the Conservation of Nature (IUCN/SSC) Conservation Breeding Specialist Group.

REFERENCES

Aveling, R., 1982, Orang utan conservation in Sumatra, by habitat protection and conservation education. Pages 299-315 in J. Schwartz (ed.) *Orang-utan Biology*. New York: Oxford University Press.

Cocks, K. and I. Baird, 1991, The role of geographical information systems in the collection, extrapolation and use of survey data. Pages 74-78 in C. Margules and M. Austin (eds.) *Nature Conservation: Cost effective biological surveys and data analysis*. Canberra, Australia: CSIRO.

Cox, R. and Collins, M., 1991, Indonesia. Pages 141-165 in M. Collins, J. Sayer, and T. Whitmore (eds.) *The conservation atlas of tropical forests: Asia and the Pacific*. New York, N.Y.: Simon Schuster Co.

Davis, F., D. Stoms, J. Estes, J. Scepan and M. Scott., 1990, An information systems approach to the preservation of biological diversity. *International Journal of Geographical Information Systems* 4(1): 55-78.

Haslett, J., 1990, Geographic information systems: a new approach to habitat definition and the study of distributions. *Trends in Ecology and Evolution* 5(7): 214-218.

Leighton, M. et al., 1993, Orangutan life history and Vortex analysis. *Orangutan Population and Habitat Viability Analysis Final Report*. Apple Valley, MN: CBSG.

Margules, C. and J. Stein., 1989, Patterns in the distributions of species and the selection of nature reserves: an example from *Eucalyptus* forests in South-eastern New South Wales. *Biological Conservation* 50: 219-238.

Rijksen, H., 1978, *A field study on Sumatran orangutans* (Pongo pygmaeus abelii *Lesson 1827)*. Wageningen: H. Veenman and Zonen.

Scott, J., F. Davis, B. Csuti, R. Noss, B. Butterfield, C. Groves, H. Anderson, S. Caicco, F. D'Erchia, T. Edwards, J. Ulliman and G. Wright, 1993, Gap Analysis: a geographic approach to protection of biological diversity. *Wildlife Monographs* 123:1-41.

Smith, T., S. Menon, J. Star, and J. Estes., 1989, Requirements and principles for the implementation and construction of large-scale geographic information systems. Pp. 19-37 in W. Ripple (ed.) *Fundamentals of Geographic Information Systems*: A compendium. Bethesda, Md.: American Congress on Surveying and Mapping.

Sugardjito, J. and C. van Schaik, 1993, Orangutans: Current population status, threats, and conservation measures. In R. Tilson, K. Traylor-Holzer and U. Seal (eds.) *Orangutan PHVA Workshop: Briefing Book*. Apple Valley, MN.: CBSG.

Tilson, R. et al., 1993, *Orangutan Population and Habitat Viability Analysis Final Report*. Apple Valley, MN: CBSG.

Yonzon, P. and M. Hunter., 1991, Conservation of the red panda *Ailurus fulgens*. *Biological Conservation* 57: 1-11.

11

ORANGUTAN LIFE HISTORY AND VORTEX ANALYSIS

M. Leighton,[1] U. S. Seal,[2] K. Soemarna,[3] Adjisasmito,[3] M. Wijaya,[3] T. Mitra Setia,[4] G. Shapiro,[5] L. Perkins,[6] K. Traylor-Holzer[7] and R. Tilson[2,7]

[1] Harvard University
[2] IUCN/SSC Conservation Breeding Specialist Group
[3] Indonesian Directorate General of Forest Protection and Nature Conservation
[4] University of Indonesia
[5] Orangutan Foundation
[6] Zoo Atlanta
[7] Minnesota Zoo

ABSTRACT

The Working Group on Life History Characteristics of the Population and Habitat Viability Analysis Workshop relied primarily on unpublished data on orangutans collected at Ketambe, Gunung Leuser National Park, Sumatra and Tanjung Puting National Park, Kalimantan. The orangutan appears to be the ultimate K-selected species, in that survivorship is high, interbirth interval is long (mean = 8 years) and the female makes a high investment in her offspring. VORTEX modelling indicated that adult females are the most valuable members of an orangutan population and that the death of an adult female has the greatest influence on increasing extinction rates of all life history variables. Infants in illegal trade may be thought of as representing dead females.

INTRODUCTION

A Population and Habitat Viability Analysis (PHVA), as developed by the Conservation Breeding Specialist Group (CBSG), provides population viability assessments for each population of a species or subspecies, as decided when a workshop is organized. The assessment for each species undertakes an in-depth analysis of information on the life history, population dynamics, ecology, and population history of each population. Information on the demography, genetics, and environmental factors pertinent to assessing the status of each population and its risk of extinction under current management scenarios and perceived

The Neglected Ape, Edited by Ronald D. Nadler et al.
Plenum Press, New York, 1995

threats are assembled in preparation for the PHVA and the data for individual populations before and during the workshop are examined.

An important feature of the workshops is the elicitation of information from the experts that is not readily available in published form, but which maybe of decisive importance in understanding the status and population dynamics of the species in the wild. This information provides the basis for constructing a simulation model of each population which can in a single model evaluate the deterministic and stochastic effects and interactions of genetic, demographic, environmental, and catastrophic factors on the population dynamics and extinction risks. The process of formulating information to put into the model requires that assumptions and the data available to support the assumptions be made explicit. This process leads to a consensus on the biology of the species, as currently known. The process also usually leads to a basic simulation model for the species that can serve as a basis for continuing discussion of management alternatives and adapting the management of the species or population in response to new information. It provides, in effect, a means for conducting management programs as scientific exercises with continuing evaluation of new information in a sufficiently timely manner to be of benefit to management practices on actual populations.

These workshop exercises assist in the formulation of management scenarios for the respective species and to evaluate their possible effects on reducing the risks of extinction. It is also possible through sensitivity analyses to search for factors whose manipulation may have the greatest effect on the survival and growth of the population(s). One can rapidly explore a wide range of values for the parameters in the model(s) to gain a picture of how the species might respond to changes in management. This approach may also assist in evaluating the information contribution of proposed and ongoing research studies to the conservation management of the species.

The CBSG PHVA Workshop process is based on biological and sociological science. Effective conservation action is best built on a synthesis of available biological information, but is dependent on actions of humans living within the range of the threatened species as well as established international interests. There are characteristic patterns of human behavior that appear to be cross-cultural; 1) in the acquisition, sharing, and analysis of information, 2) in the perception and analysis of risk, 3) in the development of trust between individuals and 4) in 'territoriality' (personal, institutional, local and national). Each of these has strong emotional links that shape human interactions. Recognition of these patterns is essential in developing processes to assist groups in reaching consensus on natural resource actions.

The motivation for organizing and participating in a PHVA workshop comes from both a fear of loss of, and a hope of recovery for, the taxon. We do not want species to go extinct on our watch. A commitment to the species is made by individuals who provide the required leadership. Effective and persistent action depends on a bottom-up approach at the level of actors which has rarely been a part of the planning of species conservation. There is a consensus among the players on a desired outcome with the goal of preventing extinction and achieving species recovery. A successful outcome depends on formulating a potentially "win-win" strategy for management scenarios or outcomes among participants and stakeholders whose interests and agendas usually differ. Institutions, groups and individuals must pool resources and information.

LIFE HISTORY VARIABLES

Orangutan life history baseline values, and lower and/or higher values around the baseline for sensitivity analysis, were estimated based on several sources of evidence and lines of reasoning. Data from wild populations were used for all variables. In a few cases, however, small sample sizes or biases from field work slightly influenced selections. In some instances, the best estimates were guided by data from captive orangutans or from other mammalian populations.

Data were examined to determine if different life history estimates for the Sumatran (*P.p. abelii*) and Bornean (*P.p. pygmaeus*) subspecies were justified. Fortunately, the field data that guided selections were from Tanjung Puting and Ketambe. These sites not only represent the two subspecies, but they also differ dramatically in ecological factors (e.g., food resources and predators) that might be expected to affect life history variables if strong variation were found. Although sample sizes were usually small for at least one and usually both studies, no variables seemed to differ between the two studies sufficiently to suggest different selections of values. For both the Tanjung Puting and Ketambe populations, better estimates for variables were obtained by using unpublished data. The review by Rodman (1988) and other original published sources, especially those from Mentoko in the Kutai National Park (East Kalimantan), also supports the conclusion that estimates for life history variables can be generalized across different orangutan populations and subspecies. No orangutan populations have been studied in mountain or hill areas, however, where individuals do not have access to a relatively rich mosaic of lowland habitats. It may be that these populations, which are apparently at the lowest densities, show different life history characteristics reflecting poorer habitat quality.

These life history variables are presented in the sequence they are encountered when running the VORTEX software model. For each variable, the values selected for the computer simulations are given and a brief description of the rationale for these selections is provided. Data from Tanjung Puting are indicated by TP and from Ketambe by KT.

1. There Is No Significant Correlation between Juvenile and Adult Survivorship

Deaths in wild populations have seldom been observed or reliably inferred. The sample indicates, however, that most deaths occur as isolated events, as expected for species of similar ecology and life history. Also, note that the "mild" catastrophe explored in the simulations assumes that high mortality is shared equally across age classes (see below).

2. The Populations Are Polygynous

Field studies provide conclusive evidence for polygyny in all populations (Mitani and Rodman, 1987). Caution should be used, however, when modeling small populations of sparse orangutans. The VORTEX model assumes that no female goes unmated when her interbirth interval expires, even in populations with very low numbers or proportions of adult males. For these cases, the model may have to be adjusted if low rates of encounter, perhaps exacerbated by the exercise of mate choice by females, lengthen the interbirth interval. This result, for example, might affect populations in mountain habitats or in logged forest.

3. Different Catastrophes

Extreme Food Shortage

Rate = 5 Times/100 yr (Range = 1,10)

P(Survival) = 0.85; P(Litter>0) = 0.

The large body mass and generalized diet of orangutans buffers them from starvation. Adults, especially adult males, store fat and can probably survive prolonged periods at negative energy balance while feeding on energy-poor, fibrous foods (leaves, bark phloem layers and pith). However, 1) this does not apply equally to juveniles and infants, 2) the phenology of foods they prefer is wildly unpredictable in space and time and 3) it is reasonable to hypothesize that orangutan carrying capacities and typical population densities are determined by fluctuations in the food supply (Leighton, pers. observ.).

Extreme food shortages causing episodes of mortality or reproductive failure have not been observed in field studies. However, the location of field studies, the presumed rarity of these events and the ranging patterns of orangutans may conspire to cause these food crisis events to go unrecorded. First, all long-term studies providing demographic information (Ketambe, Tanjung Puting, Gunung Palung and Mentoko), are probably located where resident orangutans enjoy nearly optimal habitat quality compared to other individuals in the population. Such conditions, perhaps, would bias the estimates of mean per capita demographic values. All research sites are located along rivers and consequently, contain a rich habitat mosaic, usually including productive alluvial bench and seasonally flooded forests, well-drained lowland forest and often swamp forests. This habitat mosaic probably best buffers orangutans from fluctuating food supplies, compared to orangutans with home ranges more restricted to single habitats, located solely in upland forests or swamps. Second, all four of these sites have reported episodes where individuals shift home ranges, especially adult males and subadults and adolescents of both sexes. These individuals migrate in and out of the research sites in response to local food supplies. The data from Gunung Palung strongly suggest that these habitat shifts support high overall densities. It may be that some of these individuals are most vulnerable to starvation. Because these migrants or transients are unmonitored by researchers, as they leave during periods of poor food supply, mortality on this subset of orangutans would go unrecorded. Third, comparative evidence from other large-bodied herbivores indicates that aperiodic, rare events that cause starvation influence long-term population dynamics. Therefore it seems wise to explore demographic simulations that include unpredictable extreme food shortages.

There are several possible scenarios suggesting which age and sex classes might suffer disproportionately from extreme food shortages. For this analysis the scenario was modeled in which juveniles (J) and infants (I) suffer 30% mortality in the year when the event occurs, and all older (O) individuals suffer 10% mortality. Ratios of (I+J)/O in "snapshots" of population structure at Tanjung Puting (during November, 1975, Galdikas 1978), and at Ketambe (during October, 1989, Tatuno, pers. comm.), were 0.33 and 0.23, respectively, so the proportion of I+J was estimated as 30%. The per capita risk of mortality was then calculated and converted (and rounded) to the average value of P (survivorship) = 85% for the year when the food crisis occurs. It was assumed that no females would reproduce during the year. Note that it would be difficult for field workers to detect a nutritionally-based delay of reproduction because on average only one of eight females reproduce per annum (see below). Few field studies monitor many more than eight females. The frequency of a mild catastrophe such as an extreme food shortage is, of course, completely unknown. However, judging from the frequency of extended periods when orangutan subsistence is on foods of low preference and low energy density, and assuming that the range of values in these simulations might indicate approximately 95% confidence intervals, then a range of values from 1-10 times per century seems to be a useful beginning point.

Other Types of "Natural" Catastrophes. Other ecological factors (diseases) might be rare but normal events that affect long-term (50-several hundred years) population dynamics. These could occur as part of the natural ecology of the species, unrelated to effects of habitat transformation or disturbance of the forest community. It is unlikely that field researchers would be aware of such rare events, given that field studies have been too limited in space and time to adequately understand and model the population dynamics of animals with population biologies like orangutans'. Orangutans live as sparse populations of long-lived individuals that shift in space on scales much larger than our study sites.

4. The Age at First Reproduction (AFR) for Females Is 15 Years (13 - 17 Years), for Males 20 Years

The sample used for estimating this variable is comprised of six females: three from Ketambe (12, 16 and 16 years) and three from Tanjung Puting (15, 16 and 16 years), giving a mean value of 15 years and indicating consistency between the two sites. Note that the range for the simulations is guided not by what might be the variation among females in a population, but what the true population mean might be, given our small samples. Note also that there is a tendency to underestimate this variable because later maturing females would be less likely to enter the sample population. It is not known how this affected the data from either study population (Galdikas and Wood, 1990).

The mean AFR value used for males does not rest on any field data, which would require paternity analysis. The value of 20 years was entered merely because in polygynous mammals, males typically have AFR's later than females. No effort was made to improve the estimate of AFR for males (for example, from comparative data of mammals with similar mating systems and life histories) because the population dynamics modeled for polygynous species with VORTEX is insensitive to male AFR unless the number of males is very few (less than 5-10).

5. Maximum Longevity Is 45 Years for Both Males and Females

Again, because of the structure of the model, only the value for the females is important. There are no accurate field estimates for this variable, but estimates were guided by two factors. First, in a large sample of captive orangutans for which the age at death was known (n = 1,239), the maximum age for females was 57 years and the maximum age for males was 58. Although health care and nutritional status of these individuals undoubtedly varied dramatically, it is likely that a large subset received nearly optimal care, extending longevity beyond the maximum achieved in the wild, as is typical in comparisons of zoo versus wild ranging individuals of other mammals. Since field workers reliably estimate that orangutans are as old as 40 or slightly more, an estimate of about 45 seemed warranted.

Secondly, one should err on the side of underestimating this variable because of the structure of the VORTEX life history parameters being used. Because a constant age-specific mortality for all individuals above one year of age is entered, the model does not include the typical and extreme increased age-specific mortality common to long-lived mammals as they approach senility. The effect of this approach is that under conditions of low average mortality (e.g., mortality rates of 1-1.5%), an artificially high number of females would reach a maximum longevity of 50 (if that were the estimate). This would translate into an average increased reproductive life span of five years, increasing average number of births from 3.75 (=(45-15)/8) to 4.4 (=(50-15)/8) per female surviving to maximum age (= 45 and 50, respectively, with an interbirth interval of 8 yr, see below).

6. Sex Ratio at Birth: A Sex Ratio of 0.50 (0.55) Was Used for Both Sexes

The most relevant data are large samples of births from the wild. Because there is now evidence for several primate species that the immediate social and demographic context of breeding females can influence sex ratio at birth, it is not advisable to use data from captive females to estimate the sex ratio. Therefore, a baseline value of 50% was used, derived from the large Tanjung Puting sample (n approx. = 30). In a recent (1989-92) sample from Ketambe, 7 of 9 births were males, but this degree of skewing is expected in small samples. Since males in polygynous species typically have higher age-specific mortalities, careful examination of the population structure of orangutans might indicate there are "too many" males and that, perhaps, the true sex ratio is slightly skewed towards males. Because the sex ratio can importantly influence population dynamics through the rate at which females are recruited into the breeding population, the effects of a 55% male bias may be explored with the model. However, because there are not enough samples over a large enough spatial scale to avoid confusion from migrating or transient individuals, there is no field evidence at this time to support the actual existence of a skewed sex ratio.

Although the sex ratio of individuals brought to rehabilitation centers do not influence the basic demographic model for orangutans, it should be taken into account in model scenarios that include a harvesting component. A 1970s' sample from Bohorok was reported to be 53% males (n=98), whereas, a current sample brought to a center in East Kalimantan is skewed towards males (Rijksen, pers. comm.).

7. Proportion (P) of Adult Females Producing Litters of Different Sizes Each Year

P (no litter) = 87.5%

P (litter=1) = 12.5%

SD (P(litter=1)) = 6%

or for mean interbirth interval: 8 yr (7,9)

This variable was calculated from the mean interbirth interval and was used in computer simulations as variation in the interbirth interval (this is possible because all litters are of size = 1). From Ketambe, there are seven observations from five females: 1) 2 and 12, 2) 7 and 7 (a rehabilitant) and three others with values of 11, 6 and 8-9 years. The mean value is 8 years. The estimated mean from a sample of approximately 20 events (successive births of the same female) at Tanjung Puting is also 8 yr, with a range of 6-10 years (Galdikas, pers. comm.).

At this time it is not known if the Tanjung Puting data include observations of intervals beginning or ending with early deaths of neonates or stillbirths. Because infant mortality (from age 0 - 1, see below) includes these deaths, it is important to include these in the calculation of interbirth interval. For example, if only 10% of these deaths is missed, the estimate in this model changes from 8 to 7 years. Counteracting this bias, there is a tendency of field data to underestimate this value because longer intervals are more likely to be excluded from analysis, as their endpoints (the next successive birth) are less likely to have occurred at any point in a field study.

Although the data set from Ketambe is small and variable, the mean value is similar to that from Tanjung Puting. The data suggest that the true population mean for orangutans is likely to fall between 7 and 9 years and that this range of values should probably be explored only in the basic demographic model for natural populations. However, note that in populations in which significant ecological factors change birth rates (e.g., female nutritional status) or cause higher infant or juvenile mortality (e.g., predation, diseases), the interbirth interval can be substantially affected. The variance in the mean proportion of adult females giving birth each year is unknown, but stochastic variation could be reasonably modeled using a standard deviation of approximately half the mean.

8. Infant Mortality (Age 0 - 1)

Mean = 10% (5%, 15%); SD (10%) = 6%.

Of 4 births at Ketambe from 1989-92, one was a miscarriage, while one of roughly 20 births resulted in early neonatal death at Tanjung Puting. There is a likely bias towards under-reporting these events in the wild. Females followed sporadically, a few days per month or every few months, may not be known to be pregnant with certainty and give birth to a dead or dying infant without the observer being aware of this. Further, neonatal mortality rates range from 10% - 20% in captivity for many mammals. On balance, a baseline value of 10% is suggested, with 5% - 15% capturing the range of the true population mean for sensitivity analysis. The SD of the mean (of the 10% baseline value) was generated from a hypothetical data set of 10 cohorts of 10 infants each, in which 6 cohorts suffer 1 death each and 2 cohorts each suffer 0 and 2 deaths.

9. Adult Age-Specific Mortality Is 2% (1.5%, 2.5%) (for Both Males and Females)

Recall that "adults" are defined as all individuals over one year of age. Although individual orangutans have a high probability of survival, even long-term studies of large demographic samples are challenged to distinguish age-specific mortality rates that vary by 0.5-1.0%. Single deaths among small numbers of known females, for example, may go unreported because females who "disappear" may shift ranges outside of the study site. But changing the mortality rate from 1% to 2% changes the percentage of one-year-old females that mature to reproductive age (= 15) from roughly 88% to 77%, thereby dramatically influencing the relative number of births contributed by a cohort of females. A range of reasonable mean values for orangutans was explored by examining the stable age distribution generated from initial simulations using values of 1% and 2%, paying special attention to the proportion of adult females over 30-years-old, given either assumption. If age-specific mortality is 2%, then the proportion of all adult females over 30-years-old is expected to be 40%; if mortality is 1%, this number is 35%. Among the current 10 adult females known at Ketambe, three are estimated to be over 30-years-old, whereas Galdikas (pers. comm.) estimates that half the adult females are over 30 years. The results are, therefore, not definitive. A mortality rate of 1% was rejected as too low, given comparative mammalian data, but there is no empirical basis for rejecting this value for orangutans. One can probably exclude mortality rates of 3% as unreasonably high, however, because this implies that nearly half of the females born fail to reach reproductive age.

10. The SD in *K* Due to Environmental Variation (EV) Is 10% (0%, 20%)

As occurs for all mammalian populations, the carrying capacity fluctuates over time because of changes in the effects of critical ecological factors, especially the food supply. Orangutans are slow breeding and often change diets for long intervals of time to survive periods of fruit shortage, so they are relatively immune to changes in carrying capacity caused by short-term changes in *K* (e.g., that occur every few years). However, shifts over time in the relative productivity of food plants of different taxa and stochastic changes in the densities of important key food species certainly occur. A model of orangutan demography, therefore, should include a component of stochastic change in *K*, but it is difficult to imagine what this value should be. However, the fact that orangutan densities are comparable between widely spaced sites within the same habitat (e.g., swamp/peat forest or lowland forest) suggests that the effective temporal change in *K* for any single population is probably small. Therefore, a narrow range of values (0% - 20%) should be explored in demographic simulations.

SIMULATION MODEL SCENARIOS AND RESULTS

The interactions of infant mortality, adult mortality and interbirth interval on *r* and the population size at 200 and 300 years over the ranges of values suggested by the available field data were examined systematically with step-wise changes in each of the 3 variables (see Fig. 1). The range of values were, 1) 1%, 2% and 3% mortality for all ages from 1 through adulthood, 2) 2%, 5% and 10% infant mortality and 3) interbirth intervals of 6, 8, and 10 years. The other conditions of these scenarios were a starting population of 1000 with $K = 1000$, no inbreeding, all adult males in the potential breeding pool, age of first reproduction equal to 15 years for females and 20 years for males and 45 years as the age of senescence.

The simulation results for values of *r* from this 3-way matrix of interactions yielded values ranging from 0.02 to -0.02. They indicate little sensitivity to infant (0 to 1 year) mortality over the range of 2-10% within a given interbirth interval (the word infant is used by orangutan biologists to refer to the time that the young animal is carried by the mother, i.e., about 4 years). Increasing the interbirth interval in two-year steps resulted in a 50% decline in *r* for each step. An increase in adult mortality from 1% to 2% per year produced a 50% or greater reduction in *r*. At 3% adult mortality, most values of *r* were negative under the conditions of these scenarios.

Estimates of projected surviving population sizes at 200 years (about 6.5 orangutan generations) with an age of first reproduction for females of 15 years and a starting population of 1000 and $K = 1000$ indicate that all populations decline when adult mortality is set at 3% and all will eventually become extinct (Fig. 2). When the interbirth interval is 8 years or greater, the populations decline with 2% adult mortality and this also leads to eventual extinction.

Current field data indicate that the orangutan populations studied in Sumatra and Kalimantan have an interbirth interval of about 8 years. However, it is not known whether a shortening on this interval occurs if the infant or juvenile dies. Data from breeding in captivity indicate that the interval could be shortened to 3 years. The simulations of wild populations indicate that they cannot sustain an increase in adult female mortality. The stable age structure of the population shows that about 50% of the animals are adults. Thus removal of 5 adults per year from a population of 1000 animals would increase adult mortality by 1%, which under this schedule of mortality and fecundity rates leads to a long-term decline of the population. If the primary targets of the removals are females, as might occur with the

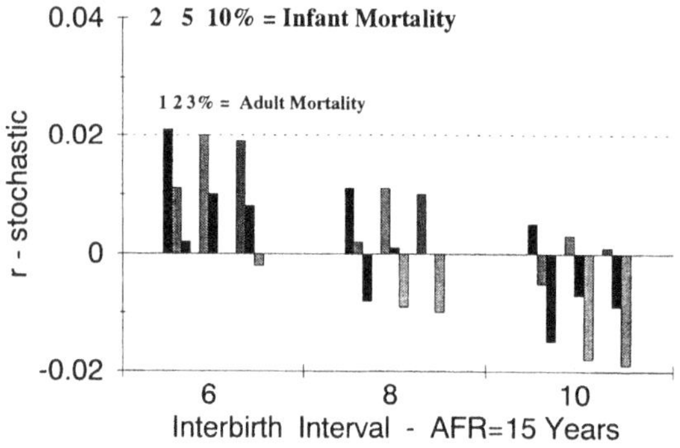

Figure 1. Effect of mortality rate and interbirth interval on the rate of increase (*r*) (AFR = 15 years).

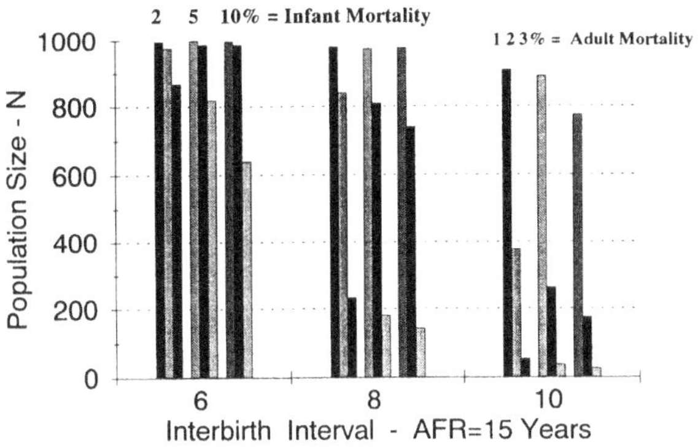

Figure 2. Effect of mortality rate and interbirth interval on population size (AFR = 15 years).

collection of infants for the pet trade, then the increase in annual adult female mortality would be 2%. This would increase the projected rate of decline and the risk of extinction. These scenarios were simulated using the harvest module of VORTEX.

These populations are thought to be at or near carrying capacity. At lower densities, orangutan populations might be able to respond by an increase in reproductive rates as can occur in captivity. This might result from a decrease in the interbirth interval or from reproduction at an earlier age or both. Both are observed in captive orangutan populations as is evident from analysis of the data in the International Studbook. In the studbook data, however, there is on average a greater likelihood of a reduction in the interbirth interval, which averaged 3-4 years for 800 births. This shortening of the interval in captivity may be partly a consequence of infant removal from the family unit, partly a consequence of good nutrition, and partly a behavioral effect of space limitations in captivity. Any of these conditions or a combination of them would allow greater resilience of wild orangutan populations to environmental variation and to occasional catastrophic losses.

A decrease of the age of first reproduction in females to 13 years yielded a small increase in mean stochastic growth rate (*r* values) with the result that all were positive, with adult mortalities of 1%, 2%, and 3% at an interbirth interval of 6 years (Fig. 3). The rate of increase was positive for 1% and 2% adult mortalities at an interbirth interval of 8 years.

Similarly, the projected population sizes at 200 years (Fig. 4) are near *K* at an interbirth interval of 6 years for all except the highest levels of infant (10%) and adult (3%) mortality.

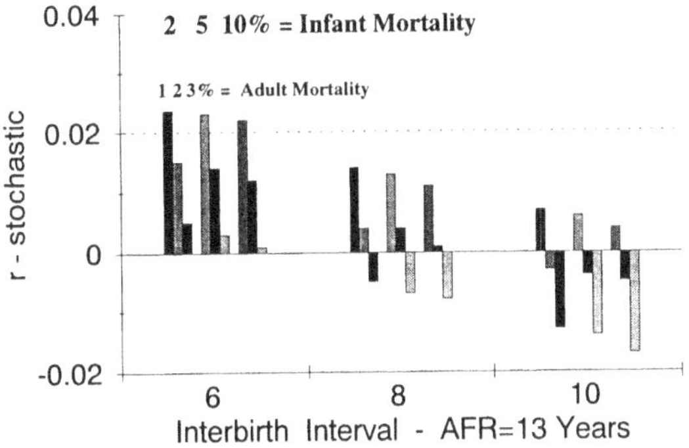

Figure 3. Effect of mortality rate and interbirth interval on *r* (AFR = 13 yrs).

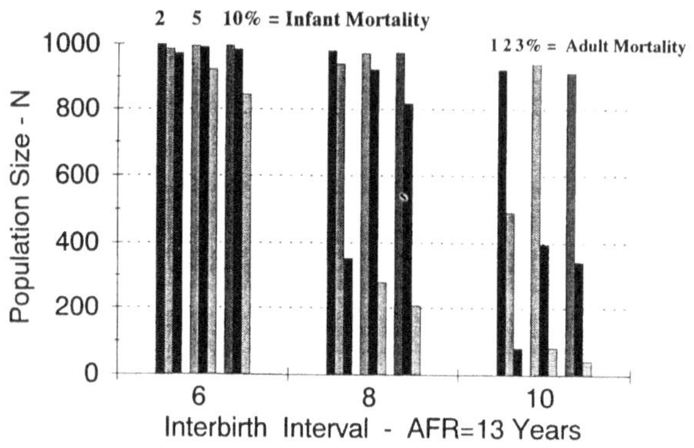

Figure 4. Effect of mortality rate and interbirth interval on population size (AFR = 13 yrs).

The simulated populations still decline at an interbirth interval of 8 years and with adult mortality of 3%. Measurement of such shifts in a wild population would require following a substantial number of marked animals. Alternatively, if a population were depleted by some event without a decline in habitat, field studies might yield information on the strategy employed by wild orangutan populations in adjusting to that event. Such information would then allow construction of a density- dependent function of changes in reproduction to include the simulations of orangutan population dynamics.

The effects of a catastrophic event on the modeled orangutan population that occurred with a probability of 5% each year, but on the average of once in 20 years.

The effects chosen were a 5% increase in mortality and the elimination of reproduction in the year of occurrence. The interaction of first year mortalities of 2%, 5%, 10%, and 15% with adult mortalities of 1%, 2%, and 3% and an interbirth interval of 7, 8, and 9 years (the 3 columns under each adult mortality level are the results for the 7-, 8-, and 9-year interbirth intervals, respectively) with this catastrophe indicate an overall decline in mean stochastic population growth rates (Fig. 5). The simulations with 1% adult mortality maintained positive mean stochastic growth rates for the 7- and 8-year interbirth intervals at levels of 2% - 15% average annual infant mortality.

Population sizes at 200 years reflected the trends indicated by the growth rates (Fig. 6). None of the populations with 3% average annual adult mortality were viable and an interbirth interval of 9 years resulted in a declining population at all levels of adult and first year mortality.

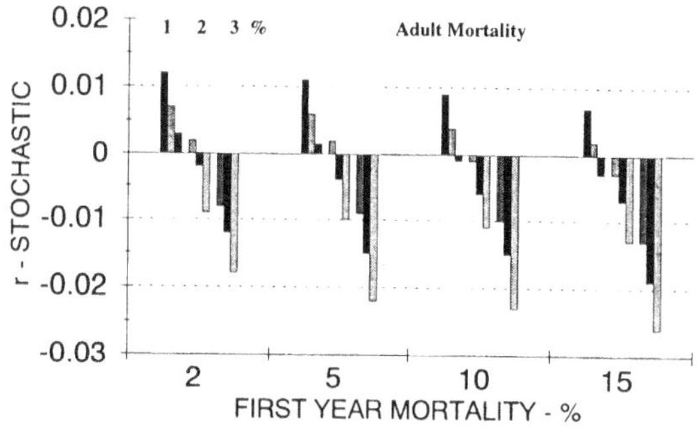

Figure 5. Effect of catastrophe and mortality rates on *r*.

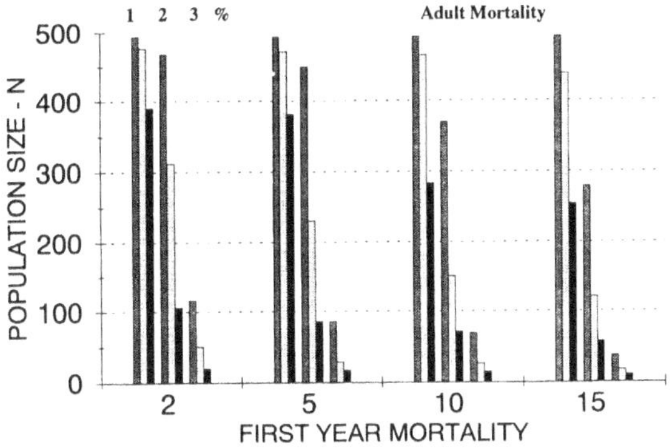

Figure 6. Effect of catastrophe and mortality rates on population size.

ACKNOWLEDGMENTS

The database, interpretations and recommendations contained in this paper were derived at the Orangutan PHVA Workshop held in Medan, North Sumatra in January 1993, which was coordinated by the Indonesian Directorate General of Forest Protection and Nature Conservation (PHPA) and the Survival Service Commission (SSC) of the International Union for the Conservation of Nature (IUCN) Conservation Breeding Specialist Group (CBSG) (see Soemarna, Ramono and Tilson, this volume, for common acknowledgements). In addition to those individuals cited in Soemarna, et al., we wish to acknowledge W. Ramono.

REFERENCES

Galdikas, B., 1978, *Orang utan adaptation at Tanjung Puting Reserve, Central Borneo*. Ph.D. Thesis, University of California, Los Angeles.

Galdikas, B.M.F. and Wood, J., 1990, Great ape and human birth intervals. *Am. J. Phys. Anthropol.* 83: 185-192.

Mitani, J. and Rodman, P., 1987, The evolution of Asian ape social systems. *Int. J. Primatol.* 8(5): 459 (abstract).

Rodman, P., 1988, Diversity and consistency in ecology and behavior. Pp. 31-51 in *Orang-utan Biology*. J.H. Schwartz, ed. New York, Oxford University Press.

Tilson, R., U. Seal, Komar Soemarna, et al., 1993, *Orangutan Population and Habitat Viability Analysis Report*. IUCN/SSC Conservation Breeding Specialist Group: Apple Valley, MN. 54 pp.

12

ESTIMATES OF ORANGUTAN DISTRIBUTION AND STATUS IN SUMATRA

C. P. van Schaik,[1] S. Poniran,[2] S. Utami,[3] M. Griffiths,[4]
S. Djojosudharmo,[2] T. Mitra Setia,[3] J. Sugardjito,[4] H. D. Rijksen,[5]
U. S. Seal,[6] T. Faust,[7] K. Traylor-Holzer,[7] and R. Tilson[6,7]

[1] Department of Biological Anthropology and Anatomy and Center for Tropical Conservation, Duke University, 3705B Erwin Road, Durham, North Carolina 27705 and
New York Zoological Society - The Wildlife Conservation Society
[2] Directorate General of Forest Protection and Nature Conservation, Indonesia (PHPA)
[3] University of Indonesia
[4] World Wide Fund for Nature
[5] Instituut voor Bosbouw en Groenbeheer
[6] IUCN/SSC Conservation Breeding Specialist Group
[7] Minnesota Zoo

ABSTRACT

At the Population and Habitat Viability Analysis (PHVA) Workshop for orangutans, estimates of habitat and population numbers for Sumatran orangutans (*Pongo pygmaeus abelii*) were derived from field biologists who studied, or are presently studying, orangutans at Ketambe in Gunung Leuser National Park. For Sumatra, the exact boundaries of orangutan distribution are not known, but there are several distinct populations, including the lesser-known Singkil population and the Sembabala-Dolok Sembelin population. The Greater Leuser orangutan population, which extends beyond the national park boundaries, is thought to cover approximately 11,710 km^2 and has two distinct populations. Using a correction or "safety" factor to derive population estimates, the western population is thought to number 5,700 individuals, and is the most important orangutan population in Sumatra; the eastern population is thought to number 3,500 individuals. Within the restricted boundaries of Gunung Leuser National Park, the area covered by the western population is 5,570 km^2 and the *corrected population size is about 3,450* individuals; the area covered by the eastern population is 2,957 km^2 and the *corrected population size is about 2,400 individuals*. The total number of orangutans in the park probably is about 5,850 individuals. The Greater Leuser populations were judged to be among the strongest in Southeast Asia, in terms of

The Neglected Ape, Edited by Ronald D. Nadler et al.
Plenum Press, New York, 1995

numbers and potential for protection. Gunung Leuser National Park was considered to be the most vital habitat to the long-term survival of the Sumatran orangutan.

INTRODUCTION

This chapter, from the Population and Habitat Viability Analysis (PHVA) Workshop for Orangutans (*Pongo pygmaeus*) (Tilson et al., 1993), focuses on the status of wild populations of orangutans in northern Sumatra *(Pongo pygmaeus abelii)*. A PHVA Workshop held previously for Sumatran tigers in 1992 had created a map-linked database that integrated vegetation types, satellite imagery, and land use patterns for the protected areas of Sumatra (Tilson et al., 1994); the map for Gunung Leuser National Park proved invaluable for developing more precise estimates of the distribution and numbers of orangutan populations in the park (Faust, Tilson and Seal, this volume). The densities used to estimate population numbers are based on the censuses of van Schaik and Azwar with more recent estimates by van Schaik (unpublished) and compared with those obtained by Rijksen (1978). Vegetation information is based on World Conservation Monitoring Centre (WCMC) maps, stored in the Geographical Information System Atlas (Faust, Tilson and Seal, this volume).

Long-term study and familiarity with Gunung Leuser National Park and surrounding areas made possible the identification of specific threats to its integrity, such as road construction and illegal encroachment and logging in lowland areas, and their effects on wild orangutan populations. This information provided a basis for the development of a number of recommendations to provide better security for Gunung Leuser National Park and its orangutans.

DISTRIBUTION

The Sumatran subspecies of orangutan (*Pongo pygmaeus abelii*) occurs only in the northern part of the island of Sumatra, where most of the forest has been uninhabited by humans since historical times. The exact boundaries of the distribution of orangutans are not known (Fig. 1). The best information is that compiled by Rijksen in the early 1970s. Orangutans that might still occur south and southeast of Danau Toba were not considered, because their numbers are likely small. This leaves several distinct populations:

1. The Singkil population — The Singkil Barat Reserve consists of extensive swamp forests. The extent of suitable habitat is not known, because there are conflicting reports on the degree of disturbance and habitat loss.
2. The Sembabala-Dolok Sembelin population — A population in production forests; the exact boundaries are not known, but the numbers are likely to be small and the prospects for maintenance are poor. Production Forest is an official term used by the Ministry of Forestry to denote forests that are meant to serve for the permanent production of timber in accordance with the officially approved selective logging system.
3. The Greater Gunung Leuser west population — This is by far the largest orangutan population anywhere. Its boundaries coincide with those of Gunung Leuser in the north and the northwest, but are wider in the west and especially the south.
4. The Greater Gunung Leuser east population — Recently, the southern and the northern sub-populations were separated because the Kutacane-Blangkejeren road has spawned development around it. The northern population consists of the national park, along with the remaining part of Tamiang (production forest) and

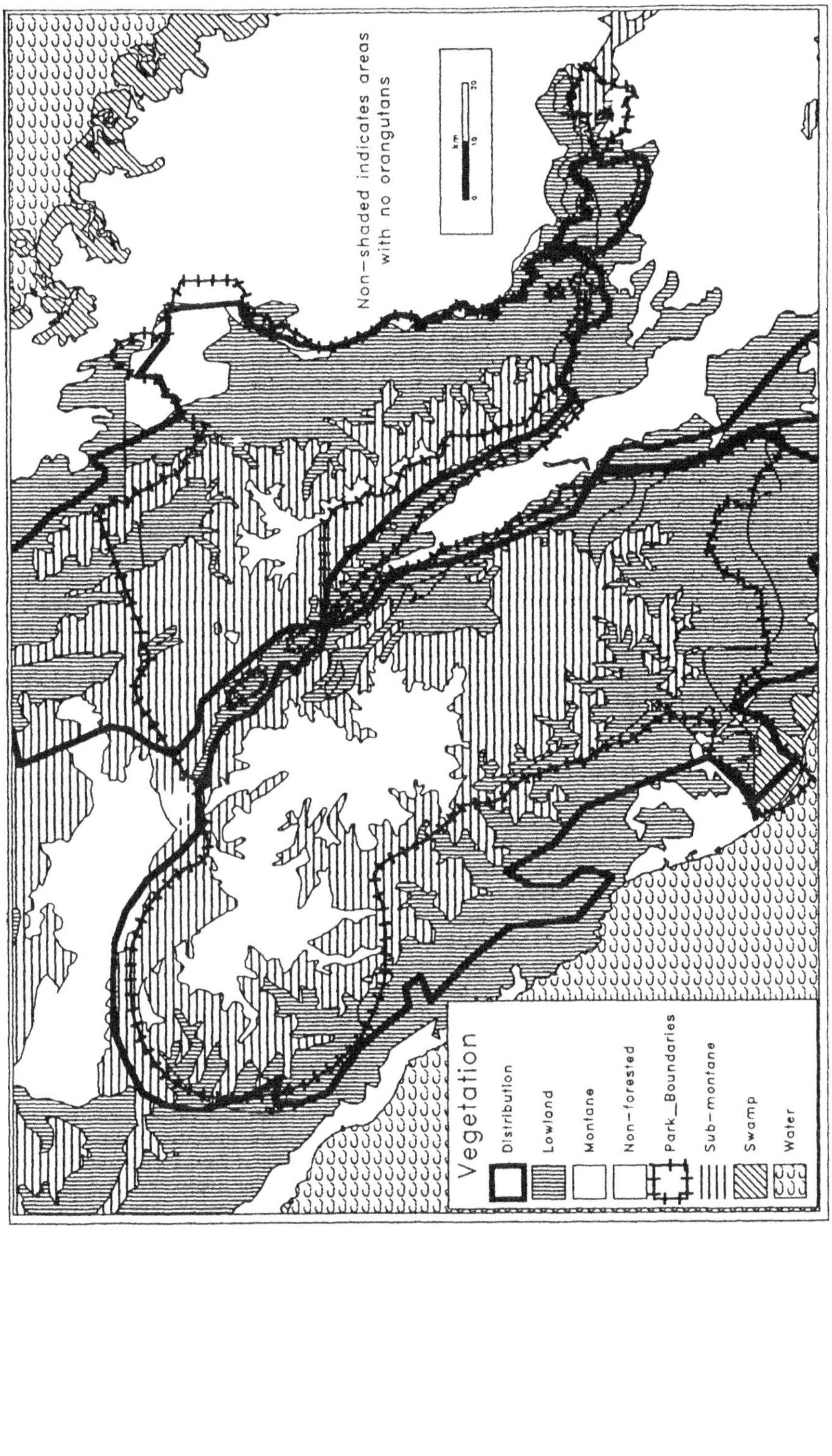

Figure 1. Current orangutan distribution around Gunung Leuser National Park.

of the Lesten area (designated production forest). To the east, there is a small sub-population separated by the Wampu River. The exact northern boundary of this population is not known, but grades uninterrupted into the forest north and northeast of Blangkejeren. However, ongoing construction of the road from Blangkejeren to Pereulak through Uring, Pinding and Lokop, will undoubtedly sever this link and create a separate northern population, of which the boundaries are only vaguely known.

5. There may possibly still be a population west of Takengon, but there are contradictory reports.

This overview demonstrates that the boundaries of the Greater Leuser orangutan population are known in detail in virtually all places. However, there is very little information on the potentially important Singkil and Takengon west populations. Field surveys will need to be conducted to establish the status of these populations. In particular, the orangutan population at Singkil is of some importance because there is a potential for connecting it to the Greater Leuser population by constructing a corridor.

THREATS AND OPPORTUNITIES

Cyclones and large-scale forest fires can be ignored as causes of habitat loss. Gunung Leuser is outside of the cyclone belt. Forest fires, when they occur, are small scale. Large-scale fires are not expected and should have occurred over the past decades given the frequent occurrence of bush fires set by local farmers. Moreover, the unpredictable long droughts as experienced on Borneo (where a major fire event occurred in 1982-83) are not found in this region.

Ecotourism could affect the orangutan population if tourists cause habitat loss or disturbances. In the case of orangutans, however, these losses are negligible with the exception of those caused by the construction of roads and the resulting fragmentation, which is discussed below.

There remain habitat loss and fragmentation as the major threats to wild orangutans. These losses can take various forms:

1. change in the legal status of land, from protected to non-protected;
2. encroachment by illegal agriculture (which, even if stopped or reversed, leaves important long-term negative impacts;
3. road construction for tourism or for connecting cities outside the park;
4. encroachment from enclaves (human settlements).

All of these threats have a component of direct habitat loss and an indirect threat of habitat fragmentation. In practice, this leads to several possible cases, each of which is examined below. In addition to losses, there also may be opportunities to gain through land acquisitions to the park. The consequences of such acquisitions will be examined as well.

BASELINE NUMBERS OF ORANGUTANS

The authors recognize the limitations of the population estimates presented here. Population estimates are usually made by estimating the size of the available habitat and then multiplying this number by the estimated densities for that species for the various habitat types. However, this figure is inevitably an overestimate. First, animals are almost never at habitat saturation across their entire range. Second, field biologists tend not to select field

Table 1. Estimated numbers of the Greater Leuser orangutan population and the Gunung Leuser National Park orangutan population

	Total area				Protected area		
	Area (km^2)	Density	Number	Corr. n	Area (km^2)	Number	Corr. n
West							
secondary	12.23	0.5	6		0	0	
swamp	50.97	5	255		50.97	255	
< 500	831.77	2.5	2079		216.38	541	
< 1000	2311.92	1.8	4161		1576.44	2838	
< 1800	2845.83	0.4	1138		2426.67	947	
> 1800	1299.47	0	0		1299.47	0	
Total	7352.19		7639	5700	5569.93	4605	3450
East							
secondary	86.63	0.5	43		0	0	
< 500	236.88	2.5	592		236.88	592	
logged lowland	176.74	1	177		176.3	176	
< 1000	1674.85	1.8	3015		1037.79	1868	
< 1800	2021.43	0.4	809		1344.04	538	
> 1800	161.56	0	0		161.56	0	
Total	4358.09		4636	3500	2956.57	3174	2400
Grand total	11710.28		12275	9200	8526.50	7779	5850

sites that have suffered recent habitat disturbance or have low population numbers of the species under study. Third, habitat distribution is almost always derived from outdated maps that do not reflect recent habitat disturbances. For all these reasons, a *correction factor* was used in all estimates of population numbers presented here. In this section of the workshop, estimates of orangutans, for all of the above reasons, were derived by employing a *correction factor of 0.75*, which all workshop participants believed to be a reasonable value.

The first two cases considered below are used as baselines (see Table 1).

The Greater Gunung Leuser populations

The area covered by these orangutan populations is approximately 11,710 km^2. The crude estimate of the populations in the entire area is 12,275 individuals; the *corrected number is 9,200* individuals. The western population of these two, covering an area of 7,352 km^2, has a total estimated number of 7,639 individuals and a *corrected number of 5,700* individuals. The eastern population has an area of 4,358 km^2 of interconnected habitat, with an estimated total population of 4,636 individuals and a *corrected total of 3,500* individuals. Hence, both of these populations are currently quite sizable.

The Gunung Leuser National Park Population

The area covered by the west population is 5,570 km^2, with an estimated 4,605 orangutans, and a *corrected estimate of 3,450* individuals. The area covered by the east population is 2,957 km^2, with an estimated number of 3,174 individuals, and a *corrected estimate of 2,400* individuals. The total number of orangutans in the GLNP is estimated at 7,779 individuals, and the *corrected estimate is 5,850* individuals.

These estimates are based on fairly crude estimates of the area, using 500 m as the boundary between lowland forest and hill forest. It neglects the area east of the Wampu River, which is slated to become part of the Tahura, and thus, will no longer be managed by national park staff.

In conclusion, despite recent attrition, the Greater Leuser orangutan populations are still the largest in the species' range. In particular, the western population, even if restricted to Gunung Leuser National Park, is the most important orangutan population of Sumatra.

THREATS

Effects of Road Construction

Road construction in this region tends to be associated with forest clearance along its edge. In some cases this actually leads to extensive conversion. At the least, the road and its associated opened strip will act as a barrier to orangutan dispersal and fragments the resident populations.

A tourist road from Bukit Lawang to near Kutacane has been proposed. This road would isolate part of the eastern orangutan population. Population estimates of this area are 767 orangutans, or a *corrected estimate of 575*. The construction of this road is not recommended from the point of view of protecting orangutans.

Another proposed road between Lubuk Keranji and Pucuk Lembang would bisect the Kluet Reserve. This small area is of great importance because it contains the only swamp forest of the park. This forest has the highest orangutan density found anywhere across their range. The estimated number of orangutans isolated by this road would be between 470 and 600, which gives a *corrected estimate of about 400 individuals*.

Loss of All Production and Protection Forest Outside of GLNP

Loss of all forests outside the park will leave two populations within the boundaries of the park as currently defined, approximately 3,450 and 2,400 animals, respectively.

Loss of All Lowland Areas Due To Attrition

If all the forests below 500 m were lost because of illegal encroachment on habitat and logging, for example, this would result in an uncorrected estimate of 3,809 individuals and a *corrected number of 2,850* individuals in the west and an uncorrected estimate of 2,406 and a *corrected estimate of 1,800* in the east.

These numbers may seem reassuring. However, in the absence of a seasonal access by orangutans to the lowland forests, the highland forests may develop a reduced carrying capacity and orangutan numbers may decline in the future. It is strongly recommended because of this risk, therefore, that no further lowland habitat be lost.

Loss Due to Encroachment and Changes in Legal Status

The Sikundur Reserve with its magnificent lowland forest has been degraded by selective logging in the late 1970s, while officially protected, and is currently recovering. Several parts of the park have been lost in the past due to changes in legal status of land use patterns. Subsequently, these areas were subjected to legal or illegal logging. Serbolangit, which was in the park initially, was reclassified to Protection Forest (and is now subject to some logging). A part of the Bengkung basin was deleted from the park recently, and has subsequently been subjected to selective logging. The eastern part of the park near Berastagi will soon be reclassified to become part of the Taman Hutan Raya.

Given these developments in the past, it is realistic to expect that some such attrition will continue in the future. A worse-case scenario was developed in which many of the accessible marginal areas of the park and some of the flatter upland parts in the Kapi will be lost. The eastern boundary is much more stable, and unlikely to change much.

As a result of this realistic worse-case scenario, the following estimates of the western and eastern park orangutan populations were produced. For the eastern population, this results in an uncorrected estimate of 1,869 individuals and a *corrected estimate of 1,400*. For the western animals, this results in an uncorrected estimate of 2,554 individuals and a *corrected estimate of 1,925*. However, the effect of the loss of lowland areas would almost certainly lead to fragmentation into two new populations. Each of these can be estimated separately; the northwestern population would be 425 (uncorrected) or *corrected to 325 individuals*; the southern population would be 2,129 (uncorrected) or *corrected to 1,600 individuals*.

In this scenario, it is impossible to restore the links among the three populations. Clearly, this situation would lead to appreciable loss of genetic variability and a significant probability of extinction of at least the smaller of the three populations.

Loss Due to Hunting And/or Pet Trade

This issue was not addressed at the workshop, but the fact that there are several hundred confiscated orangutans in Indonesia and several hundred more living in Taiwan suggests that wild orangutans have been captured for the pet trade at some time in the past (also see Eudey, this volume). It is not established where these orangutans were captured. Although past levels of removal are unknown, it is believed that it has greatly diminished in recent years.

OPPORTUNITIES

Translocating Human Inhabitants from Enclaves

The current separation of the park's two orangutan populations is due to the explosive human population growth in the enclaves and along the rest of the road between Kutacane and Blangkejeren. Plans exist to translocate the human inhabitants of the areas inside the park and, in some scenarios, those inhabiting the enclaves as well. In any case, this will in due course (less than 25 years) restore the unity of the park. All other things being equal, especially without further attrition of park habitat, this will create an orangutan population in Gunung Leuser National Park of about 5,850 individuals.

CONCLUSIONS

The two orangutan populations present in the Gunung Leuser National Park and surroundings (the Greater Leuser area) are among the strongest in the world in terms of numbers and the potential for protection. The Gunung Leuser National Park is, therefore, vital to the long-term protection of the Sumatran orangutan. Other important orangutan populations may occur south of the park (Singkil Barat) and northwest of the park (Takengon-West). Field surveys in these areas are needed to establish detailed information on the distribution of orangutans. Given the currently estimated population of 5,850 animals, the following recommendations were suggested to minimize the loss of genetic variability in the greater Gunung Leuser population:

- Add forested lowland areas to the park wherever possible because the lowland and swamp forests are optimum orangutan habitats.
- Restore the connection between the western and eastern populations to create a larger unit.
- Refrain from building roads that bisect the protected area and modify the design of established roads so that they permit arboreal passage of orangutans and other animals.
- Maintain an absolute commitment to protection of the park. Specifically, buffer zones should be established wherever possible, e.g., sago plantations around the Kluet part of the park.

It is apparent that more research is needed to assess whether and to what extent animals in the highlands can survive without seasonal access to lowlands. To support the ecological data on the forest outside of the park where orangutans occur, it is recommended that:

- Regular field surveys be conducted on orangutan densities inside and outside Gunung Leuser National Park.
- A Geographical Information System satellite imagery map of vegetation and land use patterns be constructed for the long-term monitoring of available orangutan habitat.
- These two activities together serve to monitor long-term viability of wild orangutan populations.

A final conclusion was that the Ketambe Research Station in Gunung Leuser National Park is an important site for long-term field research on orangutans and should be given priority in its role to develop an orangutan management strategy.

ACKNOWLEDGMENTS

The database, interpretations and recommendations contained in this paper were derived at the Orangutan PHVA Workshop held in Medan, North Sumatra in January 1993, which was coordinated by the Indonesian Directorate General of Forest Protection and Nature Conservation (PHPA) and the Survival Service Commission (SSC) of the International Union for the Conservation of Nature (IUCN) Conservation Breeding Specialist Group (CBSG) (See Soemarna, Ramono and Tilson, this volume, for common acknowledgements). In addition to the individuals cited in Soemarna, et al., we wish to acknowledge K. Soemarna and W. Ramono.

REFERENCES

Tilson, R., U. Seal, Komar Soemarna, et al. 1993. *Orangutan Population and Habitat Viability Analysis Report.* IUCN/SSC Conservation Breeding Specialist Group: Apple Valley, MN. 54 pp.

Tilson, R., Komar Soemarna, Widodo Ramono, Sukianto Lusli, K. Traylor-Holzer, U. Seal. 1994. *Sumatran Tiger Population and Habitat Viability Analysis Report.* IUCN/SSC Conservation Breeding Specialist Group: Apple Valley, MN. 124 pp.

13

ESTIMATES OF ORANGUTAN DISTRIBUTION AND STATUS IN BORNEO

H. D. Rijksen,[1] W. Ramono,[2] J. Sugardjito,[3] A. Lelana,[3] M. Leighton,[4] W. Karesh[5] G. Shapiro,[6] U. S. Seal,[7] K. Traylor-Holzer,[8] and R. Tilson[7,8]

[1] Instituut voor Bosbouw en Groenbeheer
[2] Indonesian Directorate General of Forest Protection and Nature Conservation
[3] World Wide Fund for Nature
[4] Harvard University
[5] NYZS/International Wildlife Conservation Park
[6] Orangutan Foundation
[7] IUCN/SSC Conservation Breeding Specialist Group
[8] Minnesota Zoo

ABSTRACT

At the Population and Habitat Viability Analysis (PHVA) Workshop for Orangutans, estimates of habitat and population numbers for orangutans (*Pongo pygmaeus pygmaeus*) in Borneo were derived from various sources. For Borneo, the known distribution of orangutans comprises eight regions with currently isolated populations. The total area of orangutan habitat on Borneo was calculated at 22,360 km^2, and the estimate of total population numbers ranged from a minimum of 10,282 to a maximum of 15,546 individuals. These figures suggest a more serious decline in the Bornean population than was previously thought.

INTRODUCTION

This paper, from the Population and Habitat Viability Analysis (PHVA) Workshop for Orangutans (*Pongo pygmaeus*) (Tilson, et al., 1993) focuses on the status of wild populations of orangutans on Borneo (*Pongo pygmaeus pygmaeus*). The known distribution of orangutans on Borneo (Kalimantan, Sabah and Sarawak) comprises eight regions with currently isolated populations. No published or unpublished data are available on population numbers for any of these areas with the exception of Gunung Palung, Kalimantan, where long-term research is being carried out by M. Leighton.

The Neglected Ape, Edited by Ronald D. Nadler et al.
Plenum Press, New York, 1995

METHODS

For each region the area of forest in square kilometers (km^2) and orangutan density based on habitat type were estimated using a correction factor to compensate for the inherent overestimation caused by lowland prime habitat. An additional 5,000 km^2 was added to account for regions where significant orangutan populations may occur in unprotected or unidentified areas. The total area of orangutan habitat on Borneo was calculated at 22,360 km^2, and the estimate of the total population ranged from a minimum of 10,282 to a maximum of 15,546 individuals. These figures suggest a more serious decline in the Bornean population than was previously thought.

A degree of specificity comparable to that for Sumatra could not be achieved for Borneo in identifying threats to known populations because of inadequate data. An "estimate of impact" of a variety of threats on population survival was calculated instead. The primary recommendation stemming from this exercise was that protection of existing national parks and other protected areas should be improved. At least 60% of the present orangutan populations on Borneo could be protected by implementation of current authority or protection laws. As an adjunct to enhanced enforcement, logging and habitat degradation should be banned in parks and proposed conservation areas and funding should be secured for boundary demarcation.

An estimate of size is given for each habitat area and an estimate of orangutan density is given, based on habitat types. Additionally, a correction factor (0.6) is used to compensate for the inherent overestimation due to prime, high density study sites and the extrapolation of these data to outdated map information. These two figures provide an upper and lower estimate of population size (Table 1).

DISTRIBUTION

The following eight areas were identified as currently isolated populations of orangutans on the island of Borneo (Kalimantan, Sabah and Sarawak).

Table 1. Estimates of population size of orangutans on Borneo

Area	Km^2	Area density	Minimum estimate	Maximumestimate
Sabah[1] 1,520	1,520	0.5	456	760
Kutai/Sanlieu	2,100	1.0	1,260	2,100
Central[2]	6,800	0.4	1,632	2,720
B. Raya/Balia	1,540	0.4	370	616
T. Puting	1,800	1.0	1,080	1,800
G. Palung[3]	700	3.0	2,100	2,100
	1,100	1.5	1,650	1,650
Kendawang	800	1.0	480	800
G. Nuit/Becapa	1,000	0.5	300	500
Additional[4]	5,000	0.5	1,500	2,500
Totals	22,360		10,830	15,546

[1]Figures for Sabah are based on current protected areas where naturally existing populations of orangutans are known to exist. Orangutans living in unprotected areas or at rehabilitation facilities are not included.

[2]This area includes the National Parks and protected areas in the border area of West Kalimantan and Sarawak (Batang Ai, Lanjak Entimau, Bentuang, Karimun).

[3]The figures for the Gunung Palung area include higher density figures for prime habitat within the park and also lower density areas within the park and the adjacent area on the west. Due to confidence levels with these data, *the 0.6 correction factor was not used.*

[4]An additional estimate of 5000 km^2 was added to account for areas where significant orangutan populations may exist in unprotected or unidentified areas.

THREATS

An estimate of impact by a variety of factors was developed (see Table 2).

GENERAL CONCLUSIONS

Indonesia has a law for the protection of orangutans, an excellent network of conservation areas which covers more than 60% of the supposed current range of orangutans in Kalimantan and a substantial organization for Protection of Forests (habitat) and Nature Conservation. Because the orangutan population of Indonesia is subject to increasing danger of persecution, habitat degradation and fragmentation, the participants of the workshop drew the following conclusions for the species and habitat:

- The major threat to the orangutan population is the loss of adult females and the low rate of population increase which is natural for this species.
- Given the life history of orangutans, continuous vigilance and strengthened enforcement of existing laws is required to protect existing populations because they are unable to withstand significant levels of removal through poaching.
- The major threat on the habitat level is habitat loss, degradation and fragmentation, especially in lowland forests.

Kalimantan

Vast areas of Borneo require thorough field surveys on a park-to-park basis to determine the status of the orangutan population and habitat types in relation to forest status. The development of a management strategy of the current metapopulations of the Bornean orangutans must also be initiated in the near future. Because of the fragmentation of populations and the need for metapopulation management, it is essential that genetic studies regarding intra-subspecific variation be completed immediately. Until this is done, it is important not to mix (through reintroductions) orangutans from widely separated populations.

Sabah

The orangutan population of Sabah is an essentially isolated population. Given the likely small number of animals and limited protected habitat, the appropriate authorities should be aware of the serious need for aggressively managing this population. Because the populations are isolated they are severely threatened.

Kutai Region

Most of the orangutans in this area (1,200 - 2,100 animals) are contained in Kutai National Park and the unprotected and protected areas to the north. A considerable gap currently exists in the center of this area. Expansion of human activities threatens to eliminate the northern half of this population (1,000 animals). The park itself is also under threat of conversion to mining activities; this would result in the loss of the other half of the population (1,000 animals).

The current status of orangutans, coupled with habitat loss and fragmentation, requires increased management efforts and immediate habitat protection to preserve this population. The northern part of this area has not been surveyed in decades. Information on

Table 2. Estimates of impact of various factors on wild orangutan populations

Cause	Sabah	Kutai	T. Puting	G.Nyiut/Becapa	*Central*	*B.Raya/B.Baka*	*G.Palung*	*Kendawangan*
Habitat destruction[1]								
Forest conversion/transmigration	0	0	0	100% of 30%	50% of 25%	50% of 25%	95% of 25%	75% of 25%
Mining	0	50% of 50%	95% of 1%	0	0	0	0	0
Recurrent fires	0	25% of 60%		0	0	0	0	0
Fragmentation (permanent)[2]	0	cut in half						
Logging	0	100% of 5%	0	0	0	0	0	0
Agricultural development	0	100% of 15%	0	0	0	0	0	0
Mining	0	100% of 5%	0	0	0	0	0	0
Roads	0	100% of 1%	0	100% of 1%	0	100% of 1%	0	100% of 1%
Degradation (temporary quality decline)[3]								
Logging	50% of 50% dr	0	100% of 5% dr	0	0	95% of 50% dr	0	0
Agricultural encroachment	0	100% of 10%	0	0	0	0	0	0
Fire	0	100% of 50% dr	100% of 20% dr	0	0	0	10% of 50%	0
Drought	10% of 50% dr	0	0	0	0	0	0	0
Illegal hunting	0.5%	2%	95% of 2%	95% of 2%	unknown	unknown	unknown	
Disease[4]	unknown	unknown	unknown	unknown	unknown	unknown	unknown	unknown

Figures expressed as % probability of occurrence in 100 years of either a % habitat loss or a % animal density reduction (dr).

[1]Permanent habitat destruction, generally set by policy decisions; Forest conversion/transmigration=intentional changes in land use practices, destroying habitat for other uses; Recurrent fires=resulting in permanent habitat destruction.

[2]Activities which separate existing continuous populations into subpopulations permanently; Agricultural development = permanent establishment of areas of agricultural use.

[3]Temporary qualitative decline of habitat, resulting in reduction of orangutan densities; Logging=illegal or legal selective logging or non-mechanized logging; Agricultural encroachment=local people clearing forest for short-term use.

[4]Introduction of a new disease or health factor that is currently not playing a role in the population dynamics of orangutans. A general figure is given that covers infectious diseases, toxicologic events or the introduction of other pathogens.

the wildlife in this area is essential for further management decisions. Given the severe reduction of orangutan populations in East Kalimantan, serious consideration should be given to establishing additional populations of East Kalimantan animals.

Central Borneo

Little good ground data are available on the Kalimantan side of this area. Surveys in this area are essential to establish the degree of fragmentation, hunting pressure and relative orangutan densities. Logging activities threaten this area; control of logging is essential to maintaining orangutan habitat. The other threat to the orangutan in this area is from hunting. Enhancing enforcement activities in the west Kalimantan and Sarawak portions would significantly enhance the possibility for survival of the orangutan populations.

Bukit Raya/Bukit Baka

This is a small orangutan population and it is not clear if connections with populations to the south at Tanjung Puting still exist. Surveys to determine the range of this population are considered essential. Control of logging and agricultural expansion in this area is critical to protecting this small population. Extension of protected habitat across permanently forested areas towards, or joining up with, the other major conservation areas (Gunung Bentuang/Karimum and/or Tanjung Puting) should be explored.

Tanjung Puting

Logging in the vicinity of the park and encroachment into the park need to be monitored and controlled to eliminate further reduction of orangutan populations in this area. Park expansion to the north could expand the habitat available and allow enlargement of this population. Elimination of mining activities and small scale hunting within the park would also significantly enhance the viability of this population.

Kendawangan

This small area being proposed for higher protection status is little known. Further studies are needed to evaluate this area and also to explore its connections with other orangutan populations. Transmigration is planned for the adjacent area to the south; threats from agricultural encroachment and habitat degradation must be controlled to prevent the loss of this small population.

Gunung Palung

There is a significant population of about 3,750 orangutans within the park and to the west of the park. The most serious threats to this population come from logging activities and agricultural development outside of the park. Expanding protection to the west of the park would save an additional 1,000 orangutans living outside of the park.

Gunung Nyuit

This is a small population of about 300-500 orangutans in a park that is particularly rich in biodiversity. The small park is severely threatened by agricultural encroachment and could be fragmented through the center by human activities. Enhancing the protection of this park is essential to maintaining the orangutan population.

ACKNOWLEDGMENTS

The database, interpretations and recommendations contained in this paper were derived at the Orangutan PHVA Workshop held in Medan, North Sumatra in January 1993, which was coordinated by the Indonesian Directorate General of Forest Protection and Nature Conservation (PHPA) and the Survival Service Commission (SSC) of the International Union for the Conservation of Nature (IUCN) Conservation Breeding Specialist Group (CBSG) (see Soemarna, Ramono and Tilson, this volume, for the common acknowledgements). In addition to the individuals cited in Soemarna, et al., we wish to acknowledge K. Soemarna.

REFERENCES

Tilson, R., U. Seal, Komar Soemarna, et al. 1993. *Orangutan Population and Habitat Viability Analysis Report.* IUCN/SSC Conservation Breeding Specialist Group: Apple Valley, MN. 54 pp.

14

CONSERVATION ACTION PLAN FOR ORANGUTANS IN INDONESIA

K. Soemarna,[1] W. Ramono,[1] S. Poniran,[1,2] C. P. van Schaik,[3] H. D. Rijksen,[4] M. Leighton,[5] D. Sajuthi,[6] A. Lelana,[6] W. Karesh,[7] M. Griffiths,[2,8] U. S. Seal,[9] K. Traylor-Holzer,[10] and R. Tilson[9,10]

[1] Indonesian Directorate General of Forest Protection and Nature Conservation
[2] Gunung Leuser National Park
[3] Duke University
[4] Instituut voor Bosbouw en Groenbeheer
[5] Harvard University
[6] Primate Research Center, Bogor Agriculture University
[7] NYZS/International Wildlife Conservation Park
[8] World Wide Fund for Nature
[9] IUCN/SSC Conservation Breeding Specialist Group
[10] Minnesota Zoo

ABSTRACT

The first Population and Habitat Viability Analysis (PHVA) Workshop for Orangutan (*Pongo pygmaeus*) focused on the status of wild populations of orangutans on Sumatra and Borneo, with major emphasis on the Sumatran population *(Pongo pygmaeus abelii).* At the Workshop, three working groups were established: Orangutan Distribution and Status in Sumatra, Orangutan Distribution and Status in Borneo, and Life History Characteristics and VORTEX Modelling. Recommendations from these working groups form the basis for the *Indonesian Orangutan Action Plan* presented here.

INTRODUCTION

Estimates before this workshop placed the number of Bornean orangutans (*P.p. pygmaeus)* between 19,000 and 30,000 and the number of Sumatran orangutans (*P.p. abelli)* at between 7,000 and 11,000. It is estimated that suitable orangutan habitat in Indonesia and Malaysia has declined by more than 80% in the last 20 years and that orangutan numbers have declined by 30-50% over the last 10 years.

The Neglected Ape, Edited by Ronald D. Nadler et al.
Plenum Press, New York, 1995

The major threat to orangutan populations is habitat conversion for timber, plantations and agriculture. Timber activities occur inside, as well as outside of protected areas. A secondary threat is capture for the pet trade. Current conservation measures for this species in the wild include increasing the effectiveness of rain forest protection efforts in Indonesia and Malaysia and research into, and implementation of, habitat restoration techniques. It is also proposed to utilize the orangutan as an umbrella species to protect rain forest areas of high biological diversity.

An international conference on the great apes, conducted in Indonesia in December 1991, identified many of the problems confronting the orangutan in the wild and proposed programs to address these problems. The more recent Workshop extends this process by conducting a Population and Habitat Viability Analysis (PHVA) on orangutans. The Workshop provided an in-depth assessment of factors that impact wild orangutan populations, explored the effects of various management options, and formulated more explicit objectives for management of this species. This process resulted in a series of recommendations for managing wild and captive populations and is referred to as the *Indonesian Orangutan Action Plan*.

INDONESIAN ORANGUTAN ACTION PLAN

Recommendations for Sumatra

On the basis of the best evidence available from the Orangutan PHVA Workshop, the population of orangutans in the Greater Leuser ecosystem should be no less than 10,000 individuals to ensure the long-term survival of the species. Specific recommendations to accomplish this are:

1. Give full protection to the South Bengkung.
1. Incorporate the Lesten valley and surrounding hills inside the Park.
3. Build no further roads through the Park to prevent fragmentation.
4. Reunite the two halves of Leuser National Park by relocating settlers in the Upper Alas valley and reforesting the vacated land.
5. Modify the road between Ketambe and Agusan so that there can be free passage of orangutans across this right of way.
6. Develop appropriate buffer zone plantations around key areas of the Park (e.g., Sago plantations around the vulnerable Kluet area of the Park).
7. Strengthen protection against the removal of orangutans, especially around the periphery of the Leuser Ecosystem; most orangutans are caught as they feed in durian orchards on the forest edge.
8. Maintain continuous forest cover over the Blangkejeren-Lokop road.

Recommendations for Borneo

At the species level:

1. Give special attention to the protection of orangutans according to the legal framework, notably, law enforcement pertaining to hunting, marketing and sale. Law enforcement can serve to diminish the killing of adult females carrying infants desired by the pet-market.

At the habitat level:

2. Give more attention and support to protective measures for existing conservation areas, especially with a focus on lowland and swamp forest.

3. Pay the utmost attention to boundary demarcation and zoning; standardize maps at the Ministry of Forestry. Seek consultation between the Directorate General for Forest Protection and Nature Conservation (PHPA) and the Directorate Generals of the Ministry of Forestry in cases where orangutan habitat outside of conservation areas is concerned.
4. Seek effective control of logging operations in relation to habitat requirements of orangutans and regulating timber concession activities.
5. Complete protected area network with permanent forested areas under special management (buffer zones, etc.) for orangutan habitat protection.
6. Restore degraded habitat and protect regeneration with respect to orangutan habitat requirements.
7. Reintroduce confiscated orangutans in permanent forested areas of good habitat quality which are isolated from other wild orangutan populations.

Recommendations for Reintroduction

The results of the PHVA demonstrate that *supplementing natural populations of orangutans has no conservation value in terms of enhancing population viability*. We expect that recent population declines in protected forests are due to habitat loss and degradation in most cases and to hunting in some cases. In addition, introduction of former captive orangutans into existing populations can have well-known negative effects, such as the damaging mixing of gene pools and the introduction of diseases. *Therefore, we strongly endorse the recommendations of the Reintroduction Group of the IUCN/SSC, emphasizing that reintroductions not occur into existing wild populations.*

In contrast, the analysis indicates that establishing a new viable population of ex-captive orangutans in habitat formerly occupied by orangutans, but where they do not now occur, would likely contribute to the viability of the metapopulations on both Borneo and Sumatra. The biological issues which should be considered in this strategy were not adequately discussed. However, if this strategy is pursued, the following guidelines are suggested:

1. Select areas to harbor a population of ex-captives on the basis of projections of population viability.
2. Population viability is maximized at maximal projected carrying capacity (*K*) of the area. The projected *K* of the selected area should be more than 500 individuals.
3. The carrying capacity can be estimated by mapping the mosaic of the habitats in the proposed area and then multiplying each habitat-specific *K* by the area of each habitat.
4. A mosaic of lowland forest habitats extending into hill forests, but more importantly, including swamp and/or peat forests, will probably provide maximum population densities. Because montane forests are well-represented in protected areas, the overall conservation value of the area from the perspective of overall biodiversity is also maximized by selection of lowland forest habitats.
5. Areas that include a large proportion of selectively logged forest, if in the lowlands, may be the best alternatives, as these forests will recover both in their overall conservation value and in their carrying capacity for orangutans, if protected from continued degradation.
6. An important value in establishing a new population is that it lowers the risk of overall extinction of the metapopulation on each island if catastrophic events or continued habitat destruction cause the extinction of local populations. Therefore, areas furthest and/or in a distinctive ecological setting from other established

populations might best contribute this benefit. However, this criteria for selecting areas for the establishment of populations of ex-captives should be secondary to the projection of the long-term viability of the population.

Recommendations for Medical Quarantine Prior to Reintroduction

Because there is already a sizable captive population of orangutans and because the long-term care and management of these captive orangutans is quickly exceeding the capabilities of organizations and individuals involved in their care, options for potential release sites and potential release candidates must be evaluated by the appropriate national and international agencies.

The confiscation of pet orangutans and their release back into a native habitat allowing contact with naturally occurring populations poses a serious threat to the health of wild populations of orangutans and other indigenous species. In order to mitigate this risk, serious efforts must be made to ensure the health of every animal prior to its entering a rehabilitation program aimed at release. At present, we cannot in good consciousness recommend the release of any orangutans into areas where viable populations of wild orangutans exist. Additionally, we cannot recommend the transfer to a rehabilitation program or the release of any ill or possibly chronically infected individuals. To ensure that only healthy animals are released, a period of quarantine and evaluation is necessary.

Before entrance into a rehabilitation program, orangutans should undergo the following:

1. A minimum quarantine period of 180 days, with records, observations, tests and treatments as outlined below. Upon admission to quarantine, a complete history should be obtained from the previous owner and a medical record initiated.
2. A one-week stabilization period to allow close scrutiny of behavioral patterns, food preferences and general condition and to allow the animal to adjust to the new environment. Observed medical problems may warrant immediate examination and attention.
3. A complete physical examination and permanent identification (tattoo and other methods, as appropriate) must be performed as soon as the animal has adjusted to quarantine or within one week of arrival (the sooner of the two). Follow-up examinations should be done any time chemical restraint is necessary for routine testing (e.g., tuberculosis (TB) or blood tests) or when the schedule concerns animals that are small enough to handle without anesthesia. Chest radiograph, Hepatitis-B surface antigen test, complete blood chemistry, chemistry panel and serum banking should be done upon initial examination and again at the end of quarantine.
4. Intradermal TB test on the upper eyelid, utilizing mammalian old tuberculin should be performed on initial examination and again on the second, third, and sixth month of quarantine.
5. The animals should be placed in separate housing quarters of sufficient space and with separate air and water circulation. Attention should be paid to maintaining psychological well-being in the face of this isolation.
6. Young animals may be housed in groups of two, though they must clear quarantine together at the end. Accommodations should be made on a case-by-case basis to provide more intimate human contact as needed with appropriate disease precautions considered.

7. Fecal bacteriological examination for salmonella, shigella, and campylobacter should be performed on samples collected at the initial examination and again at the end of the third and sixth month of quarantine.
8. Fecal parasite examinations should be done on samples collected during the initial examination and again at two, four, and six months of quarantine. The last examination should be done 14 days prior to the end of quarantine.
9. Routine anthelmintic treatment should be done every month using ivermectin, pyrantel pamoate, fenbendazole, mebendazole and/or any other appropriate medications. A rotating schedule may be used if necessary. Routine antiprotozoal treatments may be incorporated into the schedule if protozoal parasites are found in the confiscated animals as a whole.
10. Vaccination against polio, DPT, measles, hepatitis-B, and rabies should be performed as appropriate for each disease entity and only after blood sampling and adequate serum banking has occurred.
11. Genetic analysis for subspecific identification should be performed using karyotypic and electrophoretic methods at some time during the quarantine period.
12. In light of the current lack of available information regarding the definitive diagnosis, treatment, and epidemiological implications of various infectious diseases (most notably TB and Hepatitis-B), a positive test to any or all of these will disqualify an animal from release into areas with wild primates.
13. All animals that die during or following the quarantine period should undergo a full necropsy examination and histopathology. No animals in contact with the individual that died should be released from quarantine until the cause of death and all related abnormal findings have been reported in writing by a pathologist to supervising authorities for quarantine procedures.
14. Thorough and complete records should be kept at all times to facilitate proper tracking and control of animals, the study of diseases and treatments and reporting. Quarterly reports should be provided to supervising authorities and copies of all reports and records should be maintained in a central location.
15. Thorough training and health surveillance of quarantine staff should be a high priority. Poorly performing or ill staff members should not be permitted to work with the animals. Accurate records of surveillance will help to track any zoonotic episodes.
16. A manual containing all operating procedures should be prepared and kept on-site at each facility. This should delineate all quarantine procedures listed above, as well as those defining the activities of support and maintenance staff. An updated copy of this manual or these procedures must be kept on file with the Department of Forestry.
17. Procedures should be developed to deal with all situations involving animals that "drop-out" of or fail the quarantine process as described above.
18. All quarantine facilities, daily procedures and routines and staff management and procedures should meet the standards of primate quarantine and handling accepted internationally in the biomedical field. Inadequate facilities or procedural aspects will invalidate all of the efforts for disease control or surveillance. These standards must be met at any facility used for the quarantine of orangutans.
19. A special panel of veterinarians, public health experts, ecologists and others should be formed under the supervision of the Department of Forestry (PHPA) to evaluate current and ongoing research that will facilitate the development of appropriate protocols and plans for dealing with the health issues related to orangutan rehabilitation.

ACKNOWLEDGMENTS

The database, interpretations and recommendations contained in this paper were derived at the Orangutan PHVA Workshop held in Medan, North Sumatra in January 1993, which was coordinated by the Indonesian Directorate General of Forest Protection and Nature Conservation (PHPA) and the Survival Service Commission (SSC) of the International Union for the Conservation of Nature (IUCN) Conservation Breeding Specialist Group (CBSG).

REFERENCES

Sajuthi, D., Karesh, W., McManamon, R., Martin, H., Amsel, S., and Kusba, J., 1992, Recommendations to the Department of Forestry of the Republic of Indonesia on the medical quarantine of orangutans intended for reintroduction. *Proc. Great Apes Conf., Ministry of Forestry and Ministry of Tourism, Post, and Telecommunication, Jakarta.*

15

POPULATION ESTIMATES AND HABITAT PREFERENCES OF ORANGUTANS BASED ON LINE TRANSECTS OF NESTS

C. P. van Schaik,[1] A. Priatna,[2] and D. Priatna[2]

[1] Department of Biological Anthropology and Anatomy and Center for Tropical Conservation, Duke University
3705B Erwin Road, Durham, North Carolina 27705 and
New York Zoological Society - The Wildlife Conservation Society
[2] Universitas Nasional, and
Biological Sciences Club
Jl. H. Noor 10, Pasar Minggu, Jakarta Selatan, Indonesia

ABSTRACT

Effective conservation of a species requires accurate information on its geographic distribution and on its densities in the range of habitats it occupies, as well as estimates of its total numbers. For the orangutan, widely divergent estimates of total numbers have been produced over the years. In this study, we used line transects of nests to estimate population densities of Sumatran orangutans. The parameters needed for this technique were obtained from large nest samples by S. Djojosudharmo and T. Mitra Setia, and from field trials. The accuracy of the method could be validated at Ketambe, a site with known density. We found that orangutan densities are highest in forests on floodplains, that they strongly decline with altitude and that this decline is most plausibly ascribed to declines in the abundance of fruits with fleshy pulp. We note that densities in one site may vary due to movements between habitats. We also present preliminary evidence that floodplain forests may act as keystone habitats for orangutans living in the adjacent hills and thus, subsidize orangutan densities in adjacent uplands.

INTRODUCTION

Fundamental prerequisites for effective conservation of any species are accurate information on its range and on its densities in different habitats, insight into the ecological factors that determine habitat suitability, and, if the species is in decline, understanding of the causes of decline.

The Neglected Ape, Edited by Ronald D. Nadler et al.
Plenum Press, New York, 1995

There is little doubt that the orangutan (*Pongo pygmaeus*) has undergone a dramatic decline. In the Pleistocene, its range extended from Java well into Indo-China (Von Koenigswald, 1982). Currently, orangutans occur only on Sumatra and Borneo, in two morphologically distinct subspecies, but even in the early decades of this century, orangutans were still recorded far south of its current southern boundary on Sumatra, the Lake Toba area (Rijksen, 1978). Accurate information on its range has become available only in the last few years (Rijksen, this volume).

The most parsimonious explanation for this decline is hunting pressure by humans. First, hunting by the indigenous people of Borneo and Sumatra has been reported ever since the earliest western travelers reached the region (see, for example, Wallace [1869]). Second, the disproportionately high abundance of orangutan remains in caves inhabited by late Pleistocene humans indicates that these recent immigrants were already hunting orangutans (but see Harrison, in press).

Indeed, the most unusual thing about the orangutan is perhaps that it has escaped the fate that befell so many of the other large mammals in areas colonized by anatomically modern humans (Martin and Klein, 1984). It has exactly the right life history to make it extremely vulnerable to hunting. Its development is very slow, with a mean age at first reproduction for wild females of about 15 years and a mean interbirth interval of about 8 years (Galdikas and Wood, 1990), the slowest for any primate species. As Leighton, et al. show, even a slight increase in the mortality of adult females, as would be brought about by hunting, could lead to population collapse (cf. Dobson and Lyles, 1989).

The remaining range on Sumatra and Borneo coincides with areas where, for at least several centuries, people have not hunted orangutans for religious reasons (MacKinnon, 1992). Hunting is still a major threat, but in recent years, habitat loss and modification have become the major cause of their decline. Obviously, orangutans cannot survive if forest is converted into simpler habitat, even if such habitat still contains trees. In northern Sumatra, for example, long-tailed macaques (*Macaca fascicularis*) and leaf monkeys (*Presbytis cristata* and *P. thomasi*) are found in rubber plantations, but orangutans are not. More importantly, orangutans are also adversely affected by selective logging, with densities in logged areas about one-half to one-third of those in pristine habitat (Rijksen 1978; van Schaik and Azwar, 1991).

It is easy conceptually, therefore, to put an end to this decline: stop the hunting and stop habitat conversion and degradation. If enough unexploited natural forests can be set aside, there is no reason why the species could not survive, provided the best forests can be identified, within the current range and with the optimum habitat conditions for orangutans. Survey efforts should focus, therefore, on delineating the current range and on identifying the optimum habitats of orangutans and their determinants.

Until recently, confusion reigned as to the total number of orangutans in the wild. Published estimates of the total number of orangutans varied from approximately 4,000 (Harrisson, 1962) to about 180,000 (MacKinnon, 1987). The best current estimates are somewhere in between (see van Schaik, et al., this volume, for Sumatra; Rijksen et al., this volume, for Borneo). In this chapter, we present the information on densities that was used to produce these numbers for Sumatra, obtained mainly by nest survey techniques in and around the Gunung Leuser National Park, in northern Sumatra (Indonesia). We also present some data on habitat suitability and movements of orangutans that are relevant for proper management of the remaining wild populations.

Estimating Oranguan Density

Until recently, reliable density estimates were available only for the handful of sites where long-term studies were conducted, and these sites may have been selected because of their unusually high densities. Thus a rapid, accurate field method was required to cover many sites over larger areas.

Accurate estimates of primate densities can be obtained, in principle, using line transects (e.g., Brockelman and Ali, 1987). However, it is easy to see that reliable estimates based on direct sightings require unrealistically large sample sizes with animals as rare and cryptic as orangutans. For example, with a density of 1 ind.km^{-2} and an effective strip width of 20 m on either side of the transect, one will obtain one sighting per 25 km of transect. Thus, we need hundreds of kilometers of survey walks before a reasonable estimate is obtained.

All great apes make nests in which to rest during the day or to spend the night. Not surprisingly, therefore, field workers have attempted to estimate great ape densities by counting nests along transects (e.g., Ghiglieri, 1984). Nests are much more commonly encountered than the animals themselves, and being visible longer, fluctuate less in density over time at a single locality. The earliest studies of orangutans, by necessity, made untested assumptions about the area sampled and used crude estimates of the production and disappearance rates of nests (e.g., Schaller, 1961). As a consequence, nest-based estimates were dismissed as unreliable (e.g., Rijksen, 1978) and, at best, as providing an index of density.

However, line transect techniques are now available that make it possible to obtain reasonably accurate estimates of nest densities. Furthermore, long-term studies of habituated animals have made it possible to get better estimates of nest parameters, as demonstrated at Ketambe by Rijksen (1978) and more recently by Suharto Djojosudharmo (unpublished). Admittedly, the technique uses so many parameters that estimates have very wide confidence limits and thus, limited reliability. Moreover, some bias in the estimation of some crucial parameters is inevitable. Conservation is best served, however, by absolute density estimates, because that allows estimation, however crude, of absolute numbers. Even a ballpark figure, provided it is not badly biased upward, is better than no figure at all (Krebs, 1989).

This chapter, then, has two aims. First, an attempt is made to validate the line transect method in order to provide absolute estimates of orangutan densities. The method should be applicable to other nest-making apes as well. Second, the method is used to identify the ecological factors and processes that influence density. The results have been used to estimate orangutan population size in the Gunung Leuser National Park, in northern Sumatra (see Faust, et al., this volume).

METHODS

Line Transects

In principle, nest counts could be obtained by carefully searching a plot of known size, but this method has some drawbacks. First, nests tend to be spatially clumped, and the ensuing large variability between plots makes it necessary to sample many plots for a reliable estimate. Second, an unknown number of nests will inevitably be missed. Third, this method is logistically more difficult in that plots have to be delineated, which is often cumbersome in difficult terrain (swamps, steep slopes). The line transect technique is less affected by these problems.

The line transect method was developed for mobile animals in open landscapes, but it has also been successfully applied to forest primates (e.g., Brockelman and Ali, 1987). Basically, it attempts to estimate the width of the strip around a transect line in which animals are located. This width (w) is estimated from a detection function fitted to the observed distribution of perpendicular line-to-animal distances; the parameter of the detection function is efficiently estimated using the nonparametric Fourier series method (Burnham, et al., 1980). Then, density can simply be estimated as:

$$d = N/(L \times 2w) \tag{1}$$

in which:

- d = density (animals/km^2),
- N = number of animals observed along the transect,
- L = length of transect covered (km), and
- w = estimated width of the strip of habitat actually censused (km).

Although the objects to be detected usually are animals, they can also be signs produced by the animals, such as nests. The equation then, must be extended to take into account the rate at which nests are produced, the rate at which they disappear and the proportion of animals that actually build nests:

$$d = N/(L \times 2w \times p \times r \times t) \tag{2}$$

where:

- p = proportion of nest builders in the population;
- r = rate at which nests are produced (n/day/individual),
- t = time during which a nest remains visible (in days).

The parameters w and t are likely to vary across habitats (e.g., Ghiglieri, 1984). The value of w should depend on the visibility of the forest, hence, on forest structure, which is known to vary among habitats. The value of t may depend on the nature of the wood and on temperature and humidity. The values of t and w must, therefore, be estimated for each separate site. Line transects allow for the estimation of w. A new technique is presented to approximate t.

The values of p and r are not expected to vary much and were taken from previous studies. Small infants do not make nests. In both the Ketambe and the Suaq Balimbing population, around 10 % of individuals are small infants (Ketambe: 4 out of 39, around 1990 [T. Mitra Setia, personal communication]; Suaq Balimbing, 4 out of 40, around 1994). Thus, p is estimated as approximately 0.9.

Rijksen (1978) gave a figure of 1.8 for r, based on 36 orangutan days at Ketambe. However, individuals produce nests at different rates. Adult females with infants or juveniles produce the most, along with subadult males (which often associate with these females); adult males and nulliparous females produce fewer nests (Table 1). Juveniles make day nests and, as they get older, begin to make their own night nests too (near the mother's). Nests are reused only rarely. Thus, T. Mitra Setia (personal communication), based on a large sample of 437 orangutan days in the same area, arrived at r = 1.7, weighted for population composition; van Schaik, based on 134 days at Suaq Balimbing, arrived at r = 1.6. The average of 1.7 was used in the present study.

Clearly, systematic variation in the composition of the population would also affect the values of p and r. However, as shown in Table 1, this variation would affect r only if there were appreciable changes in the representation of adult males. Attempts have been

Table 1. Variation in nest production rates (nests/day) among age-sex classes for two Sumatran sites (Ketambe and Suaq Balimbing)[1]

Site and sample size (days)	Ketambe (437)	Suaq Balimbing (134)
Adult males	1.2	1.4
Adult femaleswith infant	2.0	1.5
Subadult males	1.9	2.1
Nulliparous females	—	1.4
Juveniles	—	—
Overall mean	1.7	1.6

[1]Data from Ketambe were made available by T. Mitra Setia.

made to recognize population composition from the height, size and position of nests (van Schaik and Azwar, 1991; E.A. Fox, unpublished), but they were not successful. Moreover, because there is no evidence for appreciable variation in age-sex composition across sites where populations have been studied intensively, we ignored possible between-site variation in r or p.

Sites

A total of 15 sites, all of them in or near the Gunung Leuser National Park, Sumatra, Indonesia, were selected on the basis of accessibility and habitat type and sampled for orangutans. The map of Figure 1 shows their location; Table 2 presents forest type and altitude of each. We either cut new trails or used existing wildlife trails. In one site (Kluet) we used an existing human foot path. In most places there was little human traffic. Except possibly at Lau Kawar, there is no hunting at any of these sites.

Table 2. The sites visited in this study, their habitat features, observed nest counts and parameters used to estimate densities

Site	Forest type	Altitude of nest (m)	Count[1] (N/km)	Visibility (d)	Estimated density (N/sq. km.)
Suaq Balimbing	Fresh water + peat swamp	10	33.5	69.9	6.9
Sikundur	Secondary (alluvium)	10	4.3	71.6	1.4
Sikundur[2]	Selectively logged (low hills)	40	2.7	71.6	1.2
Pucuk Lembang	Selectively logged (alluvium)	40	5.0	71.6	0.7
Manggala	Alluvial-hill	150	4.9	77.8	1.2
Ketambe	Alluvial-hill	375	24.4	92.4	5.2
Bukit Lawang	Hill dipterocarp	500	11.3	101.6	2.2
Bengkung	Hill dipterocarp	700	12.2	118.3	2.0
Ketambe-2	Submontane	1175	10.2	170.0	1.2
Mamas	Submontane + alluvium	1325	7.1	190.5	0.7
Ketambe-3	Montane	1425	7.5	205.6	0.7
Deleng Megaro	Montane	1475	4.0	213.6	0.4
Lau Kawar[2]	Montane	1500	0.4	217.7	0.04

[1]Based on the first sample collected by the author for a particular site.
[2]Sites not sampled by the author.

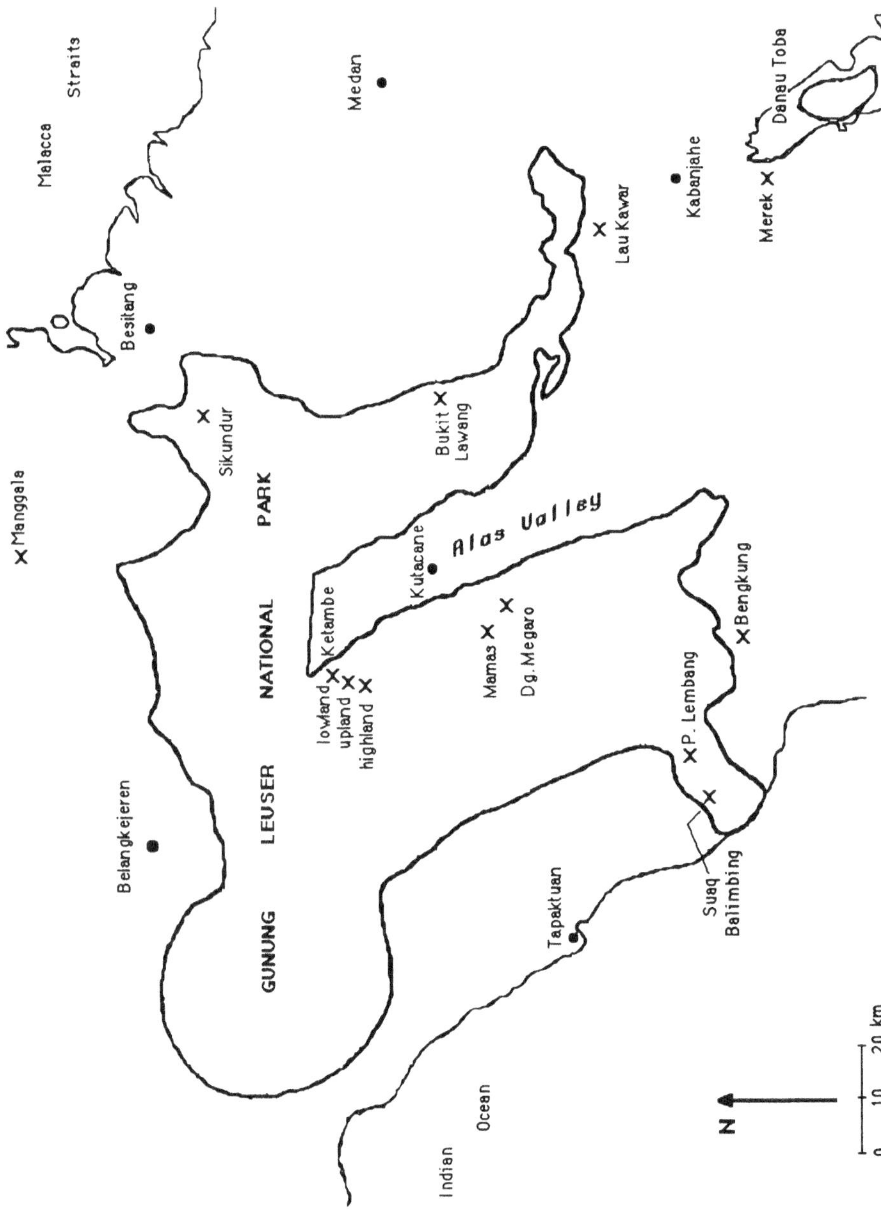

Figure 1. A sketch map of the Gunung Leuser National Park, northern Sumatra, and the location of the sites where the orangutan nests were sampled (indicated by X).

Details of the first nest samples are presented in Table 2 (for further details, see van Schaik and Azwar, 1991). Observations were made between June, 1990 and March, 1994. Many samples were collected between July and October, 1990; others were added in subsequent years and used where needed.

Field Procedure

In the field, two observers walked very slowly along the transect line, often pausing and looking back, while recording all nests observed. For this report, two features of each nest that was discovered are relevant: (i) the perpendicular distance from the transect line in m and (ii) the stage of decay of the nest, using four classes: a - fresh; leaves are still green, b - older; leaves may still be attached and the nest is still firm and solid, c - old; leaves are gone and holes are visible in the nests, d- very old; twigs and branches are still present, but no longer in the original shape of the nest. The transect line was also used to describe habitat features of the forest. Djojosudharmo and van Schaik (1992) had examined the altitudinal gradient at Ketambe and concluded that the best ecological predictor of the decline in density was the abundance of either strangling fig trees or that of fruits with fleshy pulp, in general. Both form a major component of the orangutan's diet and are eaten disproportionately. The current study makes it possible to test these hypotheses. The number of strangling fig trees visible from the trail was counted and their abundance was expressed as number/km of trail. Fruits found on the trail were classified according to their type (for definitions see Djojosudharmo and van Schaik, 1992) and their abundance. Abundance was expressed as the number of fruit sources per km. Several sites were visited only once and at different times of the year. Because there is seasonal variation in fruit abundance and this becomes more extreme at increasing altitude (van Schaik, 1986; Djojosudharmo and van Schaik, 1992), the absolute amount of fruit with fleshy pulp is not a suitable indicator of their availability. The relative abundance of fleshy-pulped fruit was adopted as the best indicator, because it is less variable than the absolute abundance, except during mast fruiting episodes (van Schaik, 1986); thus, one such event was excluded from consideration for the Manggala site.

Estimating *t*

In order to assess the variation in *t*, a technique was developed that does not require following a nest cohort and hence, long-term presence and "follows" of orangutans at a site. It relies instead on re-recording the stage of decay of nests of known initial stage of decay. This allows for the estimation of the transitions rates between decay stages. The transitions are considered to represent a Markov chain with an absorbing state (= the nest is gone). This allows one to calculate the expected number of steps it takes to reach the absorbing state (see, e.g., Kemeny, et al., 1956), and hence, the mean duration of visibility. The number of steps to disappearance can be determined by taking the transition matrix among the non-absorbing states, Q (nest stages a through d). The fundamental matrix, N, of this Markov chain, is then defined as:

$$N = (I-Q)^{-1},$$

where I is the identity matrix. The entries in N give the expected number of steps in each of the non-absorbing states from each possible non-absorbing starting state. From this matrix we can easily determine the estimated number of steps from stage *a* until absorption (nest disappearance).

To minimize the influence of information on the stage observed in the first census, the only information of the first census taken into the field was the location.

RESULTS

Methodological Aspects

Testing Assumptions. The line transect method makes a number of assumptions (Burnham, et al., 1980; Buckland, et al., 1993). In the case of nests, the most important assumptions are (i) that all objects located exactly on (or above) the trail are in fact discovered, and (ii) that the transect is located randomly in the habitat, i.e., independently of the dispersion of habitat features. These assumptions are probabaly violated, so the importance of these violations must be assessed.

Completeness of Sampling. The first assumption is easily met in habitats such as grasslands, but not in a rainforest with a multi-layered canopy, especially when the objects do not flee or deteriorate, like either orangutans or their nests, respectively. During censusing, special attention was given to the canopy immediately above the transect line. Nonetheless, if some objects right above the trail had been missed, this would produce an incorrect estimate of strip width and hence, of density. This possibility was tested by having groups of other observers carefully check the transect covered by the regular census team. In one such test, the regular team had missed 4 of 37 nests (11%); in another 4 of 25 (16%), although the regular team always discovered more nests than the checking team. None of these nests, however, were above or close to the trail. Hence, the estimates are unlikely to be far too low because of the compensating effect on w of missing nests far from the transect line (Burnham, et al., 1980).

Transect Line Location. The second assumption is that the trail is randomly located relative to landscape features such as rivers or ridges, unless one specifically intends to stratify the sample. In lowland areas, we tended to follow elephant trails or man-made trails in existing study areas. This probably does not cause major bias, although both kinds of trails tend to avoid patches of dense undergrowth or tree falls. We did not get the impression that using wildlife trails caused any difference in density, especially since the density of the trails in the lowlands is very high.

In rugged terrain, animal trails tend to follow ridges. This latter activity may produce seriously distorted orangutan densities if the apes show a preference to nest close to ridges or to avoid them. At the Ketambe highlands (mean altitude 1175 m), we checked for the presence of such a systematic bias. We cut a total of 1390 m of transects on the (steep) slopes and found 8 nests (a count of 5.8 nests/km), quite close to the 6 nests/km obtained on the 1500 m of established ridge trail. Hence, using existing animal trails is unlikely to bias the estimates of orangutan density.

Estimating the Parameters

Estimating t. For Ketambe, Rijksen's (1978) small sample indicated a median value of 81 days for t, whereas, Djojosudharmo's (unpublished) large sample produced a mean t value of around 90 days. The latter value is adopted here for Ketambe. However, we expect that time to disappearance of a nest depends on wood characteristics of the tree involved and on temperature and humidity, and thus, this number should not be used for all sites.

Table 3. Transition matrices for three samples with sufficiently large transitions between stage *d* and "nest gone" to allow for estimation of *t*

a. Ketambe lowland 1990 (n = 83 nests; interval = 31 days)

	First survey			
	a	b	c	d
Second survey				
Stage a	0	0	0	0
Stage b	6	3	0	0
Stage c	2	18	7	0
Stage d	0	3	14	6
"gone"	0	2	11	11

b. Ketambe highlands, combined 1990 (n = 35; interval = 31 days)

	First survey			
	a	b	c	d
Second survey				
Stage a	0	0	0	0
Stage b	2	5	0	0
Stage c	0	4	11	0
Stage d	0	0	5	4
"gone"	0	0	2	2

c. Suaq Balimbing, 1992 (n = 106; interval = 17 days)

	First survey			
	a	b	c	d
Second survey				
Stage a	0	0	0	0
Stage b	6	3	0	0
Stage c	9	15	14	1
Stage d	2	8	14	21
"gone"	0	0	3	10

In order to estimate *t*, we rechecked all the nests recorded in one sample after a certain interval of time and then reassigned the stages. The first requirement was to determine the optimal interval between subsequent assessments of the decay stage of each nest. It turned out that stage *d* remained visible the longest and that in order to get a reasonable estimate of the transition form, stage *d* to "nest gone", either long intervals (over 20 days) or very large nest samples were required. Four of our samples (with intervals of less than 21 days) had to be discarded because no nests had disappeared. Only three suitable samples remained (Table 3).

The effect of vegetation type could be assessed because two samples with almost identical intervals were available from Ketambe and Bukit Lawang, which have different vegetation, but are similar in altitude. Although the two particular transition matrices could not be used for estimating *t* because of the low transitions from stage *d* to "gone", they are similar in structure. This suggests that vegetation composition is not a major determinant of nest decay rates.

As expected, the remaining samples show a strong effect of altitude on disappearance time. Seasonal droughts can influence the decomposition rate of litter and hence, that of a nest as well. However, neither van Strien (1985) nor Djojosudharmo (unpublished) found

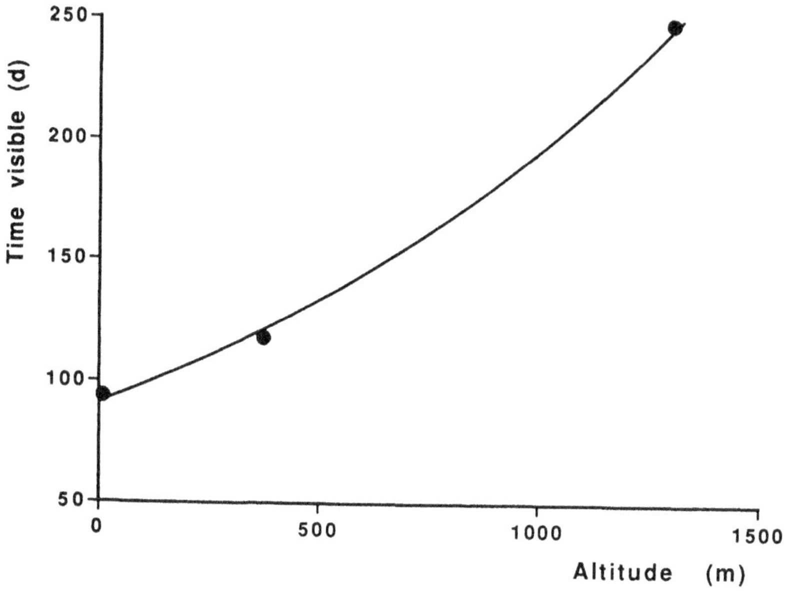

Figure 2. The estimated total duration of nest visibility in three sites of different altitude (lowest is Suaq Balimbing: n = 106 nests, interval = 17 days; intermediate is Ketambe: n = 83, interval = 31 days; highest is Ketambe submontane and montane combined; n = 35 nests, interval = 31 days). The curve is a fitted exponential function (see text).

appreciable differences in total amount or temporal distribution of rainfall between lowland (< 500 m) and nearby highland (1,000 - 1,500 m) areas. If anything, cloudiness increases with altitude, which would lead to higher average relative humidity and high rates of decomposition. None of the sites examined here have dry seasons with less than 100 mm of rain per month. Given this climatic similarity, it is probably safe to assume that the decomposition process is mainly temperature-dependent. Mean air temperature is a linear function of altitude. In this region, mean temperature decreases linearly with altitude (by about 0.625°C per 100 m increase in altitude from a sea-level value of approximately 27°C [van Beek, 1982]). The relationship between altitude and duration of visibility (expected to be exponnential; cf. Schmidt-Nielsen, 1975) for the three points was determined by a regression analysis ($r^2 = 0.997$; $P < 0.05$; Fig. 2).

The re-census method gave a value of t = 118 days for Ketambe, clearly higher than the values obtained by Rijksen and Djojosudharmo. However, a larger value is to be expected. Nests vary in firmness, depending upon who made them, whether they were made for the day or for the night, and depending upon the tree in which it was made, thus leading to a wide range of decay rates. In particular, day nests, which are quickly constructed and tend to be flimsy, are known to disappear very quickly (sometimes within days). This creates enormous variation in decay rates, and when one starts the observation of a cross-sectional sample (those nests observed during the first nest count) at a point when the most flimsy nests have already disappeared, this should produce higher estimates of t. Expressed differently, the higher a nest's t value, the more likely the nest is to be observed in a cross-sectional sample of nests. In order to remove this bias, the estimated t values were corrected by multiplying them by 0.76 (90/118). In the subsequent calculations, we used t as obtained from the regression equation in Fig. 2, multiplied by 0.76 (see Table 2).

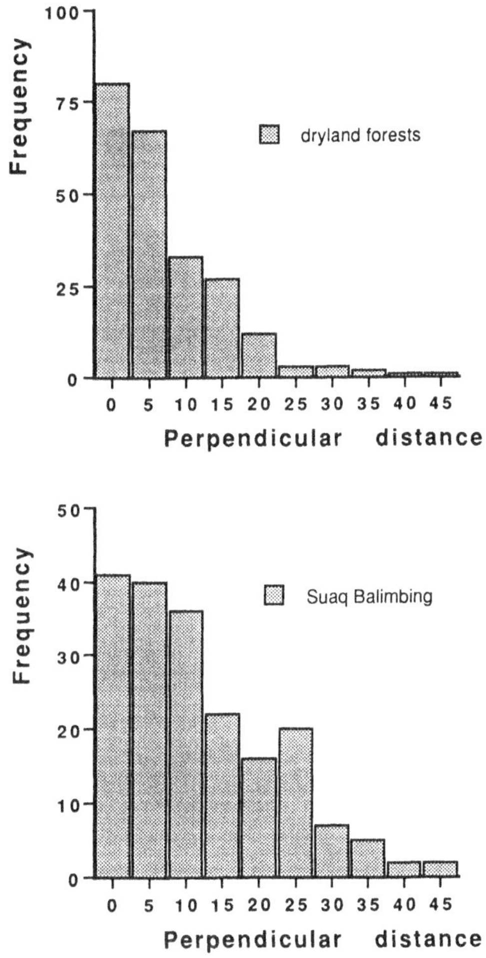

Figure 3. The distribution of perpendicular distances of nests to transect line, a) for all dryland forests sampled in 1990 and b) for the swamp forest sampled in 1992 -1993. The more open swamp forest had a significantly different distribution (see text).

Estimating w. The position of the maximum perpendicular distances included in the calculations has a small effect on the estimated *w* (cf. Whitesides, et al., 1988). Out of a total of some 745 nests in all samples, seven observed perpendicular distances were between 40 and 50 m, and four were at over 50 m. In accordance with the recommendations of Burnham, et al. (1980), it was decided to put the maximum perpendicular distance at 50 m and to discard the four observations with a greater value.

The distributions of perpendicular distances of nests to the transect line could vary with habitat, degree of disturbance and observers. The larger samples (at least 30 nests) of various sites sampled in 1990 were subjected to pairwise comparisons of shapes of the distribution of perpendicular distances using Kolmogorov-Smirnov tests. None of these comparisons approached significance. Also, no significant differences were found between the pooled samples for all undisturbed forests and selectively logged forests, or between teams of observers. Indeed, as the samples that were compared became larger, the similarity between the distributions became closer. Figure 3a shows the distribution of all observed perpendicular distances in undisturbed forests of the team containing the author in 1990. The estimated strip width, *w*, based on the iterative solution of a Fourier transformation (explained clearly in Brockelman and Ali, 1987), was 16.7 m.

In 1992 and 1993, extensive observations in swamp forest suggested that the much more open canopy would produce a higher *w* value (Fig. 3b). The estimated value of *w* here was 22.6 m. A Kolmogorov-Smirnov test indicated a significant difference with the other combined sample ($D=0.23$, $m=229$, $n=187$, $P<0.001$).

Sources of Error

Inter-Observer Differences. Observers may vary in the ability to discover nests (cf. Payne, 1987). This need not lead to large differences in density estimates, however, since lower nest counts tend to be compensated by lower values of *w* (Burnham, et al., 1980), provided that both observers record all nests above the trail (see above). In 1990, when different teams censused and one was more experienced than the other, differences were found in the number of nests recorded per km. These differences were only partly compensated for by the narrower strip width of the inexperienced team. Because some sites were visited only by the inexperienced team, their estimates were corrected on the basis of the ratio of nest counts obtained by both teams at the same sites (details are found in van Schaik and Azwar, 1991). In subsequent years, data were only collected by experienced nest counters.

Minimum L Required. How long a transect is required to obtain a realistic estimate of the orangutan density in an area? Here we will focus on the practical side of this question: after what length of time does the cumulative mean density stabilize? Figure 4 shows the results for a number of sites with long transects. Clearly, the number of nests per km of trail continues to change even after 5 km, but after about 1.5 km of transect, the rank order of estimated density remains the same across the five samples. This indicates that the minimum length of 3 km of trail per sample that we aimed for in this study, while not ideal, is a reasonable compromise between precision and efficiency.

One reason for the continuing changes in mean density is probably that long transect lines sample different habitats. Surveys aimed at providing numbers are directed, above all, toward crude density (the density in a study area as a whole [Eisenberg, 1979]), whereas, those interested in understanding a species' ecology will be most interested in estimating

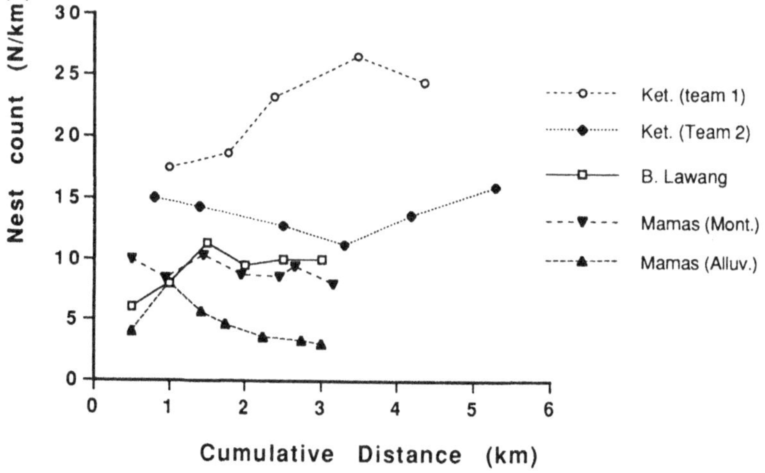

Figure 4. The change in mean nest counts in nests/km of transect with increased cumulative length of transect sampled. After about 1.5 km, the rank order of the differences between sites remains constant.

Table 4. Variation in orangutan nest densities (expressed as nests/km) among habitats at Suaq Balimbing (taken from samples between January and May)[1]

Sample[2]	River levee	Deep swamp	Transitional swamp	Peat swamp	Hills
March, 1992	14.7	11.8	52.8		
February, 1993			53.2		
January, 1994			66.1		
March, 1994	11.5			95.8	17.5
May, 1994	12.3			100.9	15.9
Mean density	12.8	11.8	57.4	98.4	16.7

[1]Deep swamps in the norther part have a high pH (over 6); transitional swamps have a pH between 5.5 and 6.5; peat swamps have a pH of less than 5.5.
[2]Sample sizes (km): March, 1992: RL (0.34 [= half[3]]), DS (1.10), TS (1.90);
February, 1993: TS (1.26);
January, 1994: TS (1.09);
March, 1994: RL (0.26 [= half[3]]), PS (1.12), H (3.09);
May, 1994: RL (0.65 [= half[3]]), PS (1.12), H (3.09).
[3]= Half of the length was sampled, because most of the forest is on one side of the trail.

ecological density (the density in the habitat types actually occupied). Where densities differ by habitat and habitat differences are not recognized, different transect lines at the same site may give very different estimated densities and low reliability will ensue.

At Ketambe, for example, patches of low or open-canopy forest related to stormthrows or riverine succession have much lower nest densities than the surrounding mature forest. At Suaq Balimbing, a breakdown of different habitats shows dramatic differences in density by habitat (Table 4), most of which became apparent only after we became more familiar with the area. Such spatial variation may affect reliability. As Table 4 shows, even short transects can achieve high reliability if they sample the same habitats. Thus, if sampling can be limited to one type of habitat, variation can be kept low.

Temporal Variation in Densities. Because nests remain visible for 3-6 months, short-term variation in the presence of orangutans at a site is attenuated. However, the existence of regular seasonal movements, as observed at Ketambe (te Boekorst, et al., 1990) and more recently at Suaq Balimbing (van Schaik, unpublished), raises the question of whether one visit to a site is enough to characterize orangutan densities. Variation among repeat surveys at the same site is presented in Table 5. While the coefficient of variation is low at Bukit Lawang and Manggala, it is fairly high at Ketambe and very high at Suaq Balimbing. The high repeatability at the first two sites indicates that 3 km of trails provides a reliable sample where there are no seasonal movements and the sample is obtained by an experienced team.

Table 5. Temporal variation in orangutan density (nests/km) as assessed by the coefficient of variation in density (CV) estimates at subsequent visits to the same site

Site	N visits	Mean density	CV (%)	Floodplain?
Bukit Lawang	2	11.4	0.8	none
Manggala	3	4.6	14.1	little
Ketambe	4	19.3	27.8	extensive
Suaq Balimbing	4	45.5	53.7	very extensive

Table 6. A comparison of methods to estimate the orangutan density at Ketambe

Period	Method	Sample size (km)	Estimated density (ind./sq. km)
ca 1990	Researchers estimate[1]	—	5 - 6
August, 1990	Line transect (nests)	4.3	5.17
April + August 1991	Line transect (A[2])	94.13	5.48
January, 1992	Line transect (T[3])	154.05	6.55

[1]J. Sugardjito and T. Mitra Setia (Personal communication).
[2]Focus was on arboreal animals.
[3]Focus was on terrestrial animals (but arboreal ones that were detected were included).

Validating the Method

The accuracy of an indirect method can be checked by comparing the density estimates obtained using different indirect methods and, of course, by comparison with the true density in an area of known density. This calibration was possible in Ketambe, where researchers estimate the density at 5 (Rijksen, 1978) or, more recently, at 6 animals/km^2 (T. Mitra Setia, personal communication). The results of differently derived estimates are compiled in Table 6. The correspondence between line transects using nests, those using direct sightings and the true density, is quite good. From these comparisons, a 10% - 15% underestimate of densities using the nest transect method may reasonably be inferred. Hence, in a flat area with known production and disappearance rates of nests, the nest survey method gives an accurate (= unbiased) and reasonably reliable estimate of orangutan density.

Habitat Patterns

Correlates of Density. Table 2 shows the best estimates of densities obtained in this study in a range of northern Sumatran sites. These estimates were used to identify the main determinants of orangutan density in pristine forests and to estimate the total number of orangutans in the National Park.

Previous studies have shown that orangutans are less abundant in disturbed habitats (e.g., Rijksen, 1978; van Schaik and Azwar, 1991). Consequently, we concentrated on the ecological correlates of pristine forests. Only 9 of the sites listed in Table 2 contained both pristine forest and at least some orangutans. One site near the margin of the range (Lau Kawar) was excluded because the orangutan density was thought to be artificially low due to hunting pressure in the past.

Altitude is a strong predictor of density (Fig. 5a), but it is interesting to identify the ecological factors underlying this variation. Earlier research implicated either the abundance of strangling figs as an important orangutan staple (e.g., Rijksen, 1978) or the abundance of fruit with fleshy pulp in general (including figs [Djojosudharmo and van Schaik, 1992]), as possible determinants of orangutan density. In the current data set, orangutan density is not correlated with strangling fig abundance ($r_s = -0.04$, $n = 9$, n.s.), but is correlated with the relative abundance of fruits with fleshy pulp ($r_s = 0.83$, $P < 0.05$; Fig. 5b).

Temporal Variation

The pattern evident in table 5 suggests that sites differ in the extent to which densities vary over time. The two sites with high temporal variability, Ketambe and Suaq Balimbing, are known for the seasonal movements of individuals. The two sites with no temporal variation are Bukit Lawang, a hill dipterocarp forest, and Manggala, a small area, partly hillside and partly alluvium, in a valley in which almost all primary forest has disappeared. Since both Ketambe

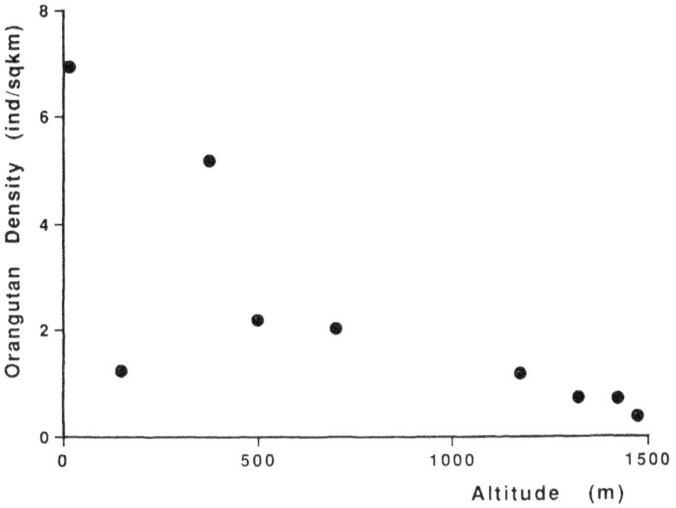

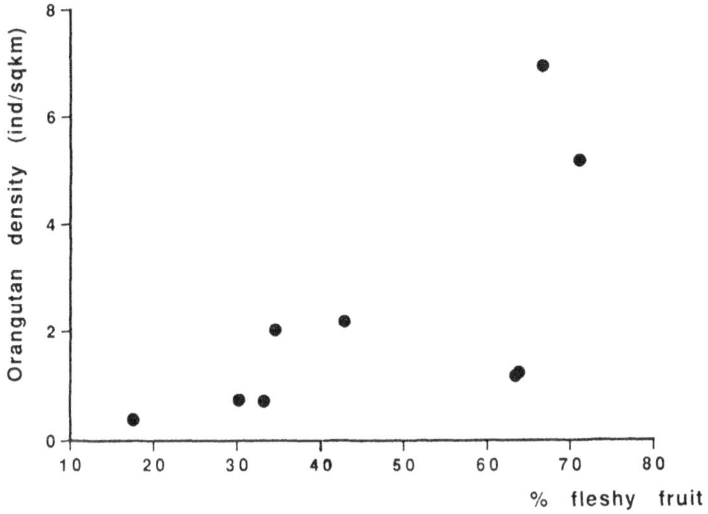

Figure 5. The relationship between orangutan density (ind./km^2) as a) a function of altitude or b) the percentage of fruits that had fleshy pulp, for the nine sites that were relatively undisturbed and presumed not to be subject to hunting.

and Suaq Balimbing are both floodplain forests adjacent to hilly areas, this contrast suggests that forests in floodplains, which are known to be more productive than hillside forests (van Schaik and Mirmanto, 1985), function as seasonal sinks for orangutans.

DISCUSSION

The Method

One would argue *a priori* that nest surveys are unlikely to be successful due to poor reliability and the possibility of bias. Reliability is bound to be low, since the estimated orangutan density is the product of many variables (see equation 2), all of which can only be measured with some error. The greatest source of error was probably variation in the time until disappearance of the nests due to variation in decay rates; more work is needed to determine the accuracy and reliability of the method developed to estimate t. It may also be possible to develop a better alternative. This method would obtain a very large sample of nests along transect lines and map their location. A repeat survey could be done after a relatively short period, e.g., 2 or 3 weeks, in which one can be confident that few nests have disappeared. All the new nests must thus have been built in the intervening period, which allows for a simple estimation of density. Although this method has been tested with success, it requires large samples to give reliable results. Nonetheless, the major patterns with altitude and fruit abundance are unlikely to be an artifact, because the estimated range of densities (about ten-fold) far exceeds the range in t values used here.

Variation in the width of the strip covered along the transect is also a source of error. However, among the undisturbed dryland forests there were no clear trends that suggested that different forest types had systematically different strip widths. The samples taken in selectively logged forest were small and a larger sample would presumably, indicate different strip width in such forests.

Bias is expected, since all parameters show true spatial (w, p, r, t) and temporal (p, r) variation, and some nests over the trail are bound to be missed. Nonetheless, as shown by the validity check at Ketambe, nest surveys are reasonably accurate, perhaps, underestimating densities by a fairly small margin (10% - 15% is likely). Fortunately, overestimation of density is not likely. Even reliability is reasonable, as shown by the high repeatability of subsequent estimates in the same habitat when seasonal variation due to movements was absent.

One should remember that the method using direct sightings along line transects takes at least five times as much time (cf. Table 2) and is hardly feasible in areas of low density. Nest surveying is, therefore, an accurate and reliable method provided it is performed by motivated and experienced observers.

One important use of line transects of nests should be highlighted. Nest surveys are especially useful for cost-effective monitoring of populations along fixed transect lines where many sources of error can be held constant and temporal variation in observed nest densities are likely to reflect true changes in absolute density. If nests are mapped at each survey, allowing identification of new nests, a reliable index of abundance can be obtained. Such an index will provide a sensitive indicator of the status of orangutans in the area and thus, can serve as an early warning system. Because of their sensitivity to logging and other forms of human disturbance, orangutans are good indicators of the level of human disturbance.

Determinants of Density

Altitude is a major predictor of density, but is not a direct determinant because there may be appreciable variation between habitats at similar altitudes (see Fig. 5a). For example, at Suaq Balimbing, densities were extremely high in the peatswamps, intermediate in the alluvial forests on the levees and the deep backwater swamps and lowest in the dipterocarp forest on the adjacent hill sides.

Altitude is correlated with various relevant ecological parameters. The decline with altitude in total forest productivity (Lonsdale, 1988; Djojosudharmo and van Schaik, 1992) is not steep enough to account for the decline in orangutan density. However, the spectrum of fruit types tends to change with altitude (Djojosudharmo and van Schaik, 1992). A good predictor of orangutan density is the abundance of fruits with fleshy pulp, which had earlier been shown to be selected preferentially by orangutans in the lowland forest at Ketambe

(Djojosudharmo and van Schaik, 1992; cf. Leighton, 1993). Especially swamp and lowland alluvium forests are rich in the protected fleshy fruits that can only be harvested by orangutans and other arboreal mammals (primates, civets, squirrels). Unfortunately, these are the habitats converted at the higest rates outside protected areas, due to their value for agriculture (Collins, et al., 1990), and under the greatest pressure inside these areas.

Seasonal Movements?

Both nest surveys and long-term observations indicate considerable seasonal variation in orangutan abundance at some floodplain sites, an observation already made centuries ago by local people in Borneo (Hornaday, 1885). Large temporal variation in the local abundance of frugivores is a hallmark of the tropical rain forests of northern Sumatra (van Schaik, unpublished). Orangutans are less mobile than hornbills and other frugivorous birds, fruit bats and primates, such as pigtailed macaques (*Macaca nemestrina*), but at Suaq Balimbing numerous animals may appear or disappear within days in an area of several km^2 (personal observation).

It is obvious that spatial variation in fruit abundance has something to do with these movements, but details are unknown. Do the floodplain forests attract visitors during their period of maximum fruit abundance or is the timing of their attractiveness based especially on poor food abundance elsewhere and the steady supply of staples in the floodplains? Another question is even more important for conservation: At what scale do these movements take place? Are they merely between different habitats that form a small-scale mosaic or are movements on the order of 10 km involved? In the latter case, the floodplains may act as seasonal refuges for upland populations and the carrying capacity of the uplands may be severely reduced when the adjacent floodplain forests, very attractive for agriculture, are cleared. These conservation questions will be addressed in subsequent studies, since the Gunung Leuser National Park consists largely of mountain areas with increasingly thin lowland fringes.

ACKNOWLEDGMENTS

We thank the L.S.B. Leakey Foundation for supporting the first survey, the NYZS - the Wildlife Conservation Society for solidly supporting all later stages of this project, the Indonesian Institute of Sciences (LIPI) for permission to conduct research in Indonesia and the Universitas Nasional and subsequently, Universitas Indonesia, for sponsoring the work. For permission to work in the Gunung Leuser National Park, I am grateful to the Director-General of Forest Protection and Nature Conservation, P.H.P.A. and the directors of National Parks and Recreation Forests and of Conservation, to the Park's former director, Mr. Poniran, for local permission and discussion, to many local staff for field support and to Mike Griffiths for his support and friendship. I am very grateful to Suharto Djojosudharmo and Tatang Mitra Setia for their unpublished data and to Idrusman and students from Universitas Nasional for considerable help in the field. L. Alterman, P. Assink, E. Bennett, B. Fox, M. van Noordwijk, H. Rijksen, T. Struhsaker, J. Sugardjito and J. Wind provided helpful comments on an earlier draft.

REFERENCES

Brockelman, W.Y. and Ali, R., 1987, Methods of surveying and sampling forest primate populations. Pp. 23-62 in (Eds. C.W. Marsh and R. A. Mittermeier), *Primate Conservation in the Tropical Rain Forest*, Alan R. Liss: New York.

Buckland, S. T., Anderson, D. R., Burnham, K. P., and Laake, J. L., 1993, *Distance Sampling; Estimating Abundance of Biological Populations*, London: Chapman and Hall.

Burnham, K. P., Anderson, D. R., and Laake, J. L., 1980, Estimation of density from line transect sampling of biological populations, *Wildl. Monogr.*, No. 72.

Collins, N.M., Sayer, J.A., and Whitmore, T.C., 1991, *The Conservation atlas of Tropical Forests; Asia and the Pacific*, London: Macmillan Press.

Davies, G. and Payne, J., 1982, *A Faunal Survey of Sabah*, Kuala Lumpur: IUCN/WWF Project No. 1692.

Djojosudharmo, S. and van Schaik, C.P., 1992, Why are orang utans so rare in the highlands? Altitudinal changes in a Sumatran forest, *Trop. Biodiv.* 1: 11-22.

Dobson, A.P. and Lyles, A.M., 1989, The population dynamics and conservation of primate populations, *Cons. Biol.* 33: 62-380.

Eisenberg, J. F., 1979, Habitat, economy and society: some correlations and hypotheses for the Neotropical primates. Pp. 215-262 in (Eds. I. S. Bernstein and E. O. Smith), *Primate Ecology and Human Origins*, New York: Garland Press.

Galdikas, B.M.F. and Wood, J., 1990. Great ape and human birth intervals. *Am. J. Phys. Anthropol.* 83: 185-192.

Ghiglieri, M. P., 1984, *The Chimpanzees of Kibale Forest; a Field Study of Ecology and Social Structure*, New York: Columbia University Press.

Harrison, T., in press, The paeloecological context at Niah cave, Sarawak: evidence from the primate fauna, *Bull. Indo-Pac. Prehist. Assoc.*

Harrisson, B., 1962, *Orang-utan*, London: Collins.

Hornaday, W. T., 1885, *The Experiences of a Hunter and Naturalist in the Malay Peninsula and Borneo*, New York: C. Scribner, Sons.

Kemeny, J.G., Snell, J.L. and Thompson, G.L., 1956, *Introduction to Finite Mathematics*, Englewood Cliffs, N.J.: Prentice-Hall, Inc.

Krebs, C. J., 1989, *Ecological Methodology*, New York: Harper Collins Publishers.

Leighton, M., 1993, Modeling dietary selectivity by Bornean orangutans: evidence for integration of multiple criteria in fruit selection. *Int. J. Primatol.*, 14, 257-313.

Lonsdale, M.W., 1988, Predicting the amount of litterfall inn forests of the world, *Ann. Bot.* 61: 319-324.

MacKinnon, K., 1987, Conservation status of primates in Malasia with special reference to Indonesia. *Prim. Cons.* 6: 175-183.

MacKinnon, J., 1992, Species survival plan for the orangutan. Pp. 209-219 in: (Eds. G. Ismail, M. Mohamed, and S. Omar), *Proceedings of the International Conference on Forest Biology and Conservation in Borneo*, Kota Kinabalu: Yayasan Sabah.

(Eds. Martin, P.S., and Klein, R.G.), 1984, *Quaternary Extinctions: a Prehistoric Revolution*, Tucson: University of Arizona Press.

Payne, J., 1987, Surveying orang-utan populations by counting nests from a helicopter: a pilot survey in Sabah, *Prim. Conserv.* 8: 92-103.

Repetto, R., 1990, Deforestation in the tropics, *Sci. Am.* 262 (4):18-24.

Rijksen, H.D., 1978, *A Field Study of Sumatran Orang Utans (Pongo pygmaeus abelii Lesson 1827)*, Wageningen: H. Veenman and Zonen B.V.

Rijksen, H.D., 1982, How to save the mysterious 'man of the rain forest'?, Pp. 317-341 in (Ed. L.E.M. de Boer), *The Orang Utan. Its Biology and Conservation*, The Hague: Dr W. Junk Publ.

Schaller, G. B., 1961, The orang-utan in Sarawak, *Zoologica* 46: 73-82.

Schmidt-Nielsen, K., 1975, *Animal Physiology, Adaptation and Environment*, New York: Cambridge University Press.

Sugardjito, J., 1983, Selecting nest sites by Sumatran orang-utans (*Pongo pygmaeus abelii*) in the Gunung Leuser National Park, Indonesia, *Primates* 24: 467-474.

Sugardjito, J., 1986, *Ecological constraints on the behaviour of Sumatran orang-utans (Pongo pygmaeus abelii) in the Gunung Leuser National Park, Indonesia, Ph.D. thesis*, University of Utrecht, Netherlands.

te Boekhorst, I.J.A., Schürmann, C.L., and Sugardjito, J., 1990, Residential status and seasonal movements of wild orang-utans in the Gunung Leuser Reserve (Sumatra, Indonesia), *Anim. Behav.* 39: 1098-1109.

Tutin, C.E.G., and Fernandez, M., 1984, Nationwide census of gorilla (*Gorilla g. gorilla*) and chimpanzee (*Pan t. troglodytes*) populations in Gabon. *Am. J. Primatol.* 6: 313-336.

van Beek, C. G. G., 1982, *Een geomorfologische bodemkundige studie van het Gunung Leuser nationale park, noord Sumatra*, Indonesië. Ph.D. thesis, University of Utrecht, Netherlands.

van Schaik, C.P., 1986, Phenological changes in a Sumatran rain forest. *J. Trop. Ecol.* 2: 327-347.

van Schaik, C.P. and Azwar, 1991, Orangutan densities in different forest types in the Gunung Leuser National Park (Sumatra), as determined by nest counts, *Report to PHPA, LIPI and L.S.B. Leakey Foundation*, Durham, NC.

van Schaik, C.P. and Mirmanto, E., 1985, Spatial variation in the structure and litterfall of a Sumatran rain forest, *Biotropica* 17: 196-205.

van Strien, N.J. , 1985, *The Sumatran rhinoceros -Dicerorhinus sumatrensis (Fisher, 1814) - in the Gunung Leuser National Park, Sumatra, Indonesia; its distribution, ecology and conservation, Ph.D thesis*, Wageningen (NL).

von Koenigswald, R., 1982, Distribution and evolution of the orang utan, *Pongo pygmaeus* (Hoppius). Pp. 1-15, in (Ed. L.E.M. de Boer), *The Orang Utan, Its Biology and Conservation*, The Hague: Dr W. Junk Publ.

Wallace, A. R., 1869, *The Malay Archipelago; the Land of the Orang-utan and the Bird of Paradise*, London: Macmillan and Co.

Whitesides, G.H., Oates, J.F., Green, S.M., and Kluberdanz, R.P., 1988, Estimating primate densities from transects in a West African rain forest: a comparison of techniques, *J. Anim. Ecol.* 57: 345-367.

SECTION FOUR – SOCIAL AND COGNITIVE BEHAVIOR

Introduction

It has been difficult to characterize the social organization of orangutans, in part, because it is unique among the diurnal primates. As was true of the chimpanzee, it was necessary with the orangutan to obtain data from several field studies before a reasonably clear picture of its social system emerged (MacKinnon, 1971, 1974; Rodman, 1973; Horr, 1975; Rijksen, 1978; Galdikas, 1985a). Some adult males become dominant animals in their area, live a relatively isolated life, and defend a territory that subsumes the ranges of several adult females. Adult and subadult males may also become migrants and range widely, presumably during the time when they are attempting to establish themselves in a territory or are developing the requisite strength and social skills to challenge a resident male and usurp his territory. Adult female orangutans and their dependent offspring occupy overlapping home ranges, but they rarely associate with one another or any other orangutans outside the mother-infant unit. Subadults of both sexes pass through a somewhat more gregarious stage in which they sometimes travel with other subadults. The subadult stage may be viewed as a bridge, an intermediate condition between the stage of constant companionship provided by the mother and the more solitary existence of adulthood. As a consequence of these findings, orangutans are generally described as "solitary" or "semi-solitary," (Galdikas, 1985 b) but this term does not apply equally to both sexes at all stages of life. (Captive orangutans, by contrast, show marked propensities to develop compatible social relationships at all ages [Zucker and Thibaut, this volume]). It is hypothesized that large body size in the orangutan coupled with its particular ecology prevents a more gregarious social system in the wild. An adult male and female essentially associate with each other only when the female is fertile and in estrus. Given the female orangutan's long interbirth interval, i.e., 8-10 years, it is clear that adult males and females rarely interact.

Orangutans are notable for the presence of two male morphs. Fully adult males possess striking secondary sexual characteristics such as fatty cheek flanges and a muscular throat pouch and they emit *long calls*. Studies on captive male orangutans suggest that the subadult male's physiological, physical, and perhaps psychological maturation are suppressed by the presence of adult males (Kingsley, 1982). The development of secondary sex characteristics may be linked to the male's hormone levels, the balance between testosterone and estrogens. Field researchers also suggest that dominant adult males inhibit the acquisition of adult features in subadult males, but relevant data are difficult to obtain in the wild.

A male's status as adult or subadult determines its mating strategy. Adult males usually establish a territory overlapping the ranges of several females. When a female comes into estrus, she usually selects a dominant resident male as a consort with which to mate.

Adult males respond positively to the sexual advances of adult estrous females, but not to nonestrous adult females or to subadults. Subadult males by contrast, initiate mating in a forcible manner and do so indiscriminately with fertile or infertile females. Their's is a "catch as catch can" strategy, since females are not attracted to them.

Van Hooff opens this section with a lively discussion of how the orangutan's semi-solitary nature is adaptive, at least in the proximate sense. He describes the evolutionary models that have been proposed to account for primate social systems as reflecting a compromise between the need for safety from predation and the rewards of competition. He notes that the orangutan's social system is atypical among diurnal primates and sets the tone for the papers that follow. Van Hooff concludes that it is difficult at this time to provide an ultimate, i.e., evolutionary, explanation for the orangutan's social system and describes the types of additional data that are required for progress in this area. He considers *female choice* to be the decisive factor for the orangutan's present social system, but regards this as only a partial solution to the larger question. The collection of long term data at sites like Tanjung Puting and Ketambe is essential for further hypothesis testing.

Galdikas' paper explores the various selective pressures that operate on individuals. She notes that reproductive strategies may change over life stages; lifetime reproductive fitness is a consequence of adopting adaptive strategies appropriate to each life stage. Forcible copulations have been described as a reproductive strategy of subadult males, but this is less easily understood and rarely discussed from the female perspective. Galdikas describes how vigorous resistance against forcible copulations is in the best reproductive interests of an *adult* female, while less resistance may actually benefit a *subadult* female. Since subadult females are unlikely to acquire an adult male mating partner, they may accrue some benefit from mating with a subadult male.

How do young orangutans learn what they need to know as adults if the only individual with which they have contact is the mother? Galdikas proposes that the gregarious subadult stage in the female provides just such an opportunity for learning. By observing and interacting with other females, a subadult female may learn about resource availability. By establishing relationships with subadult and adult male orangutans, the subadult female may forge the ties that are crucial to her later survival.

Utami and Mitra Setia provide the first field data on the "overthrow" of a resident male orangutan. They note the behavioral dynamics between the resident male and the young challenger over the time interval in question. It is noteworthy that successful takeovers by male orangutans may require the cooperation and assistance of area females. This study is particularly interesting in light of Galdikas' proposal that behavioral interaction and social knowledge of other individuals may be crucial to reproductive success, even in this "solitary" species.

Russon and Galdikas describe the keen observational and imitative skills of free-ranging wild-born ex-captive orangutans in the context of tool use. The analysis of data on spontaneous imitation provides insight into the learning process of these animals and the relationship between their cognitive abilities and those of humans. These authors point out that an essential characteristic of human cognition is the ability to interconnect cognitive skills, such as tool use and imitation, something which has never been demonstrated in monkeys. Their conclusion that orangutans have this ability is based on a careful analysis of the approach taken by the animals in response to demonstrated tool use. The orangutans focused their attention on the organization of the activity and they performed repetitions of the demonstrated components, both hallmarks of human imitation.

Section four closes with Shapiro and Galdikas' description of the only sign language experiment ever attempted with orangutans in a natural habitat. They report that the juvenile orangutans exhibited extreme variability in their learning of different signs, related in part to their attentiveness to the object during the referent identification phase. The role of referent interest was further suggested by the more rapid learning of food signs in comparison to nonfood signs.

REFERENCES

Galdikas, B.M.F., 1985a, Subadult male orangutan sociality and reproductive behavior at Tanjung Puting. *Am. J. Primatol.* 8:87-99.

Galdikas, B.M.F., 1985b, Orangutan sociality at Tanjung Puting. *Am. J. Primatol.* 9:101-119.

Horr, D.A., 1975, The Borneo orang-utan: Population structure and dynamics in relationship to ecology and reproductive strategy. In: *Primate Behavior: Developments in Field and Laboratory Research 4* (Ed. L.A. Rosenblum), New York, Academic Press, 307-323.

Kingsley, S., 1982, Causes of non-breeding and the development of the secondary sexual characteristics in the male orang utan: a hormonal study. In: *The Orang Utan Its Biology and Conservation* (Ed. L.E.M. De Boer), The Hague, Dr. W. Junk, Publishers.

MacKinnon, J.R., 1971, The orang-utan in Sabah today. *Oryx* 11:141-191.

MacKinnon, J.R., 1974, The behaviour and ecology of wild orang utans (*Pongo pygmaeus abelii*). *Anim. Behav.* 22:3-74.

Rodman, P.S., 1973, Population composition and adaptive organization among orang-utans of the Kutai Reserve. In: *Comparative Ecology and Behaviour of Primates* (Eds. R.P. Michael and J.H. Crook), London, Academic Press, 171-209.

Rijksen, H.D., 1978, *A Field Study of Sumatran Orang Utans* (*Pongo pygmaeus abelii*, Lesson 1827), *Ecology, Behaviour and Conservation*, 78-2. Wageningen, H. Veenman and Zonen B.V.

16

THE ORANGUTAN: A SOCIAL OUTSIDER

A Socio-Ecological Test Case

J. A. R. A. M. van Hooff

Ethologie and Socio-Ecologie
Universiteit Utrecht
Pb. 80.086, 3508 TB Utrecht, Netherlands

THE RED APE: A NEGLECTED LONER?

At first sight, the orangutan may well be the least spectacular of the great apes. In comparison with its African cousins, this solitary creature seems to lead a far less exciting life. Superficially, its social organization is exceptional in comparison with that of other diurnal primates. It is precisely this exceptional characteristic which makes the red ape an interesting test case for socio-ecological theorizing.

Firstly, orangutans live comparatively solitarily, that is, they move as independent units most of the time. This is true for both the Bornean (Horr, 1975; MacKinnon, 1974; Rodman, 1973, 1979; Rodman and Mitani, 1987), and the Sumatran subspecies (Rijksen, 1978). Especially in the latter case, however, occasional aggregations and even associations have been reported (Schürmann, 1982; Sugardjito, *et al.*, 1987; te Bockhorst, *et al.*, 1990). Both males and females occupy large, overlapping home ranges.

A SOLITARY AND YET SEXUALLY DIMORPHIC PRIMATE

This solitary primate, in addition, has a marked sexual dimorphism. Fully adult males have impressive secondary sexual characteristics (a coat of long hair, cheek flanges, throat sack and a far reaching and loud *long call*). More remarkable still, there is a conspicuous dimorphism also within the sexually mature males. A male may for many years retain the form of a sexually mature *subadult male*, which is distinct from that of an *adult male*. A subadult male has not yet developed the secondary sexual characteristics (SSCs) of a fully adult male, mentioned above. The subadult males roam in areas which overlap considerably with those of the fully adult males. Adult males are intolerant of other adult males they encounter in their home range. They advertise their presence with long calls (MacKinnon, 1974; Mitani, 1985a). These calls may serve to repel less powerful adult males, to intimidate subadult males and to inform females of the whereabouts of the adult males, but their precise function has yet to be determined.

The Neglected Ape, Edited by Ronald D. Nadler et al.
Plenum Press, New York, 1995

This sexual dimorphism is remarkable. Normally, sexual dimorphism in size and strength indicates the existence of strong intermale contest competition, by which dominant males achieve exclusive access to fertile females (van Hooff and van Schaik, 1992). For a number of taxa, the degree of sexual dimorphism is correlated with the number of females over whose sexual contacts a male has control in a "harem" (Alexander, *et al.* 1979). Given the size of the home ranges of orangutans, however, and given the short visual detection range in their dense arboreal habitat, it is difficult to see how one male orangutan could prevent another, if the latter behaved inconspicuously, from moving throughout its range and searching for females which make use of that same range.

To appreciate the remarkable character of the orangutan, one must consider this character in the light of recent theoretical models about the evolutionary causes of primate social systems. Across the primates there is great variation in these systems. In the last 15 years, moreover, there have been considerable advances in unraveling the underlying principles (e.g., Wrangham, 1979, 1980, 1987; van Schaik, 1983, 1989; van Schaik and van Hooff, 1983; Dunbar, 1988; van Hooff and van Schaik, 1992).

TYPES OF COMPETITION AND THE SOCIAL SYSTEM

There is a growing consensus that primate sociality is a compromise, resulting from the interaction of at least two important ultimate factors, safety and competition. Living together in groups may offer safety against predation, and also, in the case of females, safety against infanticidal threats from unfamiliar males. Group-living may also be beneficial because it makes possible cooperative defense of resources (Wrangham, 1980). There are good reasons to believe that the protection which groups offer against predation are an especially important factor (Dunbar, 1988). This safety benefit is obtained, however, at the cost of increased competition over resources among the associating animals.

It has been argued that the socio-ecological factors which were mentioned, safety through protective cooperation, on the one hand, and resource competition, on the other hand, affect the distribution of females most directly. Females are most directly affected because their fitness depends directly on a long and healthy life which allows them to produce a series of viable offspring over a long period of time with short interbirth intervals. The fitness of males, conversely, depends directly on the number of females that can be fertilized. In primates, paternal investment is usually low and is restricted largely to the protective consequences of the male's presence near its offspring. The distribution and the reproductive strategies of males, therefore, are affected largely by the distribution and accessibility of fertile females (Wrangham, 1979).

Van Schaik and van Noordwijk (1988) argued that to understand the precise form which the relationships between community members assume, it is important to distinguish between two types of competition that may exist among the members. Such competition may be of the *contest* type or of the *scramble* type.

In contest competition, one individual (or coalition of individuals) invests in attempts to exclude others from access to a resource. Contest competition occurs only if the distribution of the resource makes this possible, i.e., if the resource items occur in sizes and clustered quantities which make monopolization possible and profitable. If there is resource competition that is predominantly of the contest type, this will stimulate the evolution of physical and motivational characteristics contributing to success in contests. The relationships among female group members are expected to become strongly hierarchical and organized in coalition systems among relatives, and females are expected to remain in their natal clans, where they obtain the support of their relatives.

Scramble competition occurs when resource items are too small or dispersed to be monopolized. Where resource competition is predominantly of the scramble type, female dominance, nepotism and philopatry will be relaxed (van Schaik, 1989). There is increasing evidence that these predictions are valid (van Hooff and van Schaik, 1992).

INTERMALE CONTEST, SEXUAL DIMORPHISM AND BIMATURISM

With respect to intermale competition for access to mates, there is an analogous situation. Where the conditions favor polygyny and where the distribution of females allows their monopolization by males, investment in monopolization contests are promoted by natural selection, leading to the formation of one-male groups or harems. This is associated with the characteristics of intermale contests (Alexander, et al., 1979). If ecological constraints permit this, sexual dimorphism in size, strength and fighting disposition results. Where monopolization of access to females is not possible and where, consequently, males scramble for mates, polygynandrous systems may develop with male sperm competition (Harcourt et al. 1981; Moller, 1988). Relationships between males are expected to be more tolerant and egalitarian, and there is more room for the development of cooperative male bonds (van Hooff and van Schaik, 1994). This also implies that there will be more opportunity for female partner choice on the basis of male characteristics other than those with contest potential. For example, 'friendly services' by males (e.g., Smuts, 1985, 1987; Noe et al., 1991; Noe and Hammerstein, 1994) may determine female preferences (e.g., van Hooff, 1992).

Clearly, the orangutan occupies a place at the extreme of the spectrum of primate social variation. Its essentially solitary nature can be explained as a species-specific outcome of the safety-competition balance, namely, as a consequence of its being the largest arboreal mammal (van Hooff, 1988). Its comparatively large size and its arboreality make it relatively invulnerable to predators once it reaches adolescence. Thus, a major incentive for association is removed. That same large size, however, in combination with the consequent restraint on foraging radius, makes the competition costs of association high. There is evidence that lengthy social congregations of orangutans can take place only at times when there is an abundance of food (Sugardjito, *et al.*, 1987).

In our discussion so far, the orangutan fits the model's predictions. In other respects, however, it does not easily do so. The existence of strong sexual dimorphism was mentioned above in a socio-ecological context where effective exclusion by contest in the competition for mates seems virtually impossible. In addition, there is the remarkable dimorphism within the class of sexually mature males. Whereas the growth in size and weight of male orangutans is gradual, the change from a subadult form into an adult one can take place rapidly, in a matter of a month or two, between the ages of 9 and 20 or more years of age (Brandes, 1929, 1931; Seitz, 1969; Jones, 1976). There is evidence that this bimaturism is socially influenced. Kingsley (1982) noted that a subadult male quickly developed the adult secondary sexual characteristics once he was isolated from an adult male in whose vicinity he was living. The change was accompanied by an endocrine change, namely in the balance between testosterone and estrogen.

These observations suggest that the subadult phase represents a specific, alternative male strategy which is followed when a dominant adult male is in the area. It is of interest to note that a similar dimorphism within males was described for a solitary nocturnal prosimian, the lesser bushbaby (*Galago senegalensis*). Older and heavier so-called, *A-males*, have territories and preferential social contacts with the females living in their territories and are highly intolerant of other A-males. The A-males are much more tolerant towards the lighter B-males which roam their territories. A high-ranking B-male can quickly become an A-male once a territory becomes available (Bearder, 1987).

MACKINNON'S "RANGE GUARDIAN HYPOTHESIS"

MacKinnon (1974), who was the first to try to explain this remarkable dimorphism in mature males, also noted that subadult male orangutans were sexually active. By contrast, he never observed sexual interactions of adult males. Subadult males were seen to engage in two types of sexual contact, 1) in the form of cooperative matings, in which the female actively participated, and 2) in the form of forced matings (or "rapes") in which the male coerced a resisting female to copulate (see also, Rijksen, 1978). MacKinnon thought, therefore, that the subadults were the reproductive males and that the fully adult males were postreproductive and defending a territory in which the subadults could roam (Fig. 1). The implication was that the adult earned an inclusive-fitness benefit because the subadults were his sons (MacKinnon, 1979) and that a certain degree of male-bondedness might develop (cf. Suzuki, 1989).

This proposal is not as strange as it seems at first glance. It accords with the observations on other apes, where, in contrast to many other so-called, *female-bonded* species of primates, the female apes are not strictly philopatric while the males are. The latter implies that kin relations between males are maintained and *intermale tolerance* is promoted on the basis of its benefits to inclusive fitness. Unfortunately the data on the dispersal patterns of orangutans are scant (see van Schaik and van Hooff, 1995 for a discussion). Although females are not strictly philopatric and may wander opportunistically, a sex difference emerges with males dispersing more widely. This reduces the opportunity for male nepotism.

MacKinnon (1974, 1979) was convinced that the adult male could never prevent other adult males from entering his area and, therefore, did not regard the system as an expression of intermale mate competition. He conceived of the adult male as a "range guardian" who, with his awe-inspiring appearance, defended the area and its resources against intruders. By thus regulating the density in his area, he preserved the resources for his male offspring and thereby, also served his own inclusive fitness.

There is a logical problem in this reasoning, however. If an adult male is supposed to be unable to keep an adult male sexual competitor at bay in his vast and unsurveyable home range, how would he then be able to keep resource competitors away? In addition, there is empirical evidence which is not in accord with this model. Observations in both Borneo (Galdikas, 1979, 1985; Mitani, 1985b) and Sumatra (Schürmann and van Hooff,

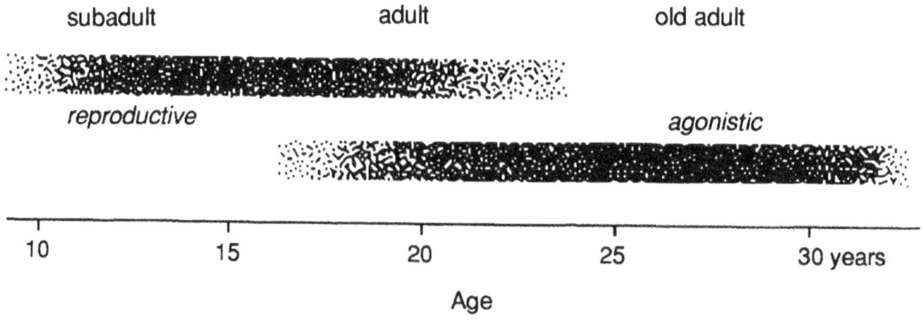

Figure 1. The distribution of reproductive and agonistic functions over the life stages of male orangutans, according to the "range-guardian hypothesis" of MacKinnon (1979).

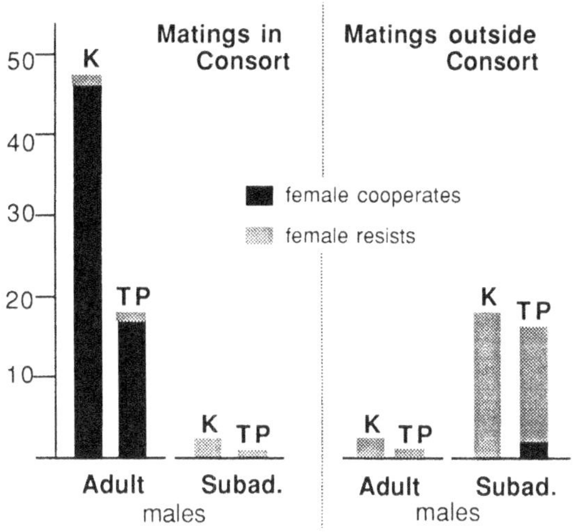

Figure 2. The distribution of matings with adult an α subadult males in which the female cooperates and in which the female resists, both in the study populations at Ketambe, Sumatra [K] and at Tanjung Puting, Borneo [TP]. After data from Schürmann and van Hooff (1986) for K, and Galdikas (1979) for TP.

1986) show that adult males are also sexually active and that they form consort relationships with a female, albeit on rare occasions (Schürmann and van Hooff, 1986; Rodman and Mitani, 1987). In contrast to the sexual contacts of subadult males, these consortships can last for many days. Observations both in Tanjung Puting, Kalimantan and in Ketambe, Sumatra suggest, moreover, that the character of the matings of subadult males and of adult males differ. The percentage of cooperative matings of adult males was higher in both locations (Fig. 2). In judging these data, however, one must realize that the samples are small and that the copulations cannot be regarded as statistically independent events, especially, the copulations which take place within a consortship.

Nevertheless, the data show that adult males are far from sexually inactive even though they may be comparatively lethargic in initial contacts. Schürmann and van Hooff (1986) suggested that the female must direct more initiative towards an adult male to awaken his sexual interest.

THE "FEMALE PREFERENCE HYPOTHESIS"

The above data created doubt over the MacKinnon model (Mitani, 1985; Schürmann and van Hooff, 1986; Rodman and Mitani, 1987). This doubt was strengthened by the proposal that the development of the SSCs seems, within certain limits, to be restrained by a social influence, namely the presence in the area of a fully adult male (see above). This suggests that the restraint in the development of SSCs represents an adaptive strategy with which subadult males avoid the agonism from adult males while staying comparatively unchallenged in the range of an adult male. It may be a "waiting room strategy": the SSCs are suppressed until their provocative effect is supported by the animal's relative power to maintain its ground against another.

Conversely, ample evidence shows that adult males engage in fierce competition (Rodman and Mitani, 1987; Utami and Mitra Setia, this volume) and that the outcome of such competition influences the behavior of both the adult males concerned and the adult females. Mitra Setia and Utami showed that the loser of a conflict stopped his long-calling, whereas, the winner increased its rate. Furthermore, the females also made more approaches to the winner than before the conflict; approaches to the winner also were greater than approaches to the loser.

These observations reduce the plausibility of MacKinnon's range guardian hypothesis that subadult reproductive males are tolerated or even protected by an adult non-reproductive male because their sexual success adds to the inclusive fitness of the resident male. The observations support the alternative hypothesis that adult males are the main reproducers, that they are engaged in strong contest competition and that they tolerate subadults simply because (1) they cannot effectively expel the subadults from their territories and (2) because females have a clear preference for associating in a *consort relation* with a successful adult male.

THE PIVOTAL ROLE OF FEMALE PREFERENCE

It should be noted that the system hinges on the condition that females have a preference for successful adult males, that is, males who make themselves known by conspicuous behavior and long-calling. If only estrous females express this preference, then the reproductive chances of subadult males would be sufficiently small to allow the resident adult male to ignore them. As a consequence, the dominant male need only concern himself about other adult males. As long as the other adult males do not make their presence known by long calls, however, neither the resident male nor the females would be able to find them easily.

Thus, the pattern can be explained in terms of male strategies maximizing *direct fitness*. Even so, this does not exclude nepotistic influences; in other words, although inclusive fitness advantages may not be the primary factor determining the proposed tolerance of subadult males, adult males may distribute their tolerance differentially in favor of subadults which were born and grew up during the residence of the adult. Long-term observations at Ketambe show that one male, John, reigned as the dominant resident from before 1972, with a short interruption, until 1990 (Utami and Mitra Setia, this volume). This would be long enough for the early male offspring of John to reach sexual maturity during his residency.

THE NEED FOR GENEALOGICAL AND ENDOCRINOLOGICAL DATA

To settle these issues and to clarify the organization of relationships and the underlying principles, the following aspects need to be investigated: Firstly, the genealogical relationships in a population must be revealed in order to determine whether the males in an area are more closely related than the females, or, more precisely, whether there is a class of natal subadults who are indeed sons of the resident male. If so, it would be possible to demonstrate whether adult males differentially tolerate subadult males on the basis of kinship.

Secondly, we must establish whether there is a cycle-dependent preference of orangutan females for adult males and their status indicators. Nadler's research on captive orangutans (1982, 1988; this volume) clearly shows that females that cannot be forced into sexual interaction by the male, but that have the possibility to make sexual advances when it suits them, do show a cycle-dependent willingness to do so. This experiment, however, leaves unanswered the question whether females also have a preference for particular males above others for which they take into account quality or status parameters. Clearly, the long call is of interest here. Two functions have been proposed for the long call: 1) to threaten and warn other males and 2) to attract females. Experimental studies in which artificially produced long calls were broadcast to feral orangutans have so far failed to demonstrate these functions convincingly, especially, an effect on female attraction to males (Mitani, 1985a). We must recognize, however, that a message may be responded to only when the message has some value for a receiver. Thus, females may selectively react only under motivational conditions which are associated with sexual proceptivity.

Only the existence of female preference can explain why two male forms exist in an ecological situation in which a dominant male is unable to prevent other males from surreptitiously contacting females. Female preference can also explain why female orangutans, which have an unadvertised ('covert') estrus, do not always avoid subadult males. If subadults never had opportunities to mate with females, as one would expect on the basis of data on other species, such as langurs and gorillas, there might be an increase in the risk of infanticide by subadults (van Schaik and Dunbar, 1990). In this context, it is also interesting to determine whether females in the first months of pregnancy actively seek out subadult males to mate with, i.e., whether they use promiscuity "to create paternity uncertainty" over the infant to be born.

RESIDENCY AND MIGRATION: IS THERE A COMMUNITY?

The composition and migration dynamics of the population are also relevant to the above issues. Is there evidence for more than a passing acquaintance between males and females? Is there some pattern of relationships between animals sharing a certain range which justifies the notion of orangutan communities? At Ketambe, certain individuals are encountered regularly whereas others are seen occasionally. Leighton (this volume) reported that the orangutans at Gunung Palung migrated between different habitats in accordance with the differential abundance of preferred foods in these habitats. Age and sex classes, moreover, varied in their spatial dynamics. These animals, however, were not recognized individually in their movements. The range guardian hypothesis implies that males must be comparatively philopatric, or, when migrating between habitats, that certain subadult males migrate in coordination with an adult male. The alternative, female-attraction hypothesis does not predict that one sex or the other would be inhibited from migration.

Analysis of the frequencies with which particular recognized individuals have been observed at Ketambe over a period of many years by te Boekhorst, *et al.* (1990) indicates that this population consists of individuals who are more or less permanently present and others which are seen only temporarily and at intervals. Preferences in relationships may well be expected, therefore, between members of the population which share home ranges and have similar migration patterns. As far as male-female relationships are concerned, however, there is no evidence of any sort of paternal investment. Still, females might profit from staying close to adult males with which they have had sexual relations. Such males might protect them against harassment by other males who might endanger their offspring (Smuts and Smuts, 1993). Although there are some indications that one might refer to as orangutan communities, the evidence is still far from conclusive (van Schaik and van Hooff, in press).

CONCLUSION

The contention here is that the social system of the orangutan can be understood as an evolutionarily stable result of individual social rules (see Table 1), only if we accept the pivotal role of female choice. This point of view contrasts with that of Rodman and Mitani (1987). They argued that the social system with its marked sexual dimorphism is explained as a result of strong intermale competition and that there is no evidence for a role of female

Table 1. Basic individual social rules leading to orangutan organization

Female	Prefer male with proven fitness prospects = seek fully adult dominant male. (Possibly, when lactating: Stay near a fully adult male to enjoy his protection)
Subadult male	Wait physiologically until there is a social opportunity = a vacancy for an adult position. Nevertheless, try to mate
Adult male	Try to remain a resident (dominant) male and broadcast your presence. Given the female preference for adult males: 1. Drive out other adult males 2. Do not be concerned about subadult males

mate choice. They noted, first of all, that a dimorphism in strength and size is also found in some other solitary mammals; in other words, that such dimorphism is not restricted to species that defend access to clusters of females, i.e., species living in one-male groups or 'harems'. This, however, is no argument in itself; it only means that one must consider these cases critically as well.

Secondly, it was noted that males do engage in contests and that larger males are able to drive away other males from females (Mitani, 1985b). That males engage in contests is clear also from the observations made elsewhere, e.g., at Ketambe (Utami and Mitra Setia, this volume). Such contests, however, can take place only if the males encounter one another in competition. In the ecological conditions in which orangutans live, however, this seems to be largely beyond the control of a dominant male. Above all, the contest model alone does not offer a satisfactory explanation for the fact that sexually mature males come in two forms. In addition, selection for contest ability does not explain certain secondary sexual characteristics, such as the cheek flanges; these are better explained as a result of epigamic selection.

We concur, therefore, with Rijksen (1978), Wrangham (1979), Schürmann (1982), Schürmann and van Hooff (1986) and Nadler (1988) that female preference must be the decisive factor that maintains the *present* system. This leaves unanswered, however, how such a system could have evolved in the first place. It is difficult to imagine that the actual strategies could have arisen as an evolutionarily stable complex in the ecological conditions in which the species is living at present. Two other types of organization are expected instead:

1. A hylobatid type of monogamous organization, with female preference for a service-rendering male. Given the consequential paternity certainty, this male would also be obliged to offer protection against the increased likelihood of infanticide by other males (cf. van Schaik and Dunbar, 1990). For the development of the latter type of organization, however, the large size of the orangutan and the consequential competition costs might be prohibitive (cf. van Schaik and van Hooff, in press).
2. Alternatively, we might expect a system arising from a situation of male scramble competition, with ensuing sperm competition, paternity uncertainty and a consequential absence of infanticidal tendencies, with intermale tolerance. In both cases, we would not expect strong sexual dimorphism and certainly not the bimaturism in males which is so remarkable in orangutans. We must assume, therefore, that the ancestral orangutans lived in a monandrous/polygynous society where male exclusion competition was more effective.

REFERENCES

Alexander, R.D., Hoogland, J.L., Howard, R.D., Noonan, K.M. and Sherman, P.W., 1979, Sexual dimorphisms and breeding systems in pinnipeds, ungulates, primates and humans. *In Evolutionary Biology and*

Human Social Behaviour: An Anthropological Perspective (Ed. N.A. Chagnon and W.A. Irons), Belmont Wadsworth, 402-603.
Bearder, S.K., 1987, Lorises, bushbabies, and tarsiers: diverse societies in solitary foragers. In *Primate Societies* (Eds. B.B. Smuts, D.L. Cheney, R.M. Seyfarth, R.W. Wrangham and T.T. Struhsaker). Chicago, University of Chicago Press, 11-24.
Boekhorst, I.J.A. te, Schürmann, C.L. and Sugardjito, J., 1990, Residential status and seasonal movements of wild orang-utans in the Gunung Leuser Reserve (Sumatra, Indonesia). *Anim. Behav.* 39:1098-1109.
Brandes, G., 1929, Die Backenwulste des Orang-mannes. *Zool. Gart. (N.F.)* 1: 365-368.
Brandes, G., 1931, Das Wachstum der Menschenaffen im Vergleich zu dem des Menschen in Kurven dargestellt. *Zool. Gart. (N.F.)* 4:339-347.
Dunbar, R.I.M., 1988, *Primate Social Systems*. London: Croom Helm.
Galdikas, B.M.F., 1979, Orangutan adaptation at Tanjung Puting Reserve: mating and ecology. In: *The Great Apes*. (Eds.: D.A. Hamburg and E.R. McCown). New York, Academic Press, 195-233.
Galdikas, B.M.F., 1985, Subadult male orangutan sociality and reproductive behavior at Tanjung Puting. *Am. J. Primatol.* 8:87-99.
Harcourt, A.H., Harvey, P.H., Larson, S.G. and Short, R.V., 1981, Testis weight, body weight, and breeding system in primates. *Nature* 293:55-57.
Horr, D.A., 1975, The Borneo orang-utan: Population structure and dynamics in relationship to ecology and reproductive strategy. In: *Primate Behavior: Developments in Field and Laboratory Research 4* (Ed. L.A. Rosenblum), New York, Academic Press, 307-323.
Jones, M.L., 1976, Fact sheet on Orang Utans in captivity (Pongo pygmaeus). *Am. Assoc. Zool. Parks Aquar. Newsl.*, 14.
Kingsley S., 1982, Causes of non-breeding and the development of secondary sexual characteristics in the male orang utan: a hormonal study. In: *The Orang Utan, Its Biology and Conservation*. (Ed. L.E.M. de Boer), den Haag, Junk, 215-229.
MacKinnon, J., 1974, The behaviour and ecology of wild orang utans (*Pongo pygmaeus abelii*). *Anim. Behav.* 22:3-74.
MacKinnon, J., 1979, Reproductive behaviour in wild orangutan populations. In: *The Great Apes* (Eds. D.A. Hamburg and E.R. McCown), Menlo Park, Benjamin/Cummings, 256-273.
Mitani, J.C., 1985a, Sexual selection and adult male orang utan long calls. *Anim. Behav.* 33:272-283.
Mitani, J.C., 1985b, Mating behaviour of male orang utans in the Kutai Game Reserve, Indonesia. *Anim. Behav.* 33:922-402.
Mitrasetia, T. and S. Utami, 1994, Male dominance in orangutan (*Pongo pygmaeus*) society. *Abstr. XVth Congr. Int. Primatol. Soc.*, Bali, Indonesia
Moller, A.P., 1988, Ejaculate quality, testes size and sperm competition in primates. *J. Hum. Evol.* 17:479-488.
Nadler, R.D. (1982) Reproductive behavior and endocrinology of orang utans. Pp. 231-248 in *The Orang Utan. Its Biology and Conservation*. L.E.M. de Boer, ed. The Hague, Dr. W. Junk Publ.
Nadler, R.D. (1988) Sexual and reproductive behavior. Pp. 105-116 in *Orang-utan Biology*. J.H. Schwartz, ed. New York, Oxford University Press.
Nadler, R.D. (1995) Sexual behavior of orangutans (*Pongo pygmaeus*): basic and applied implications (*this volume*).
Noe, R., van Schaik, C.P. and van Hooff, J.A.R.A.M., 1991, The market effect, an explanation for pay-off asymmetries among collaborating animals. *Ethology* 87:97-118.
Noe, R. and Hammerstein, P., 1994, Biological markets: supply and demand determine the effect of partner choice in cooperation, mutualism and mating. *Behav. Ecol. Sociobiol.*, in press.
Rijksen, H.D., 1978, *A Field Study of Sumatran Orang Utans* (*Pongo pygmaeus abelii*, Lesson 1827), *Ecology, Behaviour and Conservation*, 78-2. Wageningen, H. Veenman and Zonen B.V.
Rodman, P.S., 1973, Population composition and adaptive organization among orang-utans of the Kutai Reserve. In: *Comparative Ecology and Behaviour of Primates* (Eds. R.P. Michael and J.H. Crook), London, Academic Press, 171-209.
Rodman, P.S., 1979, Individual activity patterns and the solitary nature of orangutans. In: *The Great Apes* (Eds. D.A. Hamburg and E.R. McCown. Menlo Park, Benjamin/Cummings, 234-255.
Rodman, P.S. and Mitani, J.C., 1987, Orang-utans: sexual dimorphism in a solitary species. In: *Primate Societies* (Eds. B.B. Smuts, D.L. Cheney, R.M. Seyfarth, R.W. Wrangham and T.T. Struhsaker). Chicago, University of Chicago Press, 146-154.
Schürmann, C.L., 1982, Mating behaviour of wild orangutans. In: *The Orang utan: Its Biology and Conservation* (Ed. L. de Boer), Den Haag, Junk, 271-286.
Schürmann, C.L. and van Hooff, J.A.R.A.M., 1986, Reproductive strategies of the orang-utan: new data and a reconsideration of existing socio-sexual models. *Int. J. Primatol.* 7:265-287.

Seitz, A., 1969, Notes on the body weights of new-born and young orang-utans. *Int. Zoo Yb.* 9:81-84.

Smuts, B.B., 1985, *Sex and Friendship in Baboons* (Ed. Hawthorne). Aldine, New York.

Smuts, B.B., 1987, Sexual competition and mate choice. In: *Primate Societies* (Ed. B.B. Smuts, D.L. Cheney, R.M. Seyfarth, R.W. Wrangham and T.T. Struhsaker), Chicago, University of Chicago Press, 385-399.

Smuts, B.B. and Smuts, R.W., 1993, Male aggression and sexual coercion of females in nonhuman primates and other mammals: evidence and theoretical implications. *Adv. Study Behav.* 22:1-63.

Sugardjito, J., te Boekhorst, I.J.A. and van Hooff, J.A.R.A.M., 1987, Ecological constraints on the grouping of wild orang-utans (Pongo pygmaeus) in the Gunung Leuser National Park, Indonesia. *Int. J. Primatol.* 8:17-41.

Suzuki, A., 1989, Socio-ecological studies of orangutans and primates in Kutai National Park, East Kalimantan in 1988-1989. *Kyoto Univ. Overseas Res. Rep. of Studies on Asian Nonhuman Primates*, 7:1-42.

Utami, S. and Mitra Setia, T., 1995, *(this volume)*.

van Hooff, J.A.R.A.M., 1988, Sociality in primates: a compromise of ecological and social adaptation strategies. In *Perspectives in the Study of Primates* (Ed. A. Tartabini and M.L. Genta), DeRose, Cosenza, 9-23.

van Hooff, J.A.R.A.M., 1992, Competing for progeny: the socioecology of primate mating systems. In: *Sex Matters*, (Eds.: W. Bezemer, P. Cohen-Kettenis, K. Slob and N. van Son-Schoones), Amsterdam, Elsevier Science Publ., 93-96.

van Hooff, J.A.R.A.M. and C.P. van Schaik, 1992, Cooperation in competition: the ecology of primate bonds. In: *Coalitions and Alliances in Humans and Other Animals*. (Eds.: S. Harcourt and F.B.M. de Waal), Oxford, Oxford Univ. Press, 357-390.

van Hooff, J.A.R.A.M. and C.P. van Schaik, 1995, Male Bonds: Afilliative relationships among nonhuman primate males. *Behaviour* 130: 309-338.

van Schaik, C.P., 1983, Why are diurnal primates living in groups? *Behaviour* 87:120-44.

van Schaik, C.P., 1989, The ecology of social relationships amongst female primates. In: *Comparative Socioecology, the Behavioural Ecology of Humans and Other Mammals* (Ed. V. Standen and G.R.A. Foley). Oxford, Blackwell Scientific Publications, 195-218.

van Schaik, C.P. and van Hooff, J.A.R.A.M., 1983, On the ultimate causes of primate social systems. *Behaviour* 85:91-117.

van Schaik, C.P. and Dunbar, R. I. M., 1990, The evolution of monogamy in large primates: a new hypothesis and some crucial tests. *Behaviour* 115:30-62.

van Schaik, C.P. and van Noordwijk, M. A., 1988, Scramble and contest among female long-tailed macaques in a Sumatran rain forest. *Behaviour* 105: 77-98.

Wrangham, R.W., 1979, On the evolution of ape social systems. *Soc. Sci. Inform.* 18: 335-368.

Wrangham, R.W., 1980, An ecological model of female-bonded primate groups. *Behaviour* 75:262-300.

Wrangham, R.W., 1987, Evolution of social structure. In: *Primate Societies* (Ed. B.B. Smuts, D.L. Cheney, R.M. Seyfarth, R.W. Wrangham and T.T. Struhsaker), Chicago, University of Chicago Press, 282-296.

17

SOCIAL AND REPRODUCTIVE BEHAVIOR OF WILD ADOLESCENT FEMALE ORANGUTANS

B. M. F. Galdikas

Simon Fraser University
Burnaby, B.C., V5A 1S6, Canada

ABSTRACT

At Tanjung Puting, adolescent female orangutans were the most social of all age/sex classes and should not be described as solitary or even semi-solitary. Non-exclusive groupings involving adolescent females and subadult males were the associations most commonly seen at Tanjung Puting. Adolescent female interactions with adult males were almost totally restricted to consortships which the female initiated, maintained, and terminated. Adolescent female interactions with adult females were largely absent due to adult female agonism. This agonism probably relates to child-care constraints and food competition experienced by the adult female. Adolescent female interactions with subadult males involved long term associations in which no copulations took place and short term (<1 day) association in which "resisted mating" occurred. Overall, heterosexual associations were social. Adolescent females did not consort with subadult males. Adolescent female interactions with their own age/sex class were invariably social, and adolescent females always traveled together following encounters. Adolescent females were the only age/sex class observed to carry out interunit grooming. First parturition for wild orangutan females is estimated at approximately 14 or 15 years of age. Adolescent females experience a period of sterility which may exceed one year prior to first parturition. Adolescent females copulate at higher rates than do adult females. Adolescent females were the only age/sex class at Tanjung Puting to display high levels of proceptive behavior. This proceptive behavior, while not always successful, appeared crucial to inducing consorts to copulate. Adolescent females preferred adult males as sexual partners, but themselves were not preferred by adult males. A large amount of learning appears to take place during the orangutan's adolescence as a result of increased sociality. Due to their small size, rapid travel and long day ranges, sociality probably is not as costly for adolescent female orangutans as it is for adults. Adolescent females probably benefit from associating with other orangutan units because, based on the age/sex class they are associating with, grouping provides: a) a potential context in which to learn adaptive solutions to social and/or environmental problems, b) increased foraging efficiency, c)

The Neglected Ape, Edited by Ronald D. Nadler et al.
Plenum Press, New York, 1995

protection from predators and/or "enemies," and d) a potential context in which dominance and affiliative relations between individuals can be developed.

INTRODUCTION

For many years preliminary descriptions of free-ranging orangutans as solitary stood in direct contrast to the complex social behavior described for all other Old World anthropoids. Subsequent field studies of longer duration confirmed that the invariant features of orangutan social organization include highly dispersed individuals and a relative absence of social interaction. Rodman (1973) went so far as to comment that "Orangutans are hardly more social than any mammal must be." Richards (1985) refers to such systems as "social networks," wherein individuals like the orangutan interact and know each other, but do not necessarily spend most of their time near one another.

The basic units of orangutan populations include solitary adult males, adult females usually accompanied by one or two dependent offspring, and immature individuals in transition between their natal units and the establishment of a solitary existence. The amount of contact among units and the nature of such contact has been little investigated owing to the infrequency of orangutan interactions, necessitating long term and continuous research to collect such data. However, the highly dispersed nature of orangutans limits the number of individuals within one study population, thus reducing the numbers of aggregations or social interactions likely to be observed. For this reason, the solitary nature of some orangutan age/sex classes has been overemphasized.

To date, primatologists have shown little interest in immature nonhuman primate behavior, particularly as it exists outside the mother/infant bond or play interactions. Literature on the life stages of immature nonhuman primates is relatively lacking, although the situation is changing. Reasons for this scarcity are varied:

1. Erroneously, adults are seen as evolutionarily representative of a species when, in reality, natural selection acts on individuals during all developmental stages of their life histories. Some evolutionary biologists believe that ideal phylogenies should include depictions of each developmental life stage of those species represented.
2. In light of the current sway sociobiology holds over many research agendas, it is not surprising that prereproductive immatures are sometimes overlooked in a world where adaptive success is measured in terms of offspring surviving to reproduce. However, a successful reproductive career as an adult is contingent on the organism's development as an immature individual. Any understanding of the adults of a species is incomplete without a grounding of that species' developmental life stages.
3. Since individuals spend much of their lives as adults, it follows that adult members of a species would receive greater systematic scrutiny.
4. Adult male primates often possess striking secondary sexual traits which, given their spectacular nature, may detract from attention being paid to other age/sex classes. The male orangutan, who is twice the size of the adult female and possesses cheekpads, a throat pouch, and a fatty crown, is a case in point.
5. Demographic factors may influence the distribution of individuals across age/sex classes, and immatures individuals may be outnumbered by adults. This proved to be the case for this particular study, as independent immature orangutans were not numerically predominant within the Tanjung Puting study area.

6. Finally, immature animals present special problems for the field researcher. Immature orangutans, for example, are usually in the canopy so that it is often difficult to see them. Compared to adult orangutans, immatures are smaller, less noisy and more difficult to follow due to their rapid travel. Furthermore, adult primates are easier to identify than are immature animals. The difficulty in identifying immature orangutans as compared to adults is clearly evident from data presented here, as immatures account for 78% of all observation time on unrecognized targets.

The lack of attention accorded to immature nonhuman primates is witnessed by the paucity of data on immature orangutans. Both MacKinnon (1974) and Rijksen (1978) comment that adolescents were the age class most frequently found in groupings, but provide little elaboration. Galdikas (1984, 1985a,b,c) collected extensive data on various orangutan age/sex class interactions, including those of adult males, adult females, and subadult males. Based on these findings, Galdikas argues that only adult male orangutans can be characterized as truly solitary, while adult females can be described as semi-solitary and subadult males as social. The following report expands on this picture of orangutan life way by describing the rich and complex social behavior of adolescent females observed during a 4-year period—November 1971 to November 1975—at the Orangutan Research and Conservation Project study area in Tanjung Puting National Park, Kalimantan Tengah, Indonesian Borneo.

FIELD METHODS

Observational data were collected by the author, who was aided initially by Rod Brindamour, and then, during the fourth year of fieldwork, by a trained local assistant. A large study area of 40 km^2 was chosen and covered with transects so that: 1) the home ranges of at least several members of each sex would be partially or totally encompassed, and 2) data on a representative sample of the orangutan population could be gathered including interactions between individuals of different age/sex classes.

Most of the data presented here consist of whole days (i.e., individuals were followed continuously from leaving their night nests in the morning to the time they nested for the following night). Target individuals were usually observed for a period of at least 2 or 3 whole days of consecutive observation.

If two units were located while associating together, one individual from one unit was chosen as the focal individual. In such cases males were selected over females and adults over immatures. When a subadult male and an adult female were in association, target observations on the two units alternated. This was done to equalize the amount of data collected on subadult males and females owing to difficulty in locating subadult males due to their ranging patterns (Galdikas, 1979). Known offspring accompanying their mothers were always counted as belonging to their natal unit unless the offspring were fully adult. For females, this meant having their own infant. The net result of the selection procedure was that adult males were discriminated for and immature orangutans against, particularly adolescent females. Such discrimination did not take place when a target individual encountered other units. Regardless of the age/sex classes involved, observations of the original target continued.

It is probably impossible to always discriminate between true social groups and aggregations brought about by factors acting independently on each unit. A *social grouping* is defined as two or more units traveling together peacefully for a distance of more than 40m—with all members of the group seemingly aware of the others' presence and none initially showing marked aggressive or avoidance responses. *Aggregations* are defined as passive, rather than social, contacts between individuals, occurring when recognition of the unit's identity was impossible. For example, when two units were foraging, each seemingly oblivious to the other unit's presence and displaying no coordination in movement, this was termed an aggregation.

Mitani (1989) argues that associations over short distances during short periods of time do not provide a biologically relevant measure of sociality among orangutans, since they are slow moving and usually do not travel more than a mean distance of 20-30m/30 min. This does not appear to be the case for adolescent female orangutans at Tanjung Puting, who travel greater distances and at a much more rapid pace than do adults. Whenever negative responses were persistent, the associations were not considered social groupings but were termed aggregations. This definition does not render two or more orangutans in contact as a social unit. An aggregation could well involve units feeding in one tree with friendly interactions taking place but each unit leaving separately. Conversely, a social grouping does not necessarily imply an interaction other than the close coordination of travel. For example, two orangutans could travel together and nest near each other for several days on end, yet never touch, with the only discernible interaction involving occasional glances at each other.

TERMINOLOGY

Adolescent females ranged from 9-14 years old and were distinguished from the other age/sex classes at Tanjung Puting based on morphological and behavioral characteristics. Adolescent females weighed approximately 20-30 kg. Their faces were whiter than fully adult females, but they lacked the defined, white eye, mouth and body patches characteristic of juveniles and infants. They were also smaller in size than adult females. Behaviorally, they were totally independent of their mother, although they occasionally traveled with her. At this point in the life cycle the female has periods of sexual receptivity during which she behaves proceptively and consorts with males. This definition of adolescent females differs slightly from that used by other researchers (MacKinnon, 1974; Rijksen, 1978). All agree, however, that adolescent females are independent of their mothers.

RESULTS

General

Out of the total 6,804 hours during which orangutans were observed at Tanjung Puting, 1,273 hours (19%) represented two or more basic units in contact. The rest of the observation time (81%) involved single units: solitary adult males, adult female/dependent offspring units and lone immatures, as have been described in all previous studies (Schaller, 1961; Davenport, 1967; MacKinnon, 1971, 1974; Rodman, 1973).

Nine individuals was the largest number of orangutans ever observed together at Tanjung Puting. This was noted on only two occasions during the four-year study. In one instance, nine individuals from six units were together for about one minute as an arriving subadult male moved into the vicinity while another subadult male fled. The second instance of nine individuals involved at least three units, possibly four (depending on whether a strange adult female and accompanying strange adolescent female were related or not). This involved three adult females, three infants, two juveniles and one adolescent female traveling together for over one hour.

Various measures, including search data, the percentage of total observation time spent in groups and the ratio of time spent in aggregations to social groups make clear that the four major independent age/sex classes of orangutans observed at Tanjung Puting (adult male, adult female, subadult male and adolescent female) were participating in contact with other units at different rates. Adult males and females were less likely to be part of a multi-unit grouping than were immature orangutans (adolescent female or subadult male). Quantitative measures reveal that adolescent females are the most social of all orangutan age/sex classes at Tanjung Puting. Qualitative measures indicate that adolescent female orangutans at Tanjung Puting have a capacity for overt affiliative behavior rarely witnessed among the other age/sex classes in the wild.

Adolescent Female Sociality

Adolescent females were observed as focal individuals for 547 hours, accounting for only 8% of the total observation time. During this time focal adolescent females participated in groupings 223 hours, which accounted for 41% of the time they were observed. Most observations of adolescent females involved their contacting one or more orangutan units, however briefly, during the course of an observation. Comparable data for the other major age/sex classes observed show that durations of contact with other units for focal adult males, adult females and subadult males was 16.8%, 13.5% and 40.9% respectively (Table 1). If one considers contact between an independent adolescent female and her natal unit to represent two different units in contact, then adolescent females spent little time alone. Using duration of contact by focal individuals as a measure of sociability, it is clear that adolescent females rank as highly social relative to adults of the species.

The search data likewise indicate that adolescent females were the most social of the age/sex classes. They were discovered 44 times on searches: 24 times alone or with their known natal unit (55%) and 20 times (45%) in contact with other units. This contrasts sharply with adult grouping tendencies, since search data indicate mature males were encountered as members of groups only 9% of the time, while adult females and subadult males were located while associating with other units 10% and 39% of the time, respectively.

Since less than 20% of total observation time was on target immatures and because immature, particularly adolescent, females were discriminated against in favor of adults when two units were contacted together, I suspect that the marked discrepancy between adult individuals and adolescent females is probably greater than these figures indicate. Non-target adolescent females, for example, were observed in groups an additional 265 hours when other individuals were being followed as focal targets. Thus, focal and non-focal adolescent females were observed a total of 812 hours in groups at Tanjung Puting. In other words, of the total 1,273 observation hours which involved two or more units in contact, one or more adolescent females (target or non-target) were present during 64% of that time. In comparison, out of the total observation time involving two or more units in contact, adult males were present only 37% of the time, while adult female and subadult males were present 43%

Table 1. Duration of contact with other units by target animals

Age/sex class of target	Hours observed (hrs:min)	Duration of contact (hrs:min)	% contact/observation hours on class	% contact/total observation hours logged
Adult male	2,699:27	449:10	16.8	6.6
Adult female	2,872:10	387:56	13.5	5.7
Subadult male	462:50	189:21	40.9	2.8
Adolescent female	547:22	223:23	40.8	3.3
Others	252:39	23:42	9.4	0.4
Total	6,804:28	1,273:32	—	18.7

and 31% of the time, respectively. The figures for age/sex class frequency of contact during the course of independent observations with other orangutan units, regardless of the length of contact, particularly highlight the differences between adult orangutan sociality versus that of adolescent females.

It is important to delineate adolescent female associations into nonsocial aggregations and social groups in order to formulate a clear picture of the true sociability adolescent females at Tanjung Puting exhibited. Based on the 812 hours of observations when independent adolescent females were seen in associations with other orangutan units, analysis reveals that 765 (94%) hours could be termed truly social interaction, versus the remaining 47 (6%) hours, which are better characterized as nonsocial aggregations, during which some element of avoidance, aggressive or indifferent behavior was observed. Comparable data on the other major age/sex classes observed suggests, once again, that adolescent females are indeed the most social of all age/sex classes at Tanjung Puting. For example, 85% of the time adult males were in contact with another orangutan unit was social and 82% and 89% for adult females and subadult males respectively. However, these true sociality figures for adults may appear artificially high in comparison to those of the immature orangutans owing to the methodology of this study, which favored observation of consort pairs over all other individuals.

Adolescent Female Orangutan Association Patterns with Age/Sex Classes

Independent adolescent females at Tanjung Puting were observed in 32 different grouping combinations, and they spent considerable time in association with all age/sex classes. Nevertheless, for adolescent females, non-exclusive contacts (other age/sex classes may be present) with adult males predominated: 42% of all contact time involved an adult male, 39% of which entailed sexual consortship. This compared to 17% of non-exclusive contact time with adult females and 24% and 22% respectively with subadults and with adolescent females (Table 2).

In total, 56% of the time that adolescent females were in contact with other units, they were exclusively in contact with males (subadults and adults). This compares to 24% of contact time in which they were exclusively associating with females other than their mothers. Mutually exclusive contacts between adolescent females and adult males predominated: 29% of all mutually exclusive contacts with one adolescent female involved one adult male. Comparable figures for one adolescent female's mutually exclusive contact with adult females, adolescent females and subadult males were 14%, 9% and 19% respectively.

Groupings between adult males and adolescent females were largely limited to sexual interactions (consortships). In no cases did contact outside of consortship last more than part of one day. A non-receptive adolescent female might continue feeding even after a mature male entered her tree but (outside of known ex-consort partners) was never seen to begin feeding in a tree where a mature male was already foraging. Non-receptive adolescent female behavior towards adult males can be characterized as avoidance or indifference, whereas receptive adolescent females appeared highly motivated to maintain contact with the seemingly indifferent adult males.

Differences between adult female and adolescent female sociality was significant (Mann-Whitney $U = 11$, $n_1 = 7$, $n_2 = 3$, $p < 0.05$), with adolescent females spending more time in groupings than did adult females. Adult female interactions with adolescent females spanned the entire spectrum of behaviors from aggression and avoidance to relationships that may be characterized as affiliative. For example, adult females occasionally associate nonaggressively with adolescent females known not to be their daughters. In one case, a primiparous female (Hen) traveled with an adolescent female for at least three days. For the most part, however, adolescent females were wary of

Table 2. Adolescent female non-exclusive contact with age/sex classes

	Duration of contact				Independent contact observations			
Age/sex classes observed in non-exclusive groupings	Total hrs:min	Percent total time adolescent females observed in groupings (total = 812 hrs)	Social	Non-social	Total	Percent contacts involving adolescent females (n = 217)	Social	Non-social
Adolescent female/ adult male	144:30	42.4	117:02	27:28	90	41.5	45	45
Adolescent female/ adult female	139:27	17.2	124:04	15:23	58	26.7	32	26
Adolescent female/subadult male	195:48	24.1	179:14	16:34	81	37.3	46	35
Adolescent female/adolescent female	178:29	22.0	176:34	1:52	36	16.6	30	6
Adolescent female/juveniles and/or adolescent male	26:54	3.3	23:38	3:16	14	6.3	10	4

adult females with offspring and did not approach or travel with them. Adult female orangutans with infants were observed to attack adolescent females who were former travel/play companions, seemingly without provocation. As a result, non-exclusive contact involving adolescent and adult females ranked lower in terms of durations than adolescent female contacts with the other orangutan age/sex classes observed at Tanjung Puting (adult male, subadult male, adolescent female). Despite the increased intolerance of adult females for nulliparous adolescents with whom they formerly associated, data indicate that friendships observed between adolescents may continue into adulthood once each female has conceived an offspring. Observations beyond the scope of this report seem to corroborate this viewpoint: following the birth of an adolescent female's (Fern) first infant, an older primiparous female (Georgina) began initiating long periods of associations with her. Nonetheless, while adolescent friendships may endure, the quantity and quality of socialization characteristic of adolescents is absent among adult females.

As would be expected, much of the variation among relationships involving adolescent females and adult females seems to be affected by the females' age, status, personality, and kinship ties. Degrees of kinship did not appear paramount in structuring associations; nonetheless, since females usually stay in the general vicinity of their known mother's range, most adolescent females in an area are probably related. It was difficult or impossible to determine a dominance hierarchy among females, as few direct interactions of any sort were observed. Age seemed to be a good indicator of dominance relations, with nulliparous and primiparous females avoiding multiparous females.

Contact time between adolescent females with their own class and with subadult males was roughly similar and together accounted for a substantial proportion of the total. When viewed over a period of years, associations between some immatures appeared persistent, in that they tended to reoccur over a number of sightings. Immatures traveling together did not, however, constitute basic population units as the associations (although recurring) were invariably temporary and broken off after a number of days or weeks.

Most subadult male associations with adolescent females were social but excluded consortships. Subadult males frequently traveled with non-receptive adolescent females, sometimes for weeks, while adult males did not. Subadult males were responsible for maintaining this contact, as adolescent females never followed them. No matings were witnessed between adolescent females and subadult males during those associations which lasted longer than one day. When a copulation did occur, it took place shortly after the two orangutans contacted each other and was often met by active resistance on the part of the adolescent female. With a few notable exceptions, the majority of sexual interactions between subadult males and adolescent females were "resisted matings" (Estep and Bruce, 1981).

Contacts between adolescent females were invariably social, and they always traveled together following encounter. The longest association between adolescent females was observed for ten days but probably lasted longer, since the two adolescent females were initially located together. The fact that all individually recognized adolescent and large juvenile females in the southern part of the core study area were seen traveling together at one time or another indicates that, at least among adolescent females, peer groups were important. Peer relations seemed more important than uterine ties in determining associations among immatures. It appeared that relationships were strongest when the age difference was the least. This suggests that females did not associate much with siblings, at least in adolescence, since the average birth interval at Tanjung Puting was approximately eight years (Galdikas and Wood, 1990).

Grooming

While allogrooming is a characteristic feature of the behavior of most monkeys and apes, it was rarely observed among wild orangutans. MacKinnon (1974) describes orangutan grooming practices as "not very thorough." Among captive orangutans Maple (1980) found females, regardless of age, to groom others more frequently and for longer durations than did males. Social grooming bouts by captive females lasted on average of 2.5 minutes, but were observed to last up to 45 minutes (Maple, 1980).

An important measure of adolescent female sociability at Tanjung Puting was the fact that they were the only age/sex class ever observed to groom individuals outside the natal unit (n = 26 occasions, Table 3). Limited intraunit grooming bouts were observed between adult females and their offspring and, more rarely, between siblings (such as a male juvenile briefly grooming his infant sister). However, aside from their dependent offspring, adult females did not groom other individuals, nor were mature males or subadult males ever observed to do so. Adolescent females were seen on rare occasions to groom one another (n = 2 observations; 7.7%) and, more frequently, to groom the mature males with whom they consorted (n = 24 observations; 92.3%).

Of the time that adolescent females groomed, most of it was with an adult male consort. Grooming bouts averaged 2.7 minutes in duration. The longest grooming bout ever recorded for an orangutan at Tanjung Puting involved an adolescent female (Noisy) grooming her consort's (Nick) left side, shoulder and back for nine minutes.

Although hormonal data are lacking, Maple, *et al.* (1979) suggest that peaks in grooming by females are correlated with the onset of ovulation. Data from Tanjung Puting indicate that factors underlying the manifestation of grooming behavior among wild orangutans are more complex. Adolescent female social grooming was focused on adult males during consortships when females were receptive; however, a correlation between the adult female reproductive state and grooming behavior appears absent as adult females were never observed to groom consorts.

Reproductive Parameters-Age at First Parturition

Although captive orangutans have been known to conceive at seven to eight years of age, wild orangutans are probably considerably older at age of first parturition due to differences in body weight and nutritional intake. In one case, I made a reasonably accurate estimate of the minimal age of a primiparous mother. I initially encountered the adolescent female (Fern) in early 1972 at which time I estimated, based on her morphological and behavioral characteristics, that she was at least nine or ten years of age. In 1977, when she was at least 14 or 15 years old, she gave birth to her first infant. Although age at first parturition undoubtedly varies among orangutans, for this study, it was the criterion demarcating adulthood from adolescence.

Table 3. Inter-unit grooming by adolescent females

				Duration of grooming	
Class of individuals groomed	Context	Number	%	Min	%
Adolescent female	Social grouping	2	7.7	2	2.9
Mature male	Consortship	24	92.3	68	97.1
Totals		26	100	70	100

The estimate for orangutan age at first parturition is somewhat similar to the data collected during long term studies on other great apes. Estimated age at first parturition for chimpanzees at Gombe ranged from around 10 to 14.5 years of age (Goodall 1986). Observations on the mountain gorillas of the Virunga Volcanoes place the estimated age at first parturition at around 10 years (Stewart and Harcourt, 1986). Data on chimpanzees of the Mahale Mountains, however, indicate that for this population, first parturition can occur much later in life. Estimated age of first parturition for Mahale chimpanzees ranged from 12-20 years with a mean of 14.6 years (Nishida, *et al.* 1990).

Adolescent Sterility, Estrous Cycles and Copulations

Laboratory studies have indicated that female primates experience anovulatory cycles between menarche and first conception, and these cycles may or may not be interspersed with ovulatory cycles (Corner, 1923; Jolly, 1985). Observations on an adolescent female (Fern) indicated that she began approaching adult males proceptively when she was approximately ten or eleven years old, but as mentioned, her first parturition occurred at 14 or 15 years of age. Another adolescent female was observed to cycle and consorted periodically with an adult male for at least a year before becoming pregnant. Failure of these two adolescents to conceive during the lengthy period prior to first conception during which numerous copulations occurred provides indirect evidence that female orangutans go through a 12 month or longer period of adolescent sterility following menarche.

Orangutan females have not been reported to undergo any readily visible physical changes, such as genital swelling, correlated with ovulation. Although on a few occasions the genitalia of consorting females appeared pinker than usual, this was not a reliable guide to receptivity, nor was it possible to monitor menstruation. Thus the onset of female cycles at adolescence could only be judged from behavior, namely proceptive behavior directed at an adult male. Since adolescent females always initiated and terminated consortships, I assume that consort periods correlate with ovulation. While six consort periods involving one adolescent female were witnessed, I did not observe many consecutive monthly periods. This may have been due to sampling limitations or to the fact that this adolescent female was not cycling every month. When consecutive consort periods were observed, she resumed consortship at intervals of 44, 26, and 22 days; copulations were observed 51 and 31 days apart. Given the general irregularity of adolescent cycles in anthropoids and the probable existence of a period of adolescent sterility among orangutans, this is not surprising.

Age/sex classes were observed to copulate at different frequencies. The data are shown in Table 4.

Selection of Sexual Partners

The choice of sexual partner in wild orangutan consortships appears very much a female prerogative. Due to exigencies of arboreal travel, although orangutan females can sometimes be forcibly mated, they cannot be herded, led or coerced into joining a male. Hence a female cannot be guarded by a male in the expectation that he will have sexual access to her when she becomes fertile. If a male attempts to consort with an unwilling female, he cannot guide her movements but has to resign himself to either following her, sometimes straight into the path of a dominant male chosen by her, or forcibly mating her.

The preferences of individual orangutans for mating partners indicated that age was an important factor, especially in consortship. Differential response towards subadult and adult males by adolescent females was clearly evident. Adolescent females approached adult males, particularly dominant ones, to initiate consortship but did not approach subadult males. Further, while adolescent females directed proceptive responses towards adult males,

Table 4. Frequencies of attempted and completed copulations for each age/sex class

	Age/sex class of males				Age/sex class of females			
	Adult male		Subadult male		Adult female		Adolescent female	
Copulation	Number	%	Number	%	Number	%	Number	%
Attempted; no intromission	0	0	8	36.4	5	25.0	3	9.4
Attempted with intromission but no ejaculation	6	20.0	1	4.6	2	10.0	5	15.6
Completed	24	80.0	13	59.1	13	65.0	24	75.0
Totals	30	100	22	100	20	100	32	100

they never focused such responses on subadults. Of course since subadult males seemed eager to mate, female proceptive response may not have been necessary.

Adult males also showed an age preference by selecting adult females as sexual partners and largely ignoring or rebuffing adolescent females' attempts to initiate consortship. For example, after approaching and directing proceptive responses toward a large adult male (Ralph) an adolescent female (Fern) followed him in an effort to initiate consortship. His eventual negative response (chase and grab) to her presence terminated the encounter. Scott (1984) found similar results during her study of baboons at Gilgil, Kenya. For the most part, adult male baboons ignored adolescents until the females' seventh cycle.

Adolescent female orangutans were, nonetheless, determined to press their unwanted attentions on seemingly indifferent adult males, at times slapping disinterested adult males or pinching their genitals. On three occasions adolescent females initiated contact with an adult male by approaching sitting directly above him and then breaking and dropping branches on him and once by urinating on the adult male. When persuaded to consort, differential adult male behavior towards adolescents in comparison to adult consort partners was also apparent.

Copulations within a Consort Context

Immature orangutans, whatever their sex, were responsible for initiating and maintaining consortships with adults. Adult males and adolescent females in consort suffered from a lack of behavioral synchronicity, probably as a result of conflicting interests over maintaining contact and due to differences in ranging and feeding behavior (Galdikas & Teleki 1981). This was true of five recurring consorts between an adolescent female and an adult male. While female consorts were usually followed within 20m, adolescent female consorts would sometimes wander great distances from their adult partners, especially in the late afternoon. Frequently they did not resume contact with the adult male consort until the next morning. One afternoon, an adolescent female was observed to travel over 200m away from the male before nesting for the night.

When consorting with adult females, adult males were observed to cease termite foraging and terrestrial locomotion and to range shorter distances. There is no such compromise between consorting adult males and adolescent females: males continue locomoting on the ground and foraging for termites. Adolescent females experienced difficulty following males, who moved rapidly along the ground. Furthermore, adolescent females generally spend longer hours active. Although adult males increased their day lengths somewhat, adolescent females sometimes continued foraging long after male consorts nested for the night. It seems clear that adolescent females were highly motivated to initiate and maintain contact with their consort partners. In the seven cases where the end of a consort period was witnessed, however, it was invariably the female who terminated the relationship by leaving the male, frequently after he nested for the night.

During the study, two classes of females were observed 62 times directing proceptive responses towards males: adolescent females directed such responses exclusively towards adult males and juvenile females towards subadult males. Most of these instances (95%) involved adolescent females; while only a few (5%) involved juvenile females. The vast majority of proceptive responses (90%) occurred in the context of adult male/adolescent female consortships. It must be stressed, however, that no non-consort proceptive responses led to copulation.

Every instance of adult male/adolescent female copulation was preceded by female proceptive behavior. Proceptive behavior by adolescent female orangutans involved the female approaching the male and soliciting copulation through proximity, touching (body or genitals), presenting, mouthing of genitals, grooming, or a combination of these proceptive behaviors (Table 5). Presenting involved the female dangling her perineal region in the male's face from above while facing him. While most adolescent female displays were unsuccessful (61%) it seems they were an essential prelude to all copulations with an adult male.

There were no male courtship displays as such. However, the adult male's long call seemed to play a role in initiating proceptive behavior. Almost half of all adolescent female proceptive behavior (42%) directly followed the male's beginning the long call (Galdikas 1983). In some cases, females were several trees away but moved rapidly and directly towards the male as he called. This percentage of adolescent female responses seemed relatively constant over seven consort periods and two different adult male/adolescent female pairs. However, the adolescent female's rate of success in inducing the adult male to begin copulation seemed about the same whether the proceptive behavior followed a long call or not.

Typically, the adult male sat very still while the female touched, groomed or mouthed his genitalia. Occasionally, he would even begin the grumbling portion of the long call as she did so. But once the female ceased behaving proceptively the adult male resumed foraging, continued sitting or even moved away. On one occasion, the male and female sat for an hour in the same position looking at each other after the female ended soliciting. However, in only three cases (5%) did the male interrupt an adolescent female's proceptive behavior by moving away or pushing away her hands.

Copulations in a Non-consort Context

MacKinnon's (1971, 1974) initial reports of "rape" among orangutans were received with a skepticism fueled, in part, by his choice of terminology. The word "rape" carries specific legal and moral implications absent in the animal world and therefore unsuitable for describing orangutan behavior. A more appropriate term coined by Estep and Bruce (1981) is "forced copulation" in which the female displays active resistance to being positioned for intromission. It includes struggle and escape behavior. Few nonhuman primates, outside of orangutans, display forced copulation as part of their sexual repertoires. "Willing resistance" is the recommended term when the victim of the sexual aggression is willing but resists mating attempts (Estep and Bruce, 1981).

Almost all forced copulations took place in a non-consort context. During these brief sexual encounters, contact between the pair lasted less than one day. The majority of copulations between subadult males and adolescent females (84%) were forced; all were pairings that were non-recurring. (No adult male/adolescent female resisted mating occurred.) Without exception, males initiated all sexual activity whenever copulation occurred in non-consort contexts. While consortships occurred only in early and mid-1972 and then

Table 5. Proceptive behavior by adolescent females[a] directed towards males

Behavior	No copulation		Copulation		Total		Long call or grumbling precedes proceptive behavior			
	No.	%	No.	%	No.	%	No copulation		Copulation	
							Long call	Grumbling	Long call	Grumbling
Approach male and sit/stand very close (within 1/10 m)	3	8.3	1	4.4	4	6.0				
Approach & solicit (dangle above & face him)	2	5.6	4	17.4	6	10.2	1		3	1
Touch body w/hand	6	16.7	3	13.0	9	15.3	2	1	1	
Touch genitals w/hand	2	5.6	1	4.4	3	5.1	1			
Groom body	12	33.3	2	8.7	14	23.7		1		1
Mouth genitals	2	5.6	5	21.7	7	11.9	2		2	
Touch body & solicit[b]	—	—	2	8.7	2	3.4			1	1
Groom body & solicit[b]	2	5.6	1	4.4	3	5.1	1		1	
Mouth genitals & solicit[b]	—	—	4	17.4	4	6.0			1	
Groom body & touch genitals	2	5.6	—	—	2	3.4		1		
Groom body & mouth genitals	5	13.9	—	—	5	0.5	1	1		
Total	36	100	23	100	59	100	8	4	9	4

Note: Total instances proceptive behavior: 59; Total instances proceptive behavior followed by long call and/or grumbling: 25.
[a]Three instances of proceptive behavior by juvenile females directed towards subadult males were observed: 3 solicitations and 1 mouthing of genitals.
[b]Not necessarily in that order.

again after mid-1974, non-consort copulations were more evenly spaced throughout the four years and involved females at various points of the reproductive cycle, suggesting that intromission, if complete, did not occur at the female's peak of ovulations and hence, that forced copulations were a "second best" strategy. Over half of non-consort copulations were incomplete because of female struggle or avoidance behavior.

The struggles during resisted matings varied in intensity from brief tussles while males attempted to position females to protracted battles which continued throughout the length of the forced copulation with females vocalizing "rape grunts." Adolescent females did not struggle vigorously, nor did they "rape grunt" during resisted matings. Most struggles involving adolescent females appeared to take the form of "willing resistance" (Estep and Bruce, 1981) in which the female struggled only briefly. In fact, one adolescent female resume eating fruits while the subadult male copulated with her.

DISCUSSION

Among the great apes, there is heavy emphasis on individual experience and learning throughout the developmental period. Input in the form of adaptive solutions to environmental and social problems is crucial for a properly functioning great ape adult. Orangutan adaptation appears paradoxical for an anthropoid, given the apparent paucity of social interactions reported from previous studies (Horr, 1972; Rodman, 1973).

Even in the comparatively desocialized Asian great ape, however, learning through social interaction is of momentous importance. This is evidenced not only through their prolonged maturation period, but also by the fact that young rehabilitant orangutans at Tanjung Puting who have been deprived of social contacts initially appear handicapped in their ability to grasp the subtle nuances that characterize wild orangutan relationships and interactions. The initially inappropriate social behavior of male rehabilitants has caused wild adult males to attack and seriously wound them (Galdikas, 1978). In its most extreme form, such inappropriate social behavior is manifested in the killing of conspecifics, a behavior never observed among wild orangutans (Galdikas 1980). Nonetheless, it must be stressed that wild adult males do chase, attack and wound wild subadult males as well.

In light of the highly dispersed nature of orangutan units and the relative absence of social interaction, how do young orangutans accumulate the vast amount of environmental and social knowledge necessary for survival and successful reproduction? Data from this and other studies (MacKinnon, 1974; Rijksen, 1978; Galdikas, 1985b) indicate that independent adolescent orangutans in transition from their natal unit to establishing a solitary or semi-solitary existence spend a large percentage of their time in association with other orangutan units, much of which constitute true social interaction and not passive aggregations. Thus, it appears that many of the interactions deemed important among other anthropoid species also occur among orangutans. What is unique in the case of the orangutan is that not until independence from the mother do individuals outside the natal unit play a significant role in the socialization process. With independence comes entrance into the wider social world of the orangutan network. During this period, independent immature orangutans may travel together for weeks or days and may briefly associate with adults. Even longer periods of immature/adult associations may occur during consortships. During these extended periods of interaction, a great deal of environmental and social observational learning surely occurs, which helps prepare the young orangutan for adulthood.

Results of this study show that adolescent females are the most social of all orangutan age/sex classes at Tanjung Puting. In contrast to male orangutan sociality, however, adolescent female sociality is not explained fully by reproductive considerations, nor, as in the case of adult females, can child-care constraints be posited as an explanation for their behavior.

Instead, adolescent female orangutan sociality at Tanjung Puting can be best understood by separately examining the benefits derived during interactions with each age/sex class.

Adolescent females face several adaptive problems, including: 1) the acquisition of their own home range, adjacent to and overlapping their natal range, 2) formation of relationships with orangutans who inhabit adjacent or overlapping ranges, 3) learning all potential food sources and safe travel routes within their new range, 4) learning the activity rhythms of predators within their new range, and 5) commencing successful reproductive behavior. The high level of socialization witnessed among adolescent females helps them, in part, solve these adaptive problems. Benefits accrued by adolescent females while associating with other orangutans might include: 1) a potential context in which to learn adaptive solutions to social and/or environmental problems, 2) increased foraging efficiency, 3) protection from predators and/or enemies, and 4) a potential context in which dominance and affiliative relations between individuals can be fleshed out.

Females interact with adult males based on female reproductive state. Even though opportunities are limited, adolescent females seem eager to begin reproduction immediately after puberty, before full maturity is attained. Copulation rates for adolescent females are higher than for the adult females, testifying to this age/sex class' high motivation towards sexual interaction. The initiation of reproduction as early as possible maximizes reproductive success over the life span. However, adolescent females in many primate species tend to be sterile. As the data presented in this paper show, this is probably true of orangutans. This may explain why adolescent females maintain consortships with adult males, rather than vice versa. Males probably prefer adult females as consorts because their cycles are more likely to be fertile.

From the proximate standpoint, Ashley Montagu (1979) reasons that adolescent female sterility among mammals can be explained by the different developmental rates for the various organs of the reproductive system that result in maturation of certain components of the reproductive system before the system as a whole is capable of carrying a fetus to term. The period of sterility provides time during which the female develops into an efficient reproductive organism, but at which time, is neither physiologically nor behaviorally prepared for the task of maternity (Montagu, 1979). Tutin (1979) has suggested that adolescent sterility observed in female chimpanzees provides them with a period of time to investigate new communities, select suitable mates, and form new relationships without the burden of an offspring. The period of adolescent sterility among orangutans also allows a measure of time during which adult behavioral patterns crucial to reproduction can be learned. Therefore, at this point in the orangutan female's life history it is in her best interest to learn the proper behaviors associated with successful adult consort maintenance and copulation. In this regard, adult males prove the optimal sexual partners because, more so than any subadult male, they are experienced in matters of consortship. Adult males might be viewed as the unwilling tutors of adolescent females on the complexities of consorting.

If consort associations between adult males and adolescent females are recurring, then the reproductive success of both parties is likely to benefit following a successful fertilization. Due to long birth intervals, receptive orangutan females are a rarity in the tropical forest. Females within the study area were in rough reproductive synchrony, so that for at least two years there were no receptive prime adult females. Since adult and subadult males are probably capable of producing sperm at all times, female reproductive capacity becomes a limiting resource for male reproductive success (Trivers, 1972). This situation may provide adult males with the motivation to at least tolerate the proceptive advances of adolescent females.

From the adolescent's perspective, consorting with an adult male may provide some measure of protection against forced copulation attempts by subadult males. For example, observations of one consort pair revealed that when the adult male moved away from his adolescent consort's immediate vicinity, subadult males raced towards her. The adolescent female's squeals attracted the attention of her consort, who drove off the subadult males. By consorting with an adult male the female preserves her choice of with whom she copulates. Further, through repeated consortships with dominant, resident adult males, adolescent females ensure any male offspring which may result carry those genes coding for traits that make for a successful breeding adult male, namely strength and large size, throat pouch, and cheek pads. Traveling in association with any age/sex class unit would probably increase the likelihood of predator avoidance simply because it increases the number of individuals who are vigilant; however, when consorting with adult males, protection against predators must be near absolute, owing to the male's massive size and intimidating nature.

By tenaciously consorting with a dominant male, adolescent females establish an affiliative bond with him that extends beyond consortship. One nonreceptive female's behavior in the presence of her ex-consort indicated that the pair had established a different kind of relationship than observed between any other female and adult male. Although she rarely traveled with him, she entered fruit trees where he was foraging and fed on a small branch several meters away. She also touched him, once in a manner reminiscent of a caress. This was extraordinary behavior for a nonreceptive female, but it allowed her access to resources monopolized by the adult male. No other nonreceptive female was ever observed to enter a tree where this male was foraging, let alone sit a few meters from him. Access to resources monopolized by adult males might prove crucial for female survival during periods of food scarcity.

An understanding of adolescent female/subadult male association requires knowledge of orangutan ranging behavior (Galdikas, 1979). Briefly, adult females and their dependent offspring occupy stable, overlapping home ranges. Males can be divided into two groups based on their ranging patterns: residents and transients. Residents stay in definable ranges which overlap with several home ranges occupied by females, while transients range much more widely without staying long in any one area. Residency appears loosely correlated with dominance (Galdikas, 1979). It must be stressed, however, that residency is not permanent over the life span of an adult male (see Utami and Mitra Setia, this volume). Eventually, the status of resident is challenged by other adult males, who, if successful, may oust the resident male and occupy his home range themselves.

Data presented here and elsewhere (Galdikas, 1979, 1981) demonstrate that resident males tend to be the receptive females' preferred consort partners. Given the impermanent nature of residency, however, it is in the adolescent female's best interest to form affiliative relationships with those individuals who will be the residents of the future, namely, the subadult males in the area. By defining relationships during the period when she is immature and sterile, no time is wasted attempting to form bonds or assess fitness when she is a reproductively functional adult. The fact that subadult males prefer to associate with nonreceptive adolescent females rather than with other males or nonreceptive adult females may ultimately be related to reproduction, even though matings were never observed during such long lasting social interactions. The formation of affiliative relationships with adolescent females may facilitate the selection of the subadult male as a consort partner when both become adult. Similarly, Tutin (1979) has suggested that chimpanzee females prefer males with whom they have affiliative relationships. There was no way to test the validity of this hypothesis during the four year study. Nonetheless, two non-consort copulations in which an adolescent female did not struggle involved a subadult male who had previously associated with her.

Other factors may promote subadult male/adolescent female associations. Since adolescent females are smaller than adult females, are not accompanied by offspring, and have day ranges similar in length to those of subadult males, food competition between associating immature orangutans is minimized. Thus, for immatures associating together, the costs of sociality are less than those incurred during associations with or between adults.

Subadult males may, in fact, benefit when associating with adolescent females by increasing their foraging efficiency. Adolescent females are always ranging in or around areas with which they are largely familiar; subadult males, however, occasionally move through unfamiliar areas. Under these circumstances, rather than wandering aimlessly or following slow-moving adult females, subadult males optimize foraging efficiency by associating with resident adolescent females who are intimately acquainted with the locations of fruit trees. This may be why subadult males maintain associations with adolescent females.

The adolescent females may also gain immediate benefits from associations with subadult males by forming coalitions against a third party in disputes over access to fruit trees. Normally, adolescent females defer to adults when encountering then in fruit trees, but one immature heterosexual pair were successful in driving adult females out of fruit trees. Sugardjito, *et al.* (1987) reported that siamangs (*Hylobates syndactylus*) at their Ketambe field site often harassed orangutan adolescents in fruit trees and on several occasions an adolescent female formed coalitions to drive away siamangs.

Associations between adolescent and adult females, while existent, are rare due to adult female intolerance of adolescent females, often culminating in attacks directed at the immature individual. Aggressive behavior on the part of the adult female can be understood in terms of female reproductive investment. Following the birth of their first offspring, female reproductive success is contingent on successful competition for food for themselves and their offspring (Trivers, 1972; Wrangham, 1979). Associations with adolescent female orangutans increase competition for food and thus, represent "costs" which should be avoided by adult females.

The change in behavior witnessed after one female gave birth to her first infant indicates that the transition from a social adolescent to a semi-solitary adult may take place rapidly, and indicates the profound affect child-rearing constraints place on females. In the fourteen months before Georgina gave birth, she was encountered on five different searches and observed for a total of 17 days. During part of all five observation periods, she traveled with one or two adolescent females, Maud or Fern. No aggressive incidents were observed; in fact, she once groomed Fern for ten seconds.

Observations during the year after her infant's birth indicated that Georgina had become considerably more solitary. Similar results have been reported for chimpanzees and bonnet macaques; following first parturition females become less gregarious and more solitary (Clark, 1978; Nishida, 1979; Pusey, 1983; Goodall, 1986). Associations between Georgina and the two adolescents decreased markedly after Georgina gave birth; from 45 hours in year two to one hour in year four of the study. Following the birth, when either of the two adolescent females traveled with her, as was observed on four different occasions, she attacked them without overt provocation. She succeeded in biting each of the two adolescents at least once. Further, the two adolescents began responding to her differently than previously. For example, they often squealed when she merely approached them and sometimes even left food trees when she entered. One fascinating incident took place which indicated that the infant may have been a factor in this new response. When the infant was approximately one year old and leaving her mother's body for short distances, she approached within a meter of one of these adolescent females as if to initiate play. The mother immediately put herself between her infant and the adolescent, whereupon the latter moved back. The adolescent seemed hesitant and twice fled a short distance, only to again cautiously approach the mother and infant pair. During the course of this interaction, Georgina took over the freshly made day nest of the adolescent.

By the fourth year of fieldwork, contact between this female and the two adolescents had virtually ceased. The mother was observed for 36 days between November 1974 and November 1975, during which time only one contact occurred. The adult female was feeding in a large vine (*Gnetum*) from which hung masses of fruit. The adolescent female slowly approached the vine, but did not enter. Instead she sat in an adjacent tree for 25 minutes, staring into it. Finally she slowly moved away to feed. The adult female continued to forage in her rich vine for another two and a half hours.

From the adolescent female's perspective limited associations with adult females may be beneficial. For example, when an independent adolescent female sets up her own home range, it lies partially overlapping with, but not encompassing, the natal range. Consequently, adolescent females may be unfamiliar with all potential food sources and safe travel routes in those areas of their new range which fall outside of the natal range. By following adult females whose ranges overlap into the less known area of the adolescent's new range, adolescent females can monitor their neighbor's feeding and movements, thereby noting any previously unknown food sources as well as safe travel routes. Associations with adult females over limited periods of time may ultimately increase the adolescent female's foraging efficiency, which may prove crucial during periods of food scarcity or when the need for additional food occurs following the birth of their first infant.

Finally, associations between adolescent females can be seen as arising from many of the benefits of grouping already discussed, the possibilities of alliance formation against enemies, increased foraging efficiency by means of noting previously unknown food sources and travel routes, and better protection against predators. Perhaps most significantly, however, associations between adolescent females provide a context in which dominance and affiliative relations can be explored. Unlike male orangutans who probably wander widely at some time in their lives, females remain in a relatively small area and in indirect association with other resident females throughout their life spans. Because the ranges of females overlap extensively, as adults they will intermittently meet, if only by accident. Due to this reoccurring contact throughout the life span, it is important that relationships are defined to avoid aggression. Walters (1986) suggests that dominance relations between immatures are more flexible compared to adults and thus they are more motivated to interact. Bernstein and Draper (1964) found that juvenile rhesus macaques exhibited less clear dominance relations and higher activity levels than adults, lending support to Walters' (1986) hypothesis. This explanation seems particularly plausible for adolescent orangutan females who do not appear to inherit their mothers' rank, as is the case among macaques and baboons. Instead, rank relative to that of their peers emerges directly through interactions.

CONCLUSIONS

In conclusion, quantitative measures make clear that adolescent female orangutans at Tanjung Puting are best described as social, not solitary or even semi-solitary. Instead of a linear, increasing tendency towards solitary living, immature orangutans experience an intense period of sociality. Qualitative observations also indicate that adolescent females have a higher capacity for overt affiliative behavior (as witnessed by their grooming and proceptivity), then any other orangutan age/sex class. Interestingly, this capacity is further reflected by the differential use of sign language by the rehabilitant orangutans; young sign more concerning social activities, while older individuals largely restrict their signing to requests for food (Shapiro, 1982). High levels of sociality can ultimately be seen as providing overall benefits to adolescent females which outweigh the costs of grouping experienced by adults of the species. These benefits prove crucial to adolescent females later in life, during their semi-solitary adulthood (Galdikas, 1984). In this sense, natural selection can be seen to act on every stage of the female orangutan's life history, and the high levels of sociality expressed during the adolescent phase are a crucial feature of orangutan adaptation at Tanjung Puting.

REFERENCES

Altman, J., 1974, Observational study of behavior: Sampling Methods, *Behavior* 49:227-267.

Ashley Montagu, A. M., 1979, *The Reproductive Development of the Female: A Study in the Comparative Physiology of the Adolescent Organism*, Littleton, MA: PSG Publishing.

Bernstein, I. S. & Draper, W. A., 1964, The behavior or juvenile rhesus monkeys in groups, *Anim. Behav.* 12:84-91.

Clarke, M. R., 1978, Social interactions of juvenile female bonnet monkeys, *Macaca radiata*, *Primates* 19(3):517-524.

Davenport, R., 1967, The orang-utan in Sabah, *Folia Primatol.* 5:247-263.

Estep, D. Q. & Bruce, K. E., 1981, The concept of rape in non-humans: a critique, *Anim. Behav.* 29:1272-1273.

Galdikas, B. M. F., 1978, Orangutans and hominid evolution, In: *Spectrum: Essays Presented to Sultan Takdir Alisjahbana on his Seventieth Birthday*, (Ed. S. Udin), Dian Rakyat, Jakarta, Indonesia.

Galdikas, B. M. F., 1979, Orangutan adaptation at Tanjung Puting Reserve: Mating and ecology, pp. 194-233 in: *The Great Apes*, (Eds. D. A. Hamburg & E. R. McCown), Menlo Park: Benjamin/Cummings.

Galdikas, B. M. F., 1980, Indonesia's Orangutans, living with the great orange apes, *Nat. Geogr. Mag.* 157:830-853.

Galdikas, B. M. F., 1981, Orangutan reproduction in the wild, pp. 281-300 in: *The Reproductive Biology of the Great Apes*, (Ed. C. E. Graham), New York: Academic Press.

Galdikas, B. M. F., 1983, The orangutan long call and snag crashing at Tanjung Puting, *Primates*, 24:371-384.

Galdikas, B. M. F., 1984, Orangutan female sociality at Tanjung Puting, pp. 217-235 in: *Female Primates: Studies by Women Primatologists*, (Ed. M. Small), New York: Alan R. Liss.

Galdikas, B. M. F., 1985a, Orangutan Sociality at Tanjung Puting, *Amer. J. Primatol.* 9:101-119.

Galdikas, B. M. F., 1985b, Subadult male orangutan sociality and reproductive behavior at Tanjung Puting, *Amer. J. Primatol.* 8:87-99.

Galdikas, B. M. F., 1985c, Adult male sociality and reproductive tactics among orangutans at Tanjung Puting, *Folia Primatol.* 45:9-24.

Galdikas, B. M. F. & Teleki, G., 1981, Variations in the subsistence activities of female and male pongids: New perspectives on the origins of hominid labor division, *Curr. Anthropol.* 22:241-256.

Galdikas, B. M. F. & Wood, J., 1990, Great ape and human birth intervals, *Amer. J. Phys. Anthropol.* 83:185-192.

Goodall, J., 1986, *The Chimpanzees of Gombe: Patterns in Behavior*, Cambridge: Harvard University Press.

Harlow, H. F. & Harlow, M. K., 1962, Social deprivation in monkeys, *Sci. Amer.* 54:244-272.

Harrison, B., 1962, *Orangutan*, London: Collins.

Horr, D. A., 1972, The Bornean orang-utan, *Borneo Res. Bull.* 4:46-50.

Horr, D. A., 1975, The Bornean orang-utan: Population structure and dynamics in relationship to ecology and reproductive strategy, pp. 307-323 in: *Primate Behavior: Developments in Field and Laboratory Research Vol. 4*, (Ed. L. A. Rosenblum), New York: Academic Press.

Jolly, A., 1985, *The Evolution of Primate Behavior*, New York: Macmillan Publishing Co.

MacKinnon, J. R., 1971, The orang-utan in Sabah today, *Oryx* 11:141-191.

MacKinnon, J. R., 1974, The behavior and ecology of wild orangutans (*Pongo pygmaeus*), *Anim. Behav.* 22:3-74.

Maple, T., 1980, *Orang-utan Behavior*, New York: Van Nostrand-Reinhold.

Maple, T. et al., 1979, Cyclic proceptivity in a captive female orang-utan, *Behav. Proc.* 4:53-59.

Mitani, J. C., 1989, Orangutan activity budgets: Monthy variations and the effects of body size, parturition, and sociality, *Int. J. Primatol.*

Nishida, T., 1979, The social structure of chimpanzees of the Mahale Mountains, pp. 73-121, in: *The Great Apes*, (Eds. D. A. Hamburg & E. R. McCown), Menlo Park: Benjamin/Cummings.

Nishida, T. et al., 1990, Demography and reproductive profiles, pp. 63-97 in: *The Chimpanzees of the Mahale Mountains: Sexual and Life History Strategies*, Tokyo: University of Tokyo Press.

Okano, T., 1965, Preliminary survey of the orang-utan in North Borneo (Sabah), *Primates* 6:123-126.

Pusey, A. E., 1983, Mother-offspring relationships in chimpanzees after weaning, *Anim. Behav.* 31:363-377.

Richards, A. F., 1985, *Primates in Nature*, New York: W. H. Freeman & Co.

Rijksen, H. D., 1978, *A Field Study on Sumatran Orangutans (Pongo pygmaeus abelii) Lesson 1827-Ecology, Behaviour, and Conservation*, Wageningen: Veeman and Zonen.

Rodman, P. S., 1973, Population composition and adaptive organization among orangutans of the Kutai Reserve, pp. 171-209 in: *Comparative Ecology and Behavior of Primates*, (Eds. R. P. Michael & J. H. Crook), London: Academic Press.

Schaller, G. B., 1961, The orang-utan in Sarawak, *Zoologica* 46:73-82.

Scott, L. M., 1984, Reproductive behavior of adolescent female baboons (*Papio anubis*) in Kenya, pp. 77-100 in: *Female Primates: Studies by Women Primatologists*, New York: Alan R. Liss, Inc.
Shapiro, G. L., 1982, Sign acquisition in a home reared/free ranging orangutan: Comparisons with other signing apes, *Amer. J. Primatol.* 3:121-129.
Steward, K. J. & Harcourt, A. H., 1986, Gorillas: Variation in female relationships, pp. 155-164 in: *Primate Societies*, (Eds. B. Smuts, et al.), Chicago: Chicago University Press.
Stott, K. & Selsor, C. J., 1961, The orang-utan in Northern Borneo, *Oryx* 6:39-42.
Sugardito, J. et al., 1987, Ecological constraints on the grouping of wild orang-utans (*Pongo pygmaeus*) in the Gunung Leuser National Park, Sumatra, Indonesia, *Int. J. Primatol.* 8(1):17-41.
Trivers, R. L., 1972, Parental investment and sexual selection, pp. 136-179 in: *Sexual Selection and the Descent of Man*, (Ed. B. Campbell), Chicago: Aldine.
Tutin, C. E. G., 1979, Mating patterns and reproductive strategies in a community of wild chimpanzees (*Pan troglodytes schweinfurthii*), *Behav. Ecol. Sociobiol.* 6:29-38.
Walters, J. R., 1986, Transition to adulthood, pp. 358-369 in: *Primate Societies*, (Eds. B. Smuts et al.), Chicago: Chicago University Press.
Wrangham, R. W., 1979, On the evolution of ape social systems, *Soc. Sci. Info.* 18:334-368.
Yoshiba, K., 1964, Report of the preliminary survey on the orangutan in North Borneo, *Primates* 5:11-26.

18

BEHAVIORAL CHANGES IN WILD MALE AND FEMALE SUMATRAN ORANGUTANS (*PONGO PYGMAEUS ABELII*) DURING AND FOLLOWING A RESIDENT MALE TAKE-OVER

S. Utami[1] and T. Mitra Setia[2]

[1] Faculty of Biology, National University, Jakarta, Indonesia
Ethology and Socioecology, Utrecht University, The Netherlands
Ketambe Research Centre, P.O. Box 004, Kutacane 24601, Aceh Tenggara, Sumatra, Indonesia
[2] Program Pasca Sarjana Biologi UI
Depok, Indonesia

ABSTRACT

Research on wild orangutans has been carried out in the Ketambe research site, Sumatra, Indonesia, since 1972. This paper focuses on the last five years during which 48 individuals were observed in the research area. In 1991, the adult male who had been the resident male for almost 18 years (with only a short interruption) was challenged and defeated by a younger adult male. Three fights between the two males were observed in detail. During the interactions, only the challenging male made long calls. Only females with offspring older than 10 years associated with the new male before the take-over. The most dominant female in the area which often associated with the new male tended to lead him toward the resident male. These encounters resulted in the final defeat of the old male. This dominant female copulated with the new resident male only after the take-over and then other females with younger offspring also associated with him. The events related to this take-over are compared with the present situation in which a newly adult male, who has been in the area as a subadult since 1972, is challenging the resident male.

INTRODUCTION

As with many primate species, adult male orangutans probably establish their form of social organization in response to the distribution of females (Wrangham, 1979). Fertile

The Neglected Ape, Edited by Ronald D. Nadler et al.
Plenum Press, New York, 1995

females are a limited resource for males, leading to intermale competition for access to the females. The home range of adult male orangutans is larger than that of adult females. The home ranges are not actually demarcated and generally overlap each other. Females tend to stay longer in a particular area than males. Both males and females sometimes leave the area and are not seen again (Galdikas, 1978; te Boekhorst, et al., 1990).

The existence of two distinct forms of the sexually mature male, the subadult and adult forms, has been well documented both in Borneo and Sumatra. Adult males with fully developed secondary sexual characteristics are intolerant of other adult males. There are good reasons to believe that the existence of both forms of male in a single population depends on a female sexual preference for adult males (Schürmann and van Hooff, 1986; van Hooff, this volume).

During the past three years at Ketambe, a fortunate development took place for witnessing and documenting the replacement of a long established adult male by a new adult male. Immediately afterwards, this new adult male was challenged by another young adult male. This provided an excellent opportunity to study the males' relationships and the relationships of these males with females. It was assumed that among orangutans, as for other primates, male relationships are organized in response to the distribution and availability of fertile females (Wrangham, 1979).

For a male orangutan, there seem to be two main options for achieving reproductive success:

1. Travel widely and try to encounter cycling females.
2. Try to monopolize several females in a particular area.

The results presented here favor option 2. Because reproductive success cannot be measured directly, however, the results permit only an estimate of reproductive success.

The main comparisons which are made here concern the behavior of one dominant male, a challenging male and the resident females during and after an adult male take-over. Changes in the frequency of the males' long calls are described, as are the frequencies of approaches between the males and the several females. Also described is the amount of time the females and males spent in association and the frequency of copulation between the males the females.

METHODS

The research was conducted in Ketambe, a primary rain forest area in the Gunung Leuser National Park, Sumatra. All behavioral data were collected with focal animal scans (Altmann, 1974). The orangutans were followed from dawn to dusk and often several individuals were followed simultaneously by different observers. The following adult males were studied: Jon, Nur and Boris.

Four time periods were distinguished during the study (Fig. 1):

1. Adult male Jon unchallenged (January, 1989 - October, 1990): During this period, Jon had been the resident adult male in the study area since Rijksen (1978) started the project in the early seventies. There were 5-6 cycling females and 3 noncycling females in the area at this time.
2. Jon challenged by Nur (October, 1990 - September, 1991): At the beginning of this period, the adult male, Nur, entered the area. There were 3-4 cycling females and 5 noncycling females in the area at this time.
3. Nur unchallenged (January - June, 1993): During this period, Jon was not in the area; he had left the area in 1991 as a result of the challenges of Nur. No systematic

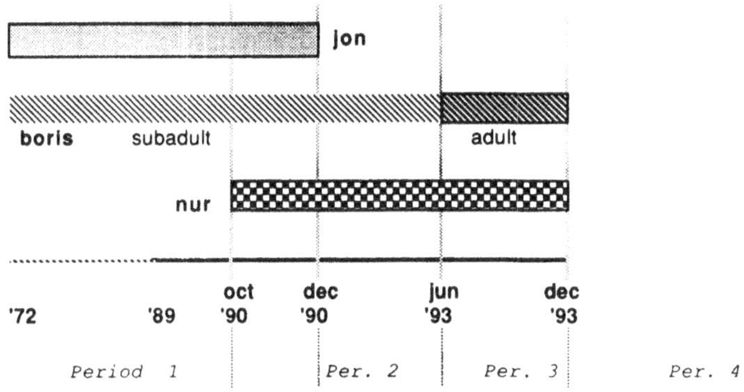

Figure 1. Residency in the Ketambe research area of the adult males present in December, 1993. Data are reported for the period 1989-1993.

observations were made in 1992. Observations were resumed in January, 1993. There were 1 cycling female and 9 noncycling females in the area at this time.

4. Nur challenged by Boris (June - December, 1993): During this period, a resident subadult male, Boris, began to develop the secondary sexual characteristics of a fully adult male and began to challenge Nur. There were 1 cycling female and 9 noncycling females in the area at this time.

The present report is preliminary, since the research is still proceeding. No detailed statistical analysis of the data is presented.

RESULTS

During the past five years, 48 individuals were sighted in the area. Among them were several adult males (Nur, Jon, Erik, I, and Mickey). For many years, until October, 1990, there was one dominant male named Jon. Another male, Nur, appeared in the area in October 1990, however, and began to challenge Jon. In December, 1990, three aggressive interactions were observed between Nur and Jon. After these interactions, Jon disappeared from the area. Subsequently, the challenging male, Nur, became the new dominant male.

Since 1972, a subadult male, Boris, was regularly sighted in the area. In the beginning of 1993, he finally began to develop the secondary sexual characteristics of a fully adult male. In June, 1993, he began to challenge Nur. Between June and December 1993, five aggressive interactions between Nur and Boris were witnessed. All of these were won by Nur. At the present time, Nur is still the dominant male.

The cycling females always remained close to the dominant male. In the period when Jon was the dominant male, they were often close to him, but after Nur took over this position, they turned to Nur.

The systematically collected data on long calls showed in all instances that the second dominant male made substantially more long calls than the nondominant males (Fig. 2). This was true for all four periods. The frequency of the long calls by Jon dropped to zero during Period 2 after Nur moved into the area. As soon as Nur entered the area, he started to make long calls; his calling reached a peak after he defeated Jon. In Period 3, Nur made fewer long calls. When Boris started challenging Nur in Period 4, however, the frequency of Nur's calling increased again. Boris hadn't yet made any long calls.

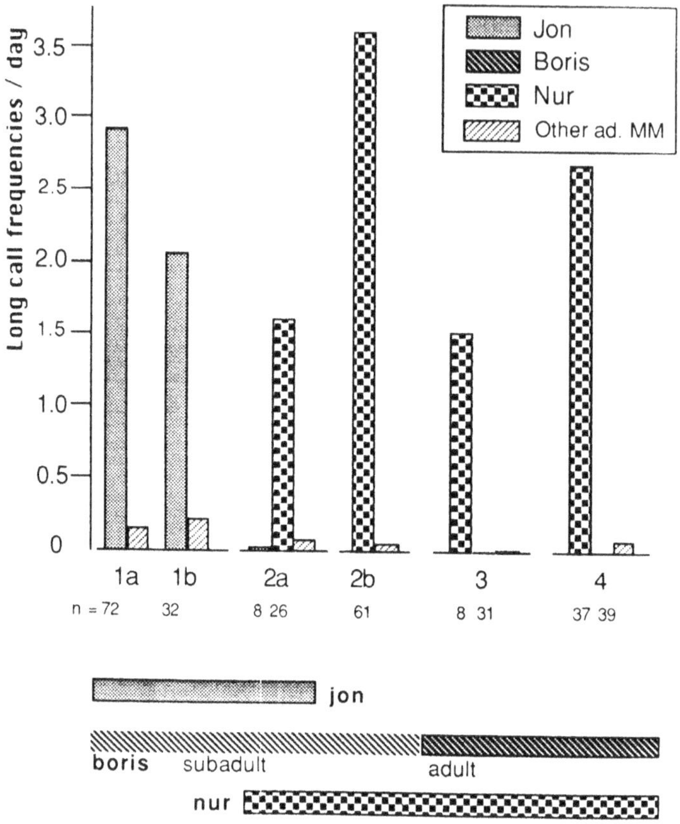

Figure 2. Frequency of long calls per day by adult males (ad. MM). Four periods are distinguished: 1) Jon is the only unchallenged adult male in th area. During Period 1, there are 3 noncycling females (NcF = 3); during Period 1a, there are 5 cycling females (CF), Period 1b, 6 CF (CF = 5/6). 2) Nur is present in the area and challenges Jon (NcF = 6; CF = 3/4). 3) Jon is absent from the area; Nur is unchallenged by Boris who is a full adult (NcF = 9; CF = 1). 4) Nur is challenged by Boris (NcF = 9; CF = 1).

Approaches of the males towards females and, conversely, of the females toward males, which occurred outside of food patches (to exclude food motivated aggregation), were analyzed (Fig 3.). In Period 1, Jon approached both cycling and noncycling females, but was approached only by cycling ones. After Nur arrived in Period 2, Jon neither approached females nor was he approached by them. Conversely, Nur began to approach cycling females as soon as he entered the area, but he was not approached by them. After the decisive interactions with Jon, Nur was approached by both cycling and noncycling females. In 1992-1993, there was a birth "boom"; most of the previously cycling females gave birth. After this, Nur approached females less frequently and the females did not approach him at all. When Boris started to challenge Nur, he both approached and was approached by females.

The association times between males and females outside the food patches were also analyzed Fig. 4). For Jon, the associations stopped altogether after Nur arrived in the area. After his arrival, Nur spent considerable time in association with cycling females; no associations with noncycling females were observed, however, until Nur established his dominance and Jon disappeared. Nur's association score decreased after most of the females

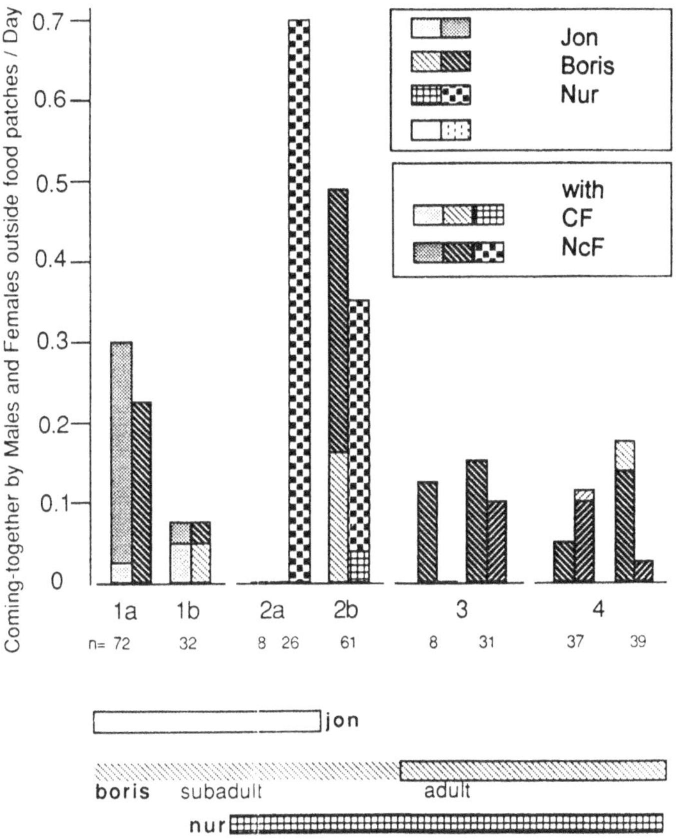

Figure 3. Frequency of approaches per day by males and females outside food patches. Four periods are distinguished: 1) Jon is the only unchallenged adult male. During Period 1, there are 3 noncycling females (NcF = 3); during Period 1a, there are 5 cycling females (CF), Period 1b, 6 CF (CF = 5/6). 2) Nur is present in the area and challenges Jon (NcF = 6; CF = 3/4). 3) Jon is absent from the area; Nur is unchallenged by Boris who is a full adult (NcF = 9; CF = 1). 4) Nur is challenged by Boris (NcF = 9; CF = 1).

had given birth; it decreased even further when Boris started challenging him. Boris spent slightly more time in association with the females than Nur.

Finally, the data on copulations outside food patches were analyzed. Jon's copulations stopped after Nur entered the area (Fig. 5).

DISCUSSION

Jon, the long established dominant male, showed a reduction in all four behavioral categories. These dropped to zero when Nur entered the area. It is important to note that at the time Nur entered the area, there were many cycling females. Nur immediately started long calling and copulating with the cycling females. They did not approach him, however, until Jon was defeated. During this period, Nur's copulations and long calls showed the highest level. After the previously receptive females had given birth, his copulations, associations, approaches and long calls decreased sharply.

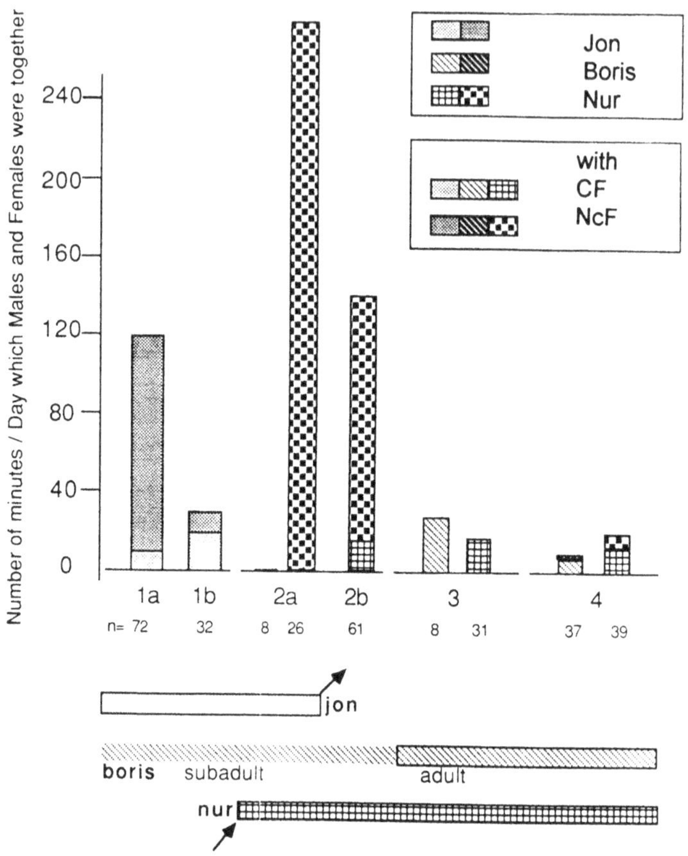

Figure 4. Number of minutes per day adult males and females spent together (association times) outside food patches. Four periods are distinguished: 1) Jon is the only unchallenged adult male. During Period 1, there are 3 noncycling females (NcF = 3); during Period 1a, there are 5 cycling females (CF), Period 1b, 6 CF (CF = 5/6). 2) Nur is present in the area and challenges Jon (NcF = 6; CF = 3/4). 3) Jon is absent from the area; Nur is unchallenged by Boris who is a full adult (NcF = 9; CF = 1). 4) Nur is challenged by Boris (NcF = 9; CF = 1).

After Boris developed secondary sexual characteristics, he both approached and was approached by females, although only by the noncycling females. It is striking that Boris was approached by these females even though he had not yet made any long calls. His only copulation was with a noncycling female.

After Boris started to challenge Nur, the frequency of Nur's long calls seemed to increase again. Both Nur and Boris continued associating with females, but less so than before. Perhaps the concern that these males had about each other's presence distracted them from contacting females.

The data suggest that the relationships between adult males are dominated by their competition for females. The dominance reversal between the adult males led to an immediate suppression of the behavioral characteristics in the nondominant male that produces sexual contacts with females. Long calls and approach initiatives were inhibited. The females, moreover, reacted with strong discrimination, shifting their attention to the winners of aggressive contests.

Finally, it may not be accidental that the challenger, Nur, arrived in the area at a time when there was a number of cycling females present. This may reflect a strategy on the part

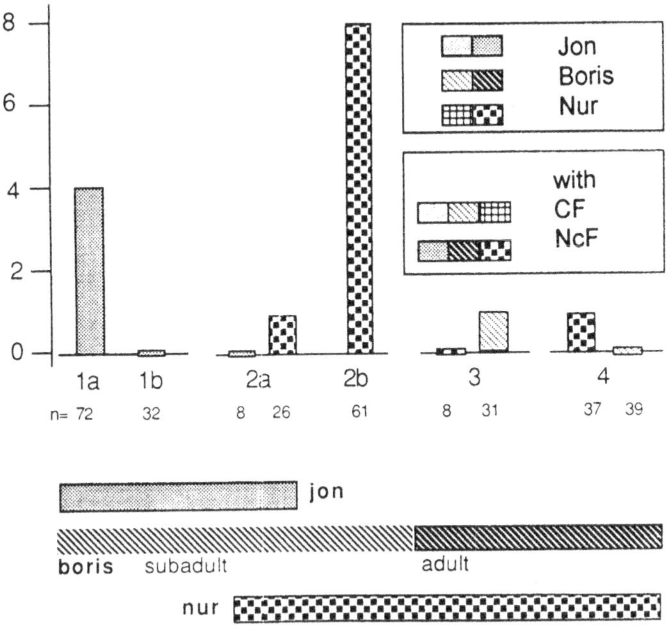

Figure 5. Number of copulations by males and females outside food patches. Four periods are distinguished: 1) Jon is the only unchallenged adult male. During Period 1, there are 3 noncycling females (NcF = 3); during Period 1a, there are 5 cycling females (CF), Period 1b, 6 CF (CF = 5/6). 2) Nur is present in the area and challenges Jon (NcF = 6; CF = 3/4). 3) Jon is absent from the area; Nur is unchallenged by Boris who is a full adult (NcF = 9; CF = 1). 4) Nur is challenged by Boris (NcF = 9; CF = 1).

of a challenger in which the risks of a take-over are offset by the benefits of a possible victory. It will be informative, therefore, to see if Boris or another adult male starts to challenge Nur more fiercely when most of the females are cycling again.

ACKNOWLEDGMENTS

This paper is based on a 4-year field study in Sumatra financed by a grant from the Netherlands Foundation for the Advancement of Tropical Research. We are grateful to the Dean of Faculty of Biology UNAS Jakarta (Dr. Kasim Moosa) and the Indonesian Nature Conservation Service for permitting this research. We are especially indebted to Prof. Dr. J.A.R.A.M. van Hooff, Romy Steenbeck, Dr. Filippo Aureli and all the students at the Ketambe Research Centre for their support and for their comments and criticisms of earlier drafts of this paper. We owe special thanks to the late Mrs. Nina Sulaiman.

REFERENCES

Altman, J., 1974, Observational study of behavior: sampling methods, *Behavior*, 49:227-267.

Galdikas, B.M.F., 1978, *Orangutan adaptation at Tanjung Puting Reserve, Central Borneo*. Unpublished doctoral dissertation, University of California, Los Angeles.

Rijksen, H.D., 1978, *A Field Study On Sumatran Orang Utans* (*Pongo pygmaeus abelii* Lesson 1827), *Ecology, Behaviour and Conservation* 78-2. Wageningen, H. Veenman and Zonen B.V..

Schürmann, C.L. and van Hooff, J.A.R.A.M., 1986, Reproductive strategies of the orangutan; new data and a reconsideration of existing socio-sexual models, *Int. J. Primatol.*, 7:265-287.

te Boekhorst, I.J.A., Schürmann, C.L., and Sugardito, J., 1990, Residential status and seasonal movements of wild orangutans in the Gunung Leuser Reserve (Sumatra, Indonesia), *Anim. Behav.*, 39:1098-1109.

Wrangham, R., 1979, On the evolution of ape social systems, *Soc. Sci. Info.*, 18:334-368.

19

IMITATION AND TOOL USE IN REHABILITANT ORANGUTANS

A. E. Russon[1] and B. M. F. Galdikas[2]

[1] Psychology Department
Glendon College, York University
2275 Bayview Ave., Toronto, Ont. M4N 3M6, Canada
[2] Archaeology Department
Simon Fraser University
Burnaby, B.C. V5A 1S6, Canada

ABSTRACT

Imitation and tool use are expressions of intelligence that have, at one time or another, played central roles in evolutionary reconstructions of human intelligence. It is fairly well established that both occur in great apes, but questions remain over the relations between the two. This relation is critical because imitation could offer a powerful process for the social transmission of tool related skills. We explored the relation between imitation and tool use with data from our observational study on spontaneous imitation in the rehabilitant orangutans in Tanjung Puting National Park, Central Indonesian Borneo. From 395 hours of observation and other reports on 26 orangutans, we identified 354 incidents of imitative behavior; 48 of these involved tool use where features of the behavior strongly suggested that the tool skills were imitatively acquired. In this paper, we discuss these incidents: the qualities of this "imitative" tool use, evidence that it was imitatively rather than experientially acquired, and the implications of these findings for models of primate and human intelligence.

INTRODUCTION

Imitation and tool use are expressions of intelligence that have both, at one time or another, been considered uniquely human. Here we use the term "imitation" in the sense of learning by observing rather than by practice (e.g., Whiten and Ham, 1992) and the term "tool use" in the sense of using detached objects to mediate attaining further goals (e.g. Beck, 1980). Each of these two abilities has played its own key role in evolutionary reconstructions of human intelligence, but the two combined have special significance: Widespread tool use in hominids is thought to have been significantly enabled by imitation.

The Neglected Ape, Edited by Ronald D. Nadler et al.
Plenum Press, New York, 1995

It is now generally accepted that great apes share complex tool capabilities with humans (e.g. McGrew, 1992), and increasingly, that they share our capacity for imitation as well (Custance and Bard, 1994; Russon and Galdikas, 1993; Tomasello, Savage-Rumbaugh and Kruger, 1993). In contrast, there has been little convincing evidence that great apes can or do use their imitative abilities to augment their tool related ones. Examples initially thought to show imitative acquisition of tool use, such as chimpanzees' termite fishing, have been discredited on the basis of evidence that direct experience or simpler forms of social influence play a large part in their normal acquisition (e.g., Galef, 1992). Since great apes seem capable of each cognitive skill separately, this may reflect their inability to link their cognitive skills; imitative acquisition of tool use requires bringing a social cognitive skill to bear on problems in the physical world. One current view is that the highly complex cognition seen in humans is possible only because humans can interconnect their different cognitive skills (tools, language, sociality) for mutual support and even enhancement (Gibson and Ingold, 1993). Some consider this interconnectness to be lacking in monkeys (Cheney and Seyfarth, 1990; Langer, 1991), who may show high level social skills but perform dismally on comparatively simple object problems.

Despite these doubts, studies continue to report examples indicating imitative acquisition of tool use in great apes, including our own studies on imitation in rehabilitant orangutans (Russon and Galdikas, 1993, in press). It is also generally accepted that great ape cognition is more complex and broad-reaching than monkey cognition. This suggests that the question of how great apes acquire their tool skills should be reconsidered. To this end, we analyzed the incidents of imitation we observed in rehabilitant orangutans for the possible role of imitation in acquiring tool related abilities.

METHODS

This was an observational study of spontaneous imitation in free-ranging wild-born ex-captive orangutans at the Orangutan Research and Conservation Project field station, Camp Leakey, in Tanjung Puting National Park, Central Kalimantan. During rehabilitation, these ex-captives experience enriched social and physical environments of the sort conducive to imitation: positive relationships (with human caregivers as well as orangutans) and free access to rich environments (the base camp as well as various forest settings). The rehabilitation procedures used in Tanjung Puting have two important characteristics: (1) the ex-captive orangutans themselves make the choice about when and how to return to the forest, and (2) the very small minority of the rehabilitants that choose to maintain some contact with the base camp seem to be "bicultural", meaning they effectively handle both human and orangutan environments rather than abnormally attaching to humans. Some of these bicultural orangutans shift to more forest-centered as they get older; since data for this study were collected, a few have abandoned the base camp but have been seen in the forest. Our work on this study focused on these bicultural individuals as those likely to show the widest range of tool use and imitation.

We collected data on orangutans in 1989 and 1990 by means of half- and full-day focal individual follows in which we described all observed cases of imitative behavior involving the focal orangutan, from complex gestures and object manipulation to simple contagious yawns and scratches. Later, we identified cases in which we could show that imitation was used in learning the imitative behavior. More detailed descriptions of our data collection, coding, and analysis procedures are available elsewhere (Russon and Galdikas, 1993, in press).

RESULTS

Our tracking produced 400 hrs of systematic observation of 26 rehabilitants; from these data, we identified 354 cases of imitative behavior. In 54/354 cases, we found evidence indicating that imitation was used in learning the imitative behavior. Therefore, we looked for tool use in these 54 cases.

In 41 of these 54 cases the imitative behavior focused on tool use, which strongly suggests that orangutans can and sometimes do use imitation in acquiring their tool skills. The range of imitative tool-use included: sweeping paths, washing things, applying insect repellent, using canoes, hammering nails, weeding paths, sharpening an axe blade, reshaping a blowgun dart, fire-making, siphoning fuel, rehanging hammocks, and sawing wood (see Russon and Galdikas, 1993, in press). One of the clearest examples is an incident Galdikas reported earlier (1982):

Rehabilitant orangutans once imitated a worker make bridges. Like him, they dragged logs to the river's edge, threw them over to the other side then scrambled across on them. They had never been observed making such bridges before they observed the worker; they acquired the technique abruptly and rapidly after observing him model it; and they used his same behavioral strategy to achieve his same goal.

The range of tool use the rehabilitant orangutans imitated is very similar to that reported for chimpanzees (Moore, 1992), suggesting that it is not simply an artifact of this setting.

Further analyses concern two facets of imitation's role in influencing orangutan tool use, (1) what the orangutans imitated and (2) how they used imitation to learn. These may suggest why it has been difficult to establish imitation of tool use in great apes.

What Orangutans Imitated

At closer look, these 41 cases of imitative tool use involved a limited repertoire of basic elements at two 'levels', actions on objects (e.g. blow, bend, throw, scrape) and simple tool relations between two objects (e.g. insert or probe one object into another). Table 1 lists some of the elements in this repertoire.

These elements were re-used throughout the 41 imitative tool use cases, but occurred in combinations, patterns, and applications which varied uniquely from one case to another. For example, we identified seven different imitative tool use routines centred around rubbing and scraping; each showed its own unique element set with an equally unique organization (see Table 2).

A good example was Supinah's attempt to reproduce a camp staff's method of sharpening an axe blade. Supinah, an adult female rehabilitant, encountered one of the male staff sharpening an axe blade with a small stone. The man dipped his stone into a pail of water he had by his side then rubbed the wet stone back and forth along the edge of the blade. Supinah watched intently, so after a minute the man gave her the stone and blade and gestured for her to do the same. She took them and immediately rubbed the stone against the face of the blade. The worker repossessed the tools and resumed his work, and Supinah resumed watching. After another five minutes she independently gestured for the stone and the blade, and the worker gave them to her. She took both but then also took the pail of water; once she had all three items, she dipped the stone into the pail of water then rubbed the wet stone against the face of the blade.

Many of these cases centred on rubbing a tool against an object and varied the pair of objects involved, substituted the tools used, or added more elements. It was this variety of combination, organization and application which produced the novelty across these routines.

Table 1. Some building blocks in the imitative tool use of orangutans

Actions on objects	Relations between two objects
blow/suck	scoop
hold	pour out (drink from)
lift	pour in
pull/push	float
tap/hit	twist (wring/wrap/wind/untie/(un)screw)
bend	open
collect/gather	close
drop	put on
throw	insert/probe/thread
wiggle	pry
scrape/rub	chop
manipulate	dig
climb on/over	scrape (draw/wipe/saw/sweep/rub)
	hammer (tap/hit)
	hook (over)
	touch together
	fan
	give

We also found that many of our cases of complex imitative tool use involved performance errors, in the sense that the orangutan's behavior did not effectively solve the problem at hand. These errors took the form of inadequate differentiation of critical elements (e.g., which direction to slide a bar lock to open rather than close it), botched performance of complex relations (e.g., awkward, unsuccessful efforts to tie knots into ropes), or

Table 2. Imitative "rub/scrape" tool use routines

Tool imitation	Elements/complexities
sweep	rub broom on path/porch surface
	strokes repeated
remove bark	scrape tool on branch
	move tool lengthwise
	substitute stick for blade
sand dart	rub dart on hairy leaf
	concentrate at rear of dart
	rub against hairy side of leaf
sharpen axe	scrape stone on blade
	repeat
	initially wet stone
paint	rub brush on surface
	first wet brush in paint
	stroke repeated
	stroke in up-down/arc patterns
brush teeth	rub brush on teeth
	first put paste on
	brush both sides
	spit out suds over the rail onto the ground
wash	rub cloth/brush on surface
	first wet cloth/brush
	perhaps rub soap on cloth/brush
	perhaps wring cloth

misordering of elements (e.g., attempting to insert a siphon tube into a fuel drum before removing the drum's cap). These errors are consistent with our earlier impression that what was acquired imitatively was not new basic elements like simple tool actions, but rather new combinations, new organizations, or new applications of known elements.

How Imitation Was Used in Learning

We assumed from the outset that the cases of imitating tool use we observed were repeat, not initial, attempts at copying a demonstration. As such, the cases we observed would have been influenced by other forms of learning than imitation; thus, they could not directly indicate how imitation is involved in acquiring tool use. However, patterns in how some of these imitative routines were performed and repeated offer interesting insights into how imitation influenced orangutan tool use.

For instance, Supinah repeated imitating sharpening an axe blade *twice* in the case described above. Supinah's second attempt at the worker's technique was a better copy than her first, but by then she had benefited from practice, so some would argue that her improvement could have derived from experience, not imitation. However, her improvement involved adding an action that she had not included in her first attempt—wetting the sharpening stone before rubbing it against the blade. This had no basis in practice—it derived not from what she experienced, but from what she observed. In this case, the nature of her improvement indicates that she did learn by imitation rather than direct experience, although she learned the tool technique a bit at a time rather than all at once.

We found even more elaborate cases in which whole bouts of imitative tool use were repeated over a period of days. For instance, we observed Supinah perform 5 bouts of hammering nails and wood over a 10-day period.

(07/11/90, 14:25) Male staff had assembled saws, nails, boards and hammers and were making a bench outside the staff quarters. Supinah joined in, hammering nails with a saw; after a while, she took a saw, nails and a board under the building. There, she held a nail properly on the board and hammered it with the saw; pushed the nail in with her mouth then her hand, and hammered it with the saw. When Siswoyo, a more dominant female arrived, Supinah left for the dining hall with her nail; there she found a new saw, reassembled her materials and resumed hammering nails into boards under the building.

(07/12/90, 8:08) The next day Supinah went under the staff quarters, retrieved boards, found a nail a few meters away, then used one board to hammer the nail into two others she placed one on top of the other. She shifted the boards around and separated them, then hammered the nail in again; she sometimes repositioned the nail, fitting it carefully back into its small indentation, before hitting it again. She finally left with the nail in her mouth.

(07/12/90, 9:08) An hour later, Supinah went under the dining hall, chopped wood with an axe she retrieved, then found a nail. She worked at putting the nail into an old stump, pushing it hard with both hands, then hammered it into the stump with the axe blade. She also pushed the nail into the ground, removed it, poked it into stump—axe handle—stump, then hammered it with the flat of the axe blade. This time she bent the nail, so she examined it closely, then hammered it gently with the edge of the axe. She finally abandoned her assembly.

(07/13/90, 11:15) The next day, Supinah went under the staff quarters, got 5 pieces of wood, placed 2 pieces together and hammered them with a 3rd.

(07/20/90) One week later, Supinah found a board with a nail in it, pulled the nail out, put it in a groove in a 2nd board, and hammered the nail with a 3rd board.

We never did observe Supinah succeed in embedding a nail into wood by hammering it, because she did not hammer hard enough.

As in the axe-blade sharpening case, Supinah's repetitions of imitative tool use seemed to be guided more by what she observed than by what she experienced. She did not simply re-encounter her carpentry kit then respond to it—in each case, she had to reassemble it by collecting its various components from scattered locations. In her repeated attempts to hammer nails into wood, she re-enacted what she had observed despite the fact that this was inconsistent with her own experience. Again, the nature of her repetitions strongly suggested that her tool using behavior was guided by observation rather than by practice.

To summarize, our data indicated that orangutan imitation does contribute to learning tool skills. However, it may not produce new tool skills in a single trial, particularly if the tool skills are complex. These orangutans seemed to acquire complex tool skills over a number of trials, one component at a time. We also found that these orangutans actively practiced re-creating routines they observed; however, even though these repeated attempts were influenced by practice and experience, a role for imitation could sometimes still be identified through the nature of the changes from one attempt to the next.

DISCUSSION AND CONCLUSION

We want to emphasize that this imitative behavior was spontaneous, not trained, and that these rehabilitants showed many normal orangutan behaviors as well. Most rehabilitants, including Supinah, spent several months a year in the forest away from human contact.

Our observations of these rehabilitant orangutans' spontaneous tool use and imitation strongly suggest that they can and do acquire some of their tool skills via imitation. They do show a certain interweaving of their cognitive skills, thus exhibiting to some degree the interconnectness found in human intelligence. We also found that the rehabilitant orangutans' imitation concentrated on the organizational features of demonstrated tool use, not on its superficial behavioral topography or basic elements; and that the process of their imitative learning may involve several attempts rather than a single one, each focusing on specific components of the demonstrated routine. In both these patterns, orangutan imitation is very like human imitation (see, for instance, Speidel and Nelson, 1989). Our findings thus support the suggestion that great apes share with humans the ability to acquire tool using skills imitatively. Consequently, we may once again require a reconsideration of the defining qualities of human intelligence.

This approach to great ape imitation and tool use has proven very fruitful. These rehabilitant subjects offer an exceptional view of orangutan intelligence. The non-interfering approach allowed us to consider spontaneous behavior in free-ranging subjects, therefore providing a high level of ecological validity. Our guess is that the difficulties researchers have encountered in eliciting imitation of tool use in great apes stem in part from ignoring the organizational-level and repetitive facets of imitation, or from treating them as impediments to research to be somehow controlled against or eliminated, rather than as integral parts of the system. Although studies like ours lack the controls extolled by experimentalists, they do offer suggestions about the form, structure and process of imitation as it is used in everyday learning. Without an understanding of such normal imitation, even the most sensitive research methods have little hope of eliciting it.

ACKNOWLEDGMENTS

We wish to acknowledge the many people in Indonesia and North America whose help and support made this specific study and the continued existence of the Orangutan Research and Conservation Project possible. There are far too many to name individually, so our thanks go collectively. Most important, we thank the many agencies of the government of Indonesia who have supported and facilitated this research. And finally, we acknowledge the very orangutans who made this study possible.

REFERENCES

Beck, B.B., 1980, *Animal Tool Behavior: the Use and Manufacture of Tools by Animals*, New York, Garland STPM Press.

Cheney, D.L. and Seyfarth, R.M., 1990, *How Monkeys See the World*, Chicago, University of Chicago Press.

Custance, D. and Bard, K.A., 1994, The comparative developmental study of self-recognition and imitation: The importance of social factors. Pp. 207-226. In: *Self-Awareness in Animals and Humans: Developmental Perspectives*, S.T. Parker; R.W. Mitchell and M.L. Boccia, eds., Cambridge, UK, Cambridge University Press.

Galdikas, B.M.F., 1982, Orang-utan tool-use at Tanjung Puting Reserve, Central Indonesian Borneo (Kalimantan Tengah), *J. Hum. Evol.*, 10:19-33.

Galef, B.G. Jr., 1992, The question of animal culture, *Hum. Nat.*, 3:157-178.

Gibson, K.R. and Ingold, T. (Eds.), 1993, *Tools, Language, and Cognition in Human Evolution*, Cambridge, UK, Cambridge University Press.

Langer, J., 1991, From Acting to Understanding, *Plenary address, 21st Ann. Meet. Jean Piaget Soc.*, Philadelphia, PA., May 30-June 1.

McGrew, W.C., 1994, *Chimpanzee Material Culture*, Cambridge, UK, Cambridge University Press.

Moore, B.R., 1992, Avian movement imitation and a new form of mimicry: Tracing the evolution of a complex form of learning, *Behaviour*, 122:231-263.

Russon, A.E. and Galdikas, B.M.F., 1993, Imitation in free-ranging rehabilitant Orangutans (*Pongo pygmaeus*), *J. Comp. Psychol.* 107:147-161.

Russon, A.E. and Galdikas, B.M.F., (in press), Constraints on great ape imitation: Model and action selectivity in rehabilitant orangutan imitation (*Pongo pygmaeus*), *J. Comp. Psychol.*

Speidel, G.E. and Nelson, K.E., (Eds.), 1989, *The Many Faces Of Imitation In Language Learning*, New York, Springer-Verlag.

Tomasello, M., Savage-Rumbaugh, E.S., and Kruger, A.C., 1993, Imitative learning of actions on objects by children, chimpanzees, and enculturated chimpanzees, *Child Devel.*, 64:1688-1705.

Whiten, A. and Ham, R., 1992, On the nature and evolution of imitation in the animal kingdom: Reappraisal of a century of research. Pp. 239-283 in *Advances In The Study Of Behavior*, Vol. 21, (Eds. P.J.B. Slater; J.S. Rosenblatt and C. Beer), San Diego, Academic Press.

20

ATTENTIVENESS IN ORANGUTANS WITHIN THE SIGN LEARNING CONTEXT

G. Shapiro and B. M. F. Galdikas

Orangutan Foundation International
822 S. Wellesley Ave., Los Angeles, California 90049

ABSTRACT

Orangutans are capable of reliably associating specific referents by producing unique gestures within a sign learning context (Shapiro, 1982). An analysis of sign performance by four juvenile orangutans indicated that sign learning varied among subjects and across signs, with some signs not learned to an acceptable criterion even after 2400 trials distributed over 15 months (Shapiro, 1985). While differences in the relative difficulty of sign movements (motor differences) could explain a portion of this failure (Fouts, 1973), individual differences in referent interest and attentiveness within the sign learning protocol could also account for differential sign performance (Shapiro, 1985).

To examine the role of visual attentiveness in sign performance, two sign experienced orangutans were given distributed sign training for five novel signs using standard molding techniques (Fouts, 1972). There was some evidence for an inverse relationship between attentiveness during the referent identification period and signing performance for both the younger and older orangutans. The combined set data indicated that during phases of referent presentation and questioning, response and molding, visual attention was averted from the target points more than half the time. During the reinforcement phase, visual attention was predominately directed towards the reward. The subjects looked at the referents of the signs less than 36% of the time during referent presentation, less than 20% during the response phase and nearly 0% of the time during molding and reinforcement. Nevertheless, there seemed to be an inverse relationship between attentiveness during the referent identification period and signing performance for both the younger and older orangutans. When data from the first trial of a set were segregated, correct signing performance and visual aversion decreased while attentiveness to the referent and trainer's face increased significantly during the question and response phases, respectively. The results are discussed within the context of such factors as method of analysis, species and subject differences, referent interest, and performance strategies.

The Neglected Ape, Edited by Ronald D. Nadler et al.
Plenum Press, New York, 1995

INTRODUCTION

More than 25 years ago, R. Allen and Beatrice Gardner published their classic report in *Science* demonstrating effective two-way propositional communication with a chimpanzee (Gardner and Gardner, 1969). They employed American Sign Language, an established linguistic system for the deaf in North America (Stokoe, et. al., 1965). The Gardeners argued that the visual-haptic modality of communication, rather than the auditory-vocal modality, was the key to productive exploration of linguistic abilities in this species. Since that time a number of studies were conducted which demonstrated linguistic abilities using different visually-based systems (Premack, 1971; Rumbaugh, 1977), different species (Patterson, 1978; Shapiro, 1982; Herman, et. al, 1984) and different aspects within the linguistic arena (Fouts, et. al., 1978; Fouts and Fouts, 1993; Patterson, 1990).

Critics of this research also asserted the phenomenon is non-linguistic and that the methods and interpretations were unreliable (Sebeok, 1980; Terrace, 1979). It is unlikely that ape researchers will completely convince the panoply of scientists and nonscientists who study language from its myriad dimensions that apes are employing signs as language. Language has been steeped, for far too long, in philosophical, religious and anthropological traditions, in addition to the more recent comparative psychological, psycholinguistic and linguistic disciplines, for any one definition of language to be acceptable to all. Perhaps, not until comparative neurophysiological studies demonstrate a correlation between the presumed linguistic behavior and neurological activity and demonstrate a homologous neurological phenomenon will the language-like behavior of productive signing by apes be considered linguistic, in the human sense, by subscribers of evolutionary theory. Nevertheless, a rich assortment of information regarding great ape cognitive and communicative abilities has emerged from ape language studies, ranging from referential communication (Gardner and Gardner, 1969), to symbolic competence (Premack, 1971), to productivity (Fouts, et.al., 1978) and to cooperative problem-solving (Savage-Rumbaugh and Rumbaugh, 1980).

Signing by apes can also be examined at a number of other levels. Signing can be seen as a paradigm to explore great ape associative and referential learning comparatively with relevance to both communicative and noncommunicative approaches to ape language research. Certainly, understanding the factors that can influence the development of a vocabulary can benefit the individual from the linguistic or communicative perspective. It would be equally beneficial to know the factors that can bias sign learning when such learning is used as a probe to explore such phenomena as functional cerebral asymmetry (Shapiro, unpublished) or developing improved language intervention therapy for linguistically impaired children (Fouts, et. al., 1978).

Orangutans are capable of learning signs (associating manual gestures and physical and nonphysical referents) and using them reliably to name referents (Shapiro, 1982). The investigation of sign learning requires measuring a unit of training effort as well as signing performance. Fouts (1973) employed the measure, "time to sign acquisition" when he taught signs to four chimpanzees. Shapiro (1985) used the measure "trials to acquisition" in a study of four orangutans. The trial was presumed to provide a more reliable unit of sign training effort than training time, since the latter does not account for the level of trainer effort or subject exposure to the referent and its gestural pairing. The trial can be broken down into four different phases:

1. Question or referent identification phase, where the object is held before the subject to identify visually;
2. Response phase, where the subject learns to produce a correct and consistent signed response with one or both hands;

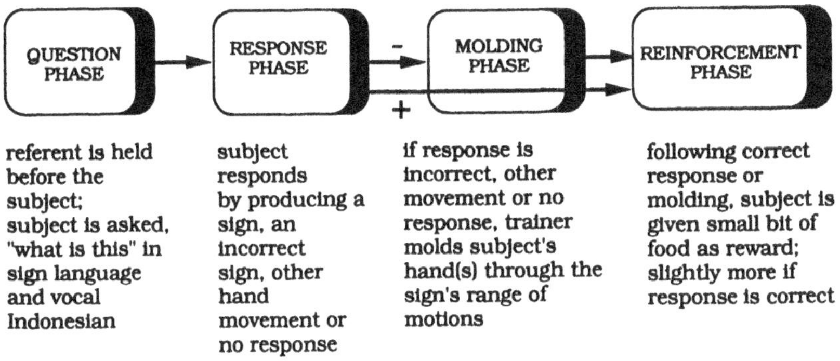

Figure 1. Phases of the unit of sign training: the trial.

3. Molding phase, where the hand(s) of the subject is(are) taken through the correct range of motions that constitute the unique gestural configuration for the sign, following an incorrect or poor gestural response;
4. Reinforcement phase, where the subject receives a social or nonsocial reward following molding or the correct signed response. The process is repeated throughout a training session, and over multiple sessions; associative learning is inferred as correct sign performance increases.

Sign learning by orangutans appeared to be influenced by gestural difficulty of the sign, previous experience with a referential concept (both seen in chimpanzees by Fouts, 1973) and referent interest (Shapiro, 1985). In a 15-month sign learning study using four juvenile orangutans (Shapiro, 1985), some signs were learned by all subjects but others were not learned to an acceptable criterion even after 2400 trials over the entire study. Performance decreased with increased gestural difficulty, but improved as previous experience and referent interest increased. For example, food signs were learned much earlier than non-food signs. This result is not surprising, since orangutans are extremely interested in food. They spend most of the day in pursuit of foods in the wild and food distribution is one strong determinant of their unique social organization (Galdikas, 1978).

Many hundreds of trials were required by the orangutans before some signs were learned to criterion. Trainers commented that the orangutan subjects did not appear to look at the referent during the question or referent identification phase of the trial. It is assumed that referent identification was required before associative and referential learning could occur. If this was the case, perhaps it could account for the large number of trials required by the orangutans before they learned the sign or before training was terminated. The present study was conducted to determine the importance of visual attention during the sign learning process.

METHODS

Project Site

These experiments were conducted in and around the camp facilities at the Orangutan Research and Conservation Project (ORCP), also known as Camp Leakey, within the Tanjung Puting Nature Reserve (now National Park), Kalimantan Tengah (Indonesian Borneo; 2° 48' S. 111° 57' E). The camp is surrounded by abandoned dry rice fields, now covered by elephant grass (*Imperata cylindrica*) and bands of lowland dipterocarp forest, incorporating limited areas of tropical heath forest, alternating with shallow peat swamp forests that deepen and interdigitate along the rivers (Galdikas, 1979).

Subjects

Two experienced signing orangutans were the subjects in the experiment. One was a home-reared, free-ranging ex-captive juvenile female (Princess, estimated to be 5-years-old), while the other was a free-ranging, ex-captive rehabilitant adult female (Rini, estimated to be 12-years-old). Each orangutan had over 18 months of intensive sign training in other sign learning and performance programs. These animals were part of an ex-captive population of orangutans undergoing rehabilitation to life in the wild in the Tanjung Puting Reserve. Although the orangutans were provisioned with rice and/or fruit twice daily, they were not caged. Between training periods, orangutans interacted with other conspecifics and their natural environment. The adult female, Rini, was part of a small population of free-ranging orangutans isolated across the river from the field station and her contact with humans was less frequent than that of the juvenile, Princess, who was being home-reared with the author. Unlike any other great ape in an experimental study, Rini was not constrained to attend the training and testing sessions. For most sessions, however, she voluntarily descended from the trees to spend up to 60 minutes with the trainer and observer.

Materials

Each subject received training on five novel signs. The signs were not part of the standard American Sign Language dictionary. Rather, they were contrived signs developed by the author to control for motoric features of the sign learning process. The signs on which the subject were trained were: tuning fork, smoking pipe, wood fungus, twig, and toy pillow. Some of the referents for the signs were natural items (wood fungus, twig) while others were human-made (pipe, tuning fork, and pillow). Table 1 illustrates the signs and their subunits or cheremes (Stokoe, et al., 1965). Cheremes designate the hand configuration, hand movement and location of the active hand. Each sign is a unique combination of these three cheremes. Referents for the signs were small objects that were relatively durable and easily transported and placed on a training stand (50 cm x 25 cm x 10 cm) for presentation. The stand was designed for the placement of referents during the administration of tests to the subject in which the observer was blind to the referent's identity (single blind).

Table 1. Signs on which two orangutans were trained, including referents and motor descriptions of the signs

Sign	Referent	Motor description
1. Pillow	A small, soft toy pillow	Extended index finger of closed right hand is drawn across the chest
2. Tuning fork	A tuning fork	Right fist contacts the right ear
3. Pipe	A smoking pipe	Extended index finger of closed left hand hooks between the 4th and 5th finger of the upturned flat right hand
4. Fungus	A wooden forest fungus	Extended index finger of closed left hand hooks the left big toe (hallux)
5. Twig	The twig of a tree	Thumb and index fingers of both curved hands repeatedly tap each other

Training Procedures

Each subject received training using standard molding techniques (Fouts, 1972) whereby, the subject's hand(s) was(were) manually positioned into the sign's proper configuration by the trainer and moved through the sign's specific range of motion. The trial was the unit of training within which the following phases were delineated:

1. Question phase: a referent is presented and the question, "what is this?" is asked of the subject in Ameslan and Indonesian languages,
2. Response phase: the subject responds to the question with a sign, a gesture or no response,
3. Molding phase: the trainer guides the subject's hand(s) through the correct range of motions following an incorrect or poor response, and
4. Reinforcement phase: the subject receives a small bit of food following a correct response or the molding phase. Subjects received a larger amount of food reward if the response was correct (i.e., differential, but continuous positive reinforcement). Distributed training was given (in contrast to massed training) in which each subject received 15 trials per sign, broken into three sets of five consecutive trials per sign during the session. Over the course of 15 months and 31 sessions, 465 trials were given to each subject for each sign. Sessions lasted from 30 minutes to nearly one hour. A sign was scored as correct when all three cheremes were performed correctly. If only two cheremes were observed, the response was scored as poor. Only correct sign performance (CSP) was used in this analysis.

Eye Contact Measures

During the last 75 trials of sign training, an observer attended the sessions to collect data on visual attentiveness by the subject during the four phases of the trial. Four possible target points were chosen: referent, trainer's face, subject's hands and the reinforcement item. The observer, in addition, graded the attentiveness into slight and prolonged eye contact. "Slight" meant the observer noticed the subject give a brief glance towards the target point during that particular phase of the trial. "Prolonged" meant the observer noticed the subject give a distinct gaze towards the target point. Eye aversion away from these four target points were also scored. The observer was positioned to the side of the trainer and subject and oriented towards the subject such that the subject's gaze could be reliably discerned. The target and relative intensity of the gaze were entered into a different databook than the sign performance data for later comparison and analysis.

RESULTS

We first examine correct signing performance (CSP) and the visual attentiveness measures over the last 75 trials of training. Although two signs (tuning fork and fungus) were performed comparably well for the two subjects, there were significant differences in signing performance between signs and subjects. Each subject had one sign that was not performed correctly even after 465 trials of training (twig for Princess; pipe for Rini). Motor difficulty may explain part of this difference. The gesture necessary to perform the sign "twig" (thumb and index finger tap repeatedly for both curved hands) may be easier to make for a physically mature orangutan than a juvenile one, as seems to be the case for chimpanzees (Fouts, 1972). Rini's poor performance of "pipe" may be due to conceptual confusion, as she frequently called it "twig" throughout training. There were no significant differences between signs

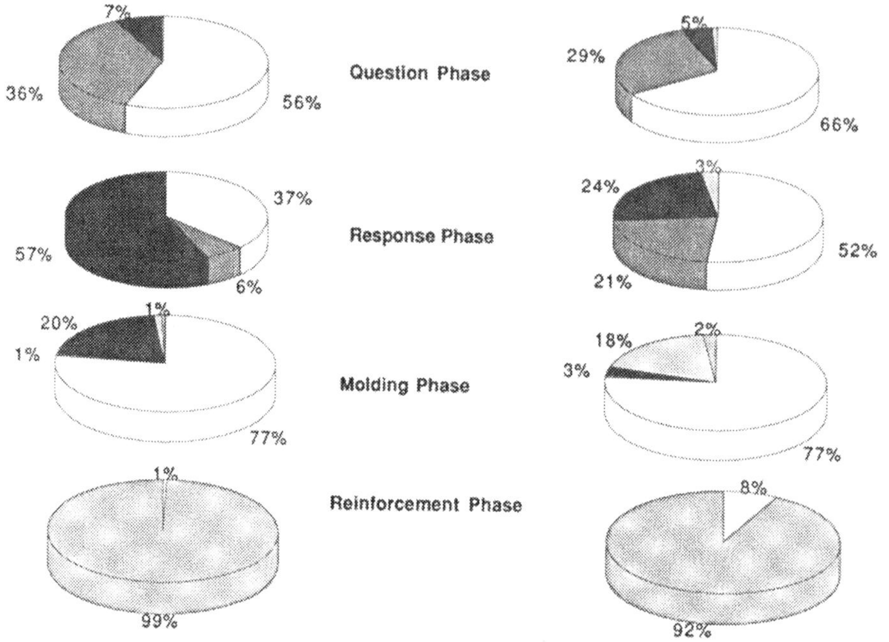

Figure 2. Relative distribution of eye contact or aversion to target points during each phase of trials for Rini (left) and Princess (right), independent of sign (n = 374/375).

with natural referents (twig and fungus) and human-made referents (smoking pipe, tuning fork and play pillow).

Aversion of the eyes from the four target points (referent, trainer's face, subject's hands, and reinforcement item) was a predominant response for both subjects during three of the four phases of the trial (question, response and molding). When the data are displayed graphically and compared, it is obvious that eye aversion is greatest during the molding phase of the trial (mean ≈ 77%) (Fig. 2). During the question or referent identification phase of the trial, the subjects showed the next highest level of eye aversion (mean ≈ 61%). Eye aversion dropped to an average of approximately 45% during the response phase of the trial while it fell to less than 10% during the reinforcement phase.

For both subjects, brief gaze seemed to dominate the type of visual attentiveness displayed towards the referent during the trial, while prolonged gaze was dominant for the type of attentiveness observed during the response and reinforcement phases of the trial. For the purpose of this paper, scores from both types of gaze were combined. While visual attentiveness towards the referent was relatively low during all phases of the trial, it was the highest during the question phase (mean ≈ 32%) followed by the response phase (mean ≈ 14%).

Visual attentiveness towards the trainer's face was highest during the response phase (mean ≈ 40%), especially for the adult female (57% vs 24%). Compared to the juvenile, the adult female continued to show higher attentiveness towards the trainer during the questioning (7.5% vs 4.8%) and molding phases (20% vs 3%) of the trials. For both subjects and their most poorly performed sign (twig for Princess; pipe for Rini), subjects showed the most attention towards the referent during the question phase and in the trainer's face during the response phase of the trials, when compared to other signs. Such attentiveness in the referent during questioning and trainer during the response phases, perhaps, indicates a strategy employed by the subjects to rely more on visual identification of the referent and then obtain nonverbal cues from the trainer when performance in the signing was poor. Finally, visual attention was directed at the reinforcement item during the reinforcement phase of the trial nearly 96% of the time.

Figure 3 shows the relationship between correct signing performance (CSP) and various measures of visual attention during the various phases of the trial. There appear to be several significant correlations between correct signing performance and these visual measures. There is a generally positive correlation between CSP and eye aversion for both question ($r = 0.38$; $p < 0.05$) and response phases ($r = 0.55$; $p < 0.001$), that is, when sign performance improves, eye aversion increases. Conversely, there is a negative correlation between CSP and attentiveness towards the referent during the question phase of the trial ($r = -0.29$; $p < 0.05$) and between CSP and attentiveness towards the trainer's face during the response phase of the trial ($r = -0.413$; $p < 0.01$).

Such significant correlations and the general observation that CSP appeared to increase following the first trial, suggested that there might be differential performance and attentive measures depending upon where within the set the trial took place. To examine the effect of trial placement, CSP and visual measurements from the first trial of the 5-trial sets were segregated for each sign for each subject. A total of 15 first trial signing and visual measures were obtained for each sign (1 trial/set x 3 sets/session x 5 sessions). Data were converted to percentage figures for ease of comparison. These data show that, with the exception of those signs never performed correctly, first trial correct signing performance was significantly lower for all signs when compared to the original 5-trial CPM scores (Fig. 4 and 5). Interestingly, the CSP for Princess's "fungus" dropped from 65% to 0%, indicating that first trial molding was necessary for subsequent correct signing performance.

Visual measures also showed a significant shift for several measures. Figure 6 displays the relative distribution of eye contact or aversion to target points during each phase of the trials and compares average percent scores for all signs and both subjects for the first trial and all five trials during the course of the last five sessions. During the question and response phases, aversion measures dropped significantly. That decrease was compensated for by a significant increase in eye contact towards the referent during the question phase and the trainer's face during the response phase. When first trial CSP and visual attentiveness measures were correlated, there were none of the significant effects seen earlier, although the direction of the best fit line was similar for CSP versus attentiveness towards the referent during the question phase of the trial ($r = -0.35$; $p > 0.31$), for CSP vs attentiveness towards the trainer's face during the response phase of the trial ($r = -0.31$; $p > 0.38$), and for CSP versus eye aversion during the response phases ($r = 0.45$; $p > 0.19$).

DISCUSSION

Gaining a better understanding of the visual dynamics of the subject during the sign learning experience may provide clues as to strategies employed by the subjects which may enhance or detract from the signing performance. Shapiro (1985) showed that selective attention (Mackintosh, 1975) could account for the differences in sign learning among four juvenile orangutans. Accordingly, the speed with which a sign is learned depends on the degree to which the subject visually and hapticly attends to the referent and its corresponding gestural configuration (all other factors being equal). One would expect a positive correlation between correct sign performance and attentiveness (visual and haptic) with the referent and gesture. Such attentiveness might be related to interest in the referent, such as food. In fact, there was a significant difference in correct sign performance between food and non-food signs in the four orangutan study. Yet, from the data presented above, it seems that the role of attention in the sign learning environment is complex and ambiguous.

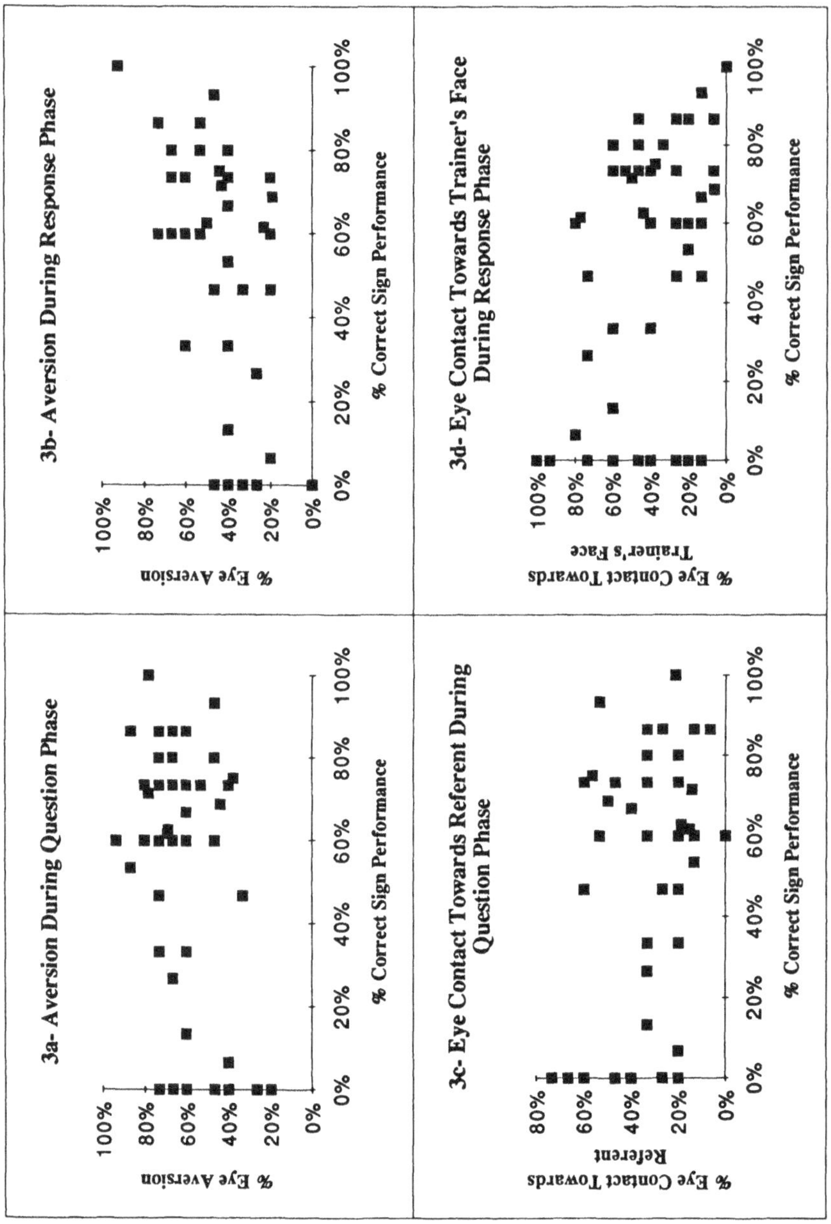

Figure 3. Percentage of trials orangutans (n = 2) averted or displayed eye contact towards target points as a function of percent correct sign performance during questioning and response phase of 15 trial sessions for each sign.

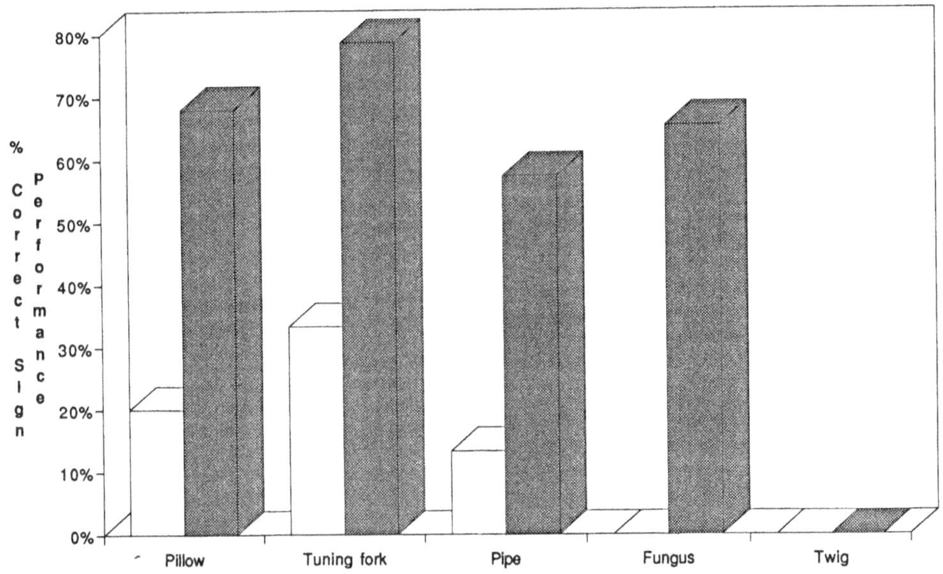

Figure 4. Correct sign performance by Princess for five signs, comparing first trial performance of 5-trial set with all trials during last 75 trials.

First Trial versus Subsequent Trial Performance

First trial performance provides a more accurate assessment of associational learning than CSP for the entire set of five trials, as there is little chance for any residual post-molding effect to influence the performance of a subsequent trial as there would be following the first

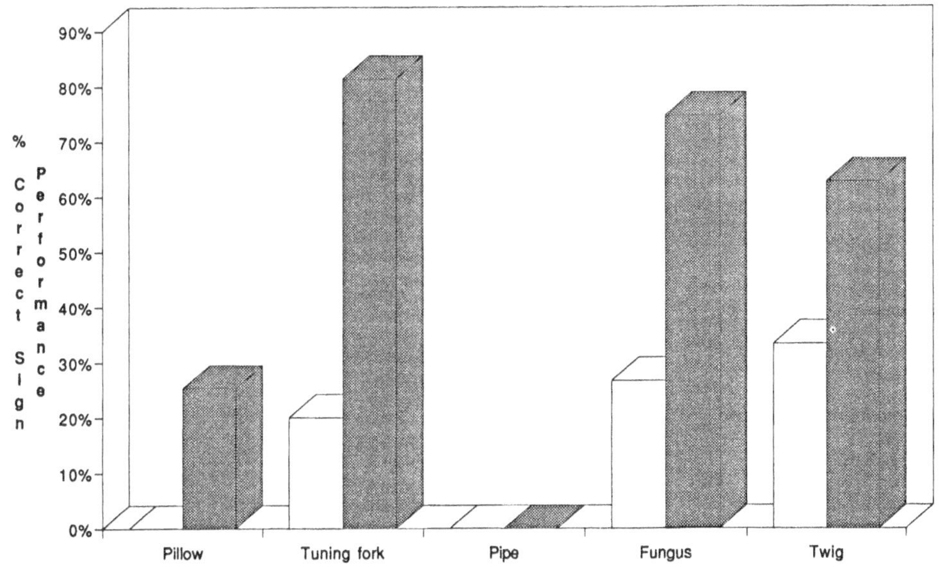

Figure 5. Correct sign performance by Rini for five signs, comparing first trial performance of 5-trial set with all trials during last 75 trials.

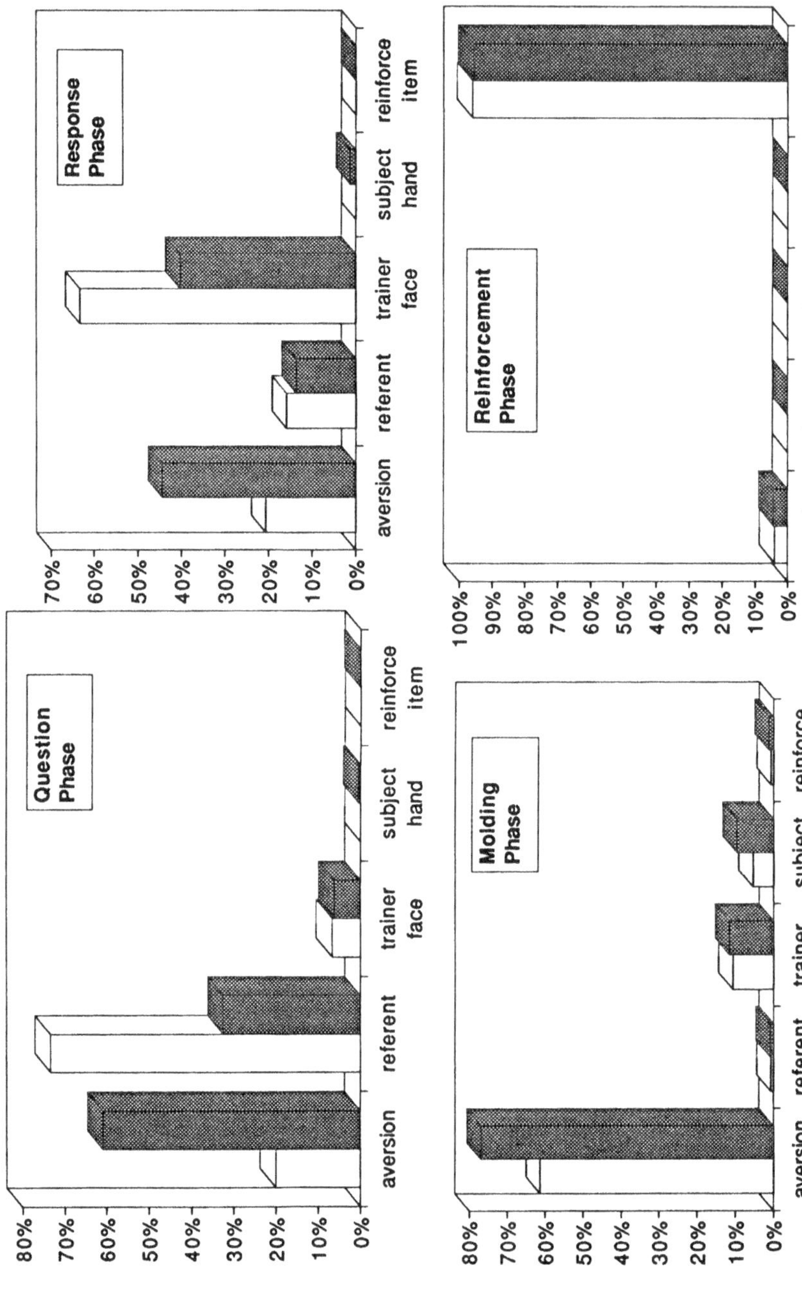

Figure 6. Relative distribution of eye contact or aversion to target points during each phase of trials for first trial performance of 5-trial sets (open) and all trials (shaded) for all signs during the last five sessions for both subjects.

trial. Such an effect would make it possible for the subject to perform the sign learning only the rule, "repeat last molded gesture when you hear the question, 'what is that?'" Alternatively, the subjects could be merely producing their "favorite" sign, regardless as to its correctness, then allowing the trainer to mold her hands when incorrect, in preparation of receiving the small reinforcement. It appears that for some of the signs, the subjects were using such a strategy, since, of the five signs trained, only three showed first trial performances greater than 0%. The discrepancy between first trial and total trial CSP scores was significant and suggests that assessments of vocabulary abilities based on sequential trial measures may not be accurate.

Aversion of Eye Contact

Following the initial trial, aversion of eye contact from the various target points was a significant component of both subjects' visual behavior. It would be tempting to suggest that this might be a species-specific behavior, as orangutans are not as social as the other great apes (Galdikas, 1978) and species-specific conventions in eye contact behavior (e.g., staring) have been long documented in social primates (Altmann, 1967). A couple of issues, however, caution against such a straightforward explanation:

1. Only two orangutans were studied. With a sample size of two, it is difficult to draw conclusions when the dependent measures vary parametrically.
2. No comparable study has been performed with chimpanzees or gorillas. Personal experience suggests that chimpanzees are much more visually attentive than orangutans in the sign learning environment, but no systematic, comparative study has been conducted, to our knowledge, in the area of great ape visual attention.
3. The eye contact measures were only taken during the last 75 trials of a much longer sign training program. Ideally, these measures should be taken during the beginning and in the middle of the training program to measure longitudinal effects of training on visual behavior. The last five sessions were not long enough to discern any significant difference in visual measures throughout the observation period. Additionally, there seemed to be a tendency for the orangutans to become distracted during the training, especially once the referent was identified by the subject during trial one of the set. For the free-ranging adult, this was frequent when branches rustled or snags crashed, signaling the appearance of a potential orangutan threat. The juvenile, on the other hand, seemed to be relatively disinterested in the sessions and easily distracted by outside stimuli. Thus, we may have a situation where the various eye contact measures following first trial referent identification are affected by individual factors, such as boredom and fear as well as species-specific factors.

The data from figures 3a and 3c suggest that during the question phase, eye aversion increases and attention towards the referent decreases as the sign performance increases. This relationship may indicate that for the two orangutans, correct signing performance was not a function of attention as selective attention theory would suggest. Rather, increased levels of sign performance and referent familiarity slowly reduced the need for the subjects to spend time visually attending to the referent during the question phase. Additionally, following the initial identification of the referent, there was apparently little motivation to visually attend to the referent during the second through fifth trials in the 5-trial set. This was borne out by the fact that eye contact towards the referent during the question phase was significantly higher (75%) than the overall average (32%) when eye contact measures from the first trial of the 5-trial sets are averaged. As noted for the poorly performed signs, "pipe" (for the adult) and "twig" (for the juvenile), visual attention towards the referent and trainer's face was high during the question and response phase, respectively, suggesting that the orangutans were employing a strategy to acquire information (referent identity; trainer approval).

Referent Interest during Sign Performance

Shapiro (1985) showed that for one subject (Princess), referent interest, as measured by pair-wise preference tests, correlated significantly and positively with correct sign performance for a set of different signs ($n = 9$, $r = 0.74$, $p < 0.01$). While visual attentiveness and interest may seem to be related cognitive states, the data in the present study indicate that visual attention towards a target point during the sign training does not necessarily correlate positively with interest in that target point. However, baseline interest and attention measures or pair-wise preference tests were not conducted with the referents in the two subject study, so it is not possible to know what the differential interest levels were for the referents of the trained signs. We only have the visual attention measures which seem to suggest that the subjects' interest in the referents were confined to their initial identification during trial one when sign performance was poorest. Interest in the referents during the remaining trials in the set waned significantly.

It is interesting to note that for both subjects, each of whom were experienced signers, spontaneous sign production outside the five sign study involved requests for referents of high interest such as food, drink, grooming, scratching and playing. These signs were learned within an informal setting, seemingly with less effort than in the formal five sign program. The rigidity of the formal program as well as the topic of study may have been a factor in reducing overall subject motivation and interest in the referents.

Subject Differences

Rini, the adult, displayed a higher amount of visual attentiveness towards the trainer's face than did Princess, the juvenile, during the question, response and molding phases. Such a difference could have been related to age, but more likely, to the differing relationships shared between the trainer and the subjects. Princess was home-reared by the trainer and interacted frequently with him, while Rini was a free-ranging individual who interacted with the trainer once a day for about one hour. Because of the adult's size and tendency to "steal" food items, the trainer occasionally restrained and verbally reprimanded the subject, especially during the other training programs which began a year and a half earlier. While the difference in attentiveness towards the trainer's face occurred during the first three phases, it was strongest during the response phase when the subject may have been seeking trainer approval or other nonverbal cues during sign production. Princess, on the other hand, did not seem to solicit as much approval, instead spending more occasions looking at the referent during the response phase.

Significance of Findings

Apart from its descriptive ethological value, an analysis of the visual attentiveness within the sign language or sign learning environment can identify factors which may lead to improved methods for teaching signs to apes. By identifying where the subjects are looking or not looking during important phases of the training protocol, adjustments can be made to increase subject attentiveness and interest and eventual rates of learning. Additionally, sign teaching methods have been called into question inasmuch as repetitive training trials and food reinforcement following each trial may have led the subjects to exploit non-associative learning strategies in order to obtain the food reward, albeit a lesser amount.

It would be of value to understand differences between species better, regarding visual attentiveness within the sign learning environment. While trainers and observers have commented anecdotally on such differences, a comparative study would permit quantification and perhaps establish species-specific conventions within this experimental domain. Such conventions might serve to formulate better *a priori* sign training procedures and choices of vocabulary items based upon species or training environment differences.

SUMMARY

Orangutans displayed differential visual attentiveness towards selected target points during the course of a trial; attentiveness towards the referent was greatest during referent presentation at the beginning of the trial (question phase), greatest towards the trainer's face during the time a signed response was to be made (response phase), greatest towards the subject's hands when the correct gestural configuration was molded by the trainer (molding phase), and greatest towards the reinforcement item when the subject was given a small bit of food following molding or a correct response (reinforcement phase).

Aversion of eye contact towards the target points was significant when all five trials of the training set were averaged. Additionally, there were significant correlations which suggested that aversion increased while attention to the referent during the question phase decreased as correct signing performance increased - opposite to what the selective attention hypothesis would predict.

Partitioning the training set into first trial measures and comparing the scores to the 5-trial averages revealed that correct signing performance was greatest during trials 2 through 5 of the set and that aversion was greatest during that same period; aversion decreased significantly during the first trial while attentiveness towards the referent during the question phase more than doubled and attentiveness towards the trainer's face also increased significantly. Significant correlations observed earlier were no longer significant for first trial measures though there were similar tendencies in the data.

Since independent referent interest tests or baseline attention measures were not conducted in this study, it is difficult to determine whether the results observed invalidate the selective attention hypothesis. Referents were non-food items which are generally of less interest to the subjects than food items; the signs of the latter are typically learned much faster than that of the former. In an independent study, one of the subjects did show a significant positive correlation between referent interest, as measured by pair-wise preference tests, and correct sign performance. Perhaps boredom strategies that favored acquiring reinforcement items, and other factors during the latter part of the training program were responsible for the ambiguous results.

The study suggests that other training procedures might be preferable in enhancing associational learning of signs whose referents are not highly preferred. Such changes could include varying the referent order between each trial, thereby requiring the subject to attend visually to the referent for accurate identification. Referent quality itself might be better characterized in advance, especially along the domains of subject interest, familiarity, and meaningfulness. Additionally, the reinforcement schedule might be adjusted to provide a greater difference in food reward following correct responses versus molding.

It should be pointed out that orangutans are not slow sign-learning automatons, as this study might suggest. Outside the formal sign training and testing environment, orangutans learn to use signs spontaneously in appropriate contexts to request food, drink and social activities such as grooming, scratching and tickling. Occasionally, they have been observed to produce novel combinations to name or describe referents for which they have no name (productivity) and to form seemingly logical multi-signed combinations. Without training, Rini invented some of her own signs (you, there/that, groom and scratch) and integrated them into appropriate multisigned combinations. In experimenter-paced studies, however, orangutans provide a unique challenge.

Only innovative future studies will elucidate methods and factors which will help to enhance and account for associational learning in a little studied species that seems resistant to formal sign learning with referents of questionable significance.

ACKNOWLEDGMENTS

The authors thank the following individuals for their moral and academic support of the presented research which took place in early 1980: Mr. and Mrs. Dwight Jenkins, Mr. and Mrs. Antanas Galdikas, Inggriani Shapiro, M.A., Bapak Bohap bin Jalan, Dr. Nancy Briggs, Dr. Roger Fouts. For permission and support in Indonesia, we thank the students of the Universitas Nasional in Jakarta and Gadjamata University in Yojakarta for counterpart support, the Indonesian Institute of Sciences (LIPI), the Indonesian Ministry of Forestry, the Indonesian Nature Protection and Forest Conservation (PHPA). The authors also acknowledge the staff of the Orangutan Research and Conservation Program who helped maintain the Camp Leakey facilities and provided assistance with many of the signing programs. Financial support was provided by grants from the Mr. and Mrs. Djura Postma of Holland, the Leakey Foundation, the van Tienhoven Foundation and The National Geographic Society, who also provided photographic support. Finally, we would like to thank Rini and Princess, the orangutan subjects who agreed to participate in the study and who tolerated our intrusion into their lives.

REFERENCES

Altmann, S.A. (ed.), 1967, *Social Communication Among Primates*. The University of Chicago Press, Chicago.

Fouts, R.S., 1972, The use of guidance in teaching sign language to a chimpanzee. *J. Comp. Physiol. Psychol.* 80: 515-522.

Fouts, R.S., 1973, Acquisition and the testing of gestural signs in four young chimpanzees. *Science* 180: 978-980.

Fouts, R.S., Shapiro, G.L., and C. O'Neil, 1978, Studies of linguistic behavior in apes and children. In: *Understanding Language Through Sign Language Research*. P. Siple, ed. Academic Press, New York.

Fouts, R.S. and D.H. Fouts, 1993, Chimpanzees' use of sign language. In: *The Great Ape Project*. P. Cavalieri and P. Singer, eds. pp 29-41. St. Martin's Press, New York.

Galdikas, B.M.F., 1978, *Orangutan Adaptation at Tanjung Puting Reserve, Central Borneo*. Ph.D. dissertation, University Microfilms International, Ann Arbor, Michigan.

Gardner, R. and Gardner, B., 1969, Teaching sign language to a chimpanzee. *Science* 165: 664-672.

Herman, L.M., Richards, D.G. and Wolz, J. P., 1984, Comprehension of sentences by bottlenosed dolphins. *Cognition* 16: 129-217.

Mackintosh, N.J., 1975, A theory of attention: variations in the associability of stimuli with reinforcement. *Psychol. Rev.* 84(4): 276-298.

Patterson, F., 1978, The gestures of a gorilla: Language acquisition in another pongid. *Brain Lang.* 5: 72-97.

Patterson, F. and Cohn, R., 1990, Language acquisition by a lowland gorilla: Koko's first ten years of vocabulary development. *Word* 41 (2): 97-143.

Premack, D., 1971, Language in chimpanzee? *Science* 172: 808-822.

Savage-Rumbaugh, E.S. and Rumbaugh, D. M., 1978, Symbolization, language, and chimpanzees: A theoretical reevaluation based on initial language acquisition processes in four young Pan troglodytes. *Brain Lang.*, 6: 265-300.

Sebeok, T. and Umiker-Sebeok, A., 1980, *Speaking of Apes*. Plenum Press.

Shapiro, G.L., 1982, Sign acquisition in a home-reared/free-ranging orangutan: comparisons with other signing apes. *Am. J. Primatol.* 3(1-4): 121-129.

Shapiro, G.L., 1985, *Factors influencing the variance in sign learning performance by four juvenile orangutans (Pongo pygmaeus)*. Ph.D. dissertation, University Microfilms International, Ann Arbor, Michigan.

Stokoe, W.C., Casterline, D., and Croneberg, C., 1965, *A Dictionary of American Sign Language*. Gallaudet College Press, Washington, D.C.

Terrace, H.S., 1979, *Nim*. Knopf, N.Y.

SECTION FIVE – BASIC SCIENCE AND CAPTIVE MAINTENANCE

Introduction

Population and habitat viability analyses demonstrate that captive populations must be considered as potential sources of stock for reintroduction to the wild when developing conservation plans for endangered species. "Perhaps 3000 vertebrate species and subspecies will require a captive propagation programme during the next 50 years to avoid extinction." (Seal, 1991, p. 40). In some cases, wild populations are so small that the inclusion of individuals and genes from captivity is crucial for the species' survival. Unfortunately, many captive populations are also quite small, so every individual is a vital part of species survival plans. Since the destruction of habitat is currently the greatest peril threatening orangutans, there is no need at the present time for using captive members of this species to supplement the wild populations. The future of wild orangutans is unknown, however, suggesting that the most expedient course is to maintain as many options available as is feasible.

Asa and her colleagues describe the techniques of gamete intra-fallopian transfer (GIFT) in orangutans, which have the potential to enhance the captive breeding of the species. This approach is particularly valuable when a female that has desirable genetic characteristics is subfertile or does not mate appropriately. The use of GIFT also permits the use of subspecies hybrid females, which are not recommended for breeding, as recipients of gametes from animals of a pure subspecies line. These investigators conclude that further research must be carried out in this area of primate reproductive biology to develop an adequately reliable methodology for orangutans.

Studies conducted in captive settings enable researchers to test hypotheses scientifically, yielding data that are impossible to obtain, or not easily gathered, in the wild. Behavioral experiments conducted in captivity frequently reveal the *capacities* of a species, which may vary *quantitatively* from the behavior observed in the wild. Qualitative differences in behavior between captive and wild settings, however, are rare. The paper of Nadler and that of Zucker and Thibaut explore several aspects of orangutan behavior about which field researchers have proposed hypotheses but were unable to test adequately. Nadler conducted laboratory pair-tests with oppositely sexed orangutans to elucidate the role female orangutans play in mate choice. When given the opportunity to control their access to males in a specially designed test, females sought out the males to mate only when they were in estrus, i.e., near the time of ovulation. Males, by contrast, forcibly initiated copulation with the females outside of estrus when the females were unable to regulate their access to the males. These experiments have implications for understanding the regulation of reproductive behavior under natural conditions, including the influence of the female's sex hormone levels during the menstrual cycle. They also suggest that the relatively low incidence of forcible copulation by fully adult males in the wild is likely due to the male's physical limitations rather than an absence of motivation. Zucker and Thibaut demonstrate the importance of social behavior in the lives of captive orangutans, at a level that is perhaps surprising for a solitary or semi-solitary species. In captive environments, both sexes are more sociable than they are in the wild; females are somewhat more so than males. The authors argue that several factors, such as release from time otherwise spent foraging, spatial proximity, and paternal certainty may all contribute to the levels of social interaction observed.

Van Hooff (this volume) suggests that it is difficult to test hypotheses about the ultimate, i.e., evolutionary, pressures that produced the orangutan's current behavior and biology. Winkler tackles this issue with the methodology of the comparative anatomist. Through an extensive examination of orangutan development, particularly through the exploration of those features that are unique to the orangutan, she presents data of particular value for interpreting the fossil record of early apes and humans. The appearance or persistence of these characteristics throughout evolutionary time may provide clues regarding their selection and the time when they were selected. As in so many other areas of research on orangutans, much more data are required; in this case, to clarify the factors which contribute to the growth and development of its morphological systems, especially under natural conditions.

Genetic variation in wild populations is of major concern to the continued survival of orangutans. Declining population size is one problem that leads to the loss of genetic variation; loss of genetic variation means that the remaining members of the population are less adaptable to environmental change. Another threat to wild orangutans is population subdivision. The fragmentation of wild populations may have a devastating impact on the propagation of a species even in the absence of an overall population decline. Muir and his collaborators conducted a pilot study related to this latter problem using polymerase chain reaction amplification of mitochondrial DNA in hair follicles and fecal samples of orangutans. Their results suggest that this noninvasive approach provided adequate samples of DNA for genetic characterization and for the identification of genomic regions that may be useful in the analysis of population subdivision.

Five hundred and seventy-three (65%) zoos and aquaria are in the "developed" world (Magin, Johnson, Groombridge, Jenkins, and Smith, 1994). These facilities serve as reservoirs of the genetic and behavioral diversity of populations from all over the world. Shepherdson (1994) notes that "Failure to reproduce an environment that is at least functionally equivalent to that of the wild will inevitably result in the loss of many forms and patterns of natural behavior." (p. 168). This could be important to orangutans even if a particular individual never leaves captivity. Since much of primate behavior is learned, the loss of a behavior means that descendants would be unable to benefit from the experience of others and hence would be required to discover anew those same lessons of the past. Box (1991) stresses the importance of the captive environment in the individual's development of species-appropriate orientation and locomotion skills, feeding and foraging behaviors, and the acquisition of socially appropriate behaviors. Hence, it is important to create captive environments which foster naturalistic behavior. Two papers in this volume recount such efforts on behalf of captive orangutans.

Markham compares the data on reproduction and development of orangutans in captivity and in the wild, noting similarities as well as differences. She emphasizes the importance of maintaining natural groupings of the species in captive collections, of systematically collecting data on captive individuals and of applying those data to develop a rational approach to their maintenance.

Mallinson (1991) earlier described the Jersey Wildlife Preservation Trust's long history of assistance with species conservation. In this volume, he describes the Trusts's recent efforts to improve one facility for captive orangutans. The changes described all increase the likelihood of eliciting naturalistic behavior from the orangutans, of considerable educational value for the human observers as well as beneficial to the orangutans.

REFERENCES

Box, H.O., 1991, Training for life after release: simian primates as examples. Pp. 111-123 in: *Beyond Captive Breeding: Re-introducing Endangered Mammals to the Wild. Symposia of the Zoological Society of London 62.* (Ed. J.H.W. Gipps), Oxford, Clarendon Press.

Magin, C.D., Johnson, T.H., Groombridge, B., Jenkins, M., Smith, H. 1994, Species extinctions, endangerment and captive breeding. Pp. 3-31 in: *Creative Conservation: Interactive Management of Wild and Captive Animals*, (Eds. P.J.S. Olney, G.M. Mace, A.T.C. Feistner), New York, Chapman and Hall.

Mallinson, J.J.C., 1991, Partnerships for conservation between zoos, local governments and non-governmental organizations Pp. 57-74 in: *Beyond Captive Breeding: Re-introducing Endangered Mammals to the Wild. Symposia of the Zoological Society of London 62.* (Ed. J.H.W. Gipps), Oxford, Clarendon Press.

Seal, U.S., 1991, Life after extinction. Pp. 39-55 in: *Beyond Captive Breeding: Re-introducing Endangered Mammals to the Wild, Symposia of the Zoological Society of London 62*, (Ed. J.H.W. Gipps), Oxford, Clarendon Press.

Shepherdson, D. 1994, The role of environmental enrichment in the captive breeding and reintroduction of endangered species. Pp. 167-177 in: *Creative Conservation: Interactive Management of Wild and Captive Animals*, (Eds. P.J.S. Olney, G.M. Mace, A.T.C. Feistner), New York, Chapman and Hall.

21

FOLLICLE STIMULATION AND OVUM COLLECTION IN THE ORANGUTAN

C. S. Asa,[1] S. J. Silber,[2] I. Porton,[1] F. Fischer,[1] K. Lenehan,[2] M. Deters,[2] R. Junge,[1] J. Hicks,[2] and R. C. Cohen[2]

[1] St. Louis Zoological Park
St. Louis, Missouri 63110
[2] St. Luke's Hospital
St. Louis, Missouri 63017

INTRODUCTION

In spite of the many years that assisted reproductive technologies have been applied to humans, these procedures have not met with success in the great apes. The advantages of artificial insemination and *in vitro* fertilization for genetically valuable animals that have not produced offspring are apparent. There also is a potential application for embryo transfer, however, which has not been widely recognized, that results from the current taxonomic circumstance of orangutans (*Pongo pygmaeus*). Two subspecies, Sumatran and Bornean, have been identified, leading the American Association of Zoos and Aquarium's Orangutan Species Survival Plan to recommend breeding only within each of the subspecies lines. Therefore, animals that represent hybridization between the two subspecies are considered genetically "surplus" and are recommended not to be bred.

Many of the orangutans that have been documented as pure subspecies have not reproduced, often for behavioral reasons. For example, animals that have been raised by humans often lack the social skills required to mate or to properly care for offspring. Artificial insemination or a variation of *in vitro* fertilization might be an effective approach for valuable females that have not reproduced due to a failure to mate or even due to a Fallopian tube blockage, as is common for humans. However, the inability to respond maternally in the species-typical fashion perpetuates the problem by requiring that offspring be hand-raised if they are to survive. These youngsters will also lack the necessary social skills to function properly as an orangutan.

A potential solution to this cycle involves a novel application for assisted reproductive technology. Maternally experienced hybrid females may serve as gestational surrogates for genetically valuable but behaviorally incompetent females from a pure subspecies line. This approach makes use of the hybrid females that now are not allowed to breed, but that have successfully reared young in the past and have proven to be competent mothers. If successful, the cycle of hand-rearing can be broken in one generation by allowing an

The Neglected Ape, Edited by Ronald D. Nadler et al.
Plenum Press, New York, 1995

experienced mother to gestate, give birth to and raise the offspring of females that are designated to breed, but are not socially adept.

A pilot study was conducted to evaluate the suitability of a protocol for oocyte stimulation and retrieval, previously shown to be successful for human females. The technique, Gamete Intra-Fallopian Transfer, is the most effective method to date for producing pregnancies in women [Silber et al., 1990]. In this procedure, the donor female is stimulated with hormones to produce multiple eggs, which are retrieved and transferred to the Fallopian tubes of the recipient, along with sperm. The current study used a Sumatran female and male as donors of ova and sperm, respectively, and a hybrid female as the potential gamete recipient.

MATERIALS AND METHODS

The donor female, Sumatran orangutan (#921216), and the potential female recipient, hybrid (#911006), that were 25- and 23-years-old, respectively, at the time of treatment, had each produced at least one infant, so they were assumed to be reproductively competent. They were housed together in an indoor enclosure (approximately 62 m X 30 m) during the day (0900 h to 1645 h) and overnight in separate sleeping quarters (8 m X 12 m or 6 m X 10 m). Overnight urine was aspirated by syringe each morning from the drains and tested for menstrual blood (Ames Hemastix, Miles, Inc., Elkhart, IN).

In addition, temperature telemetry was used to aid in monitoring ovarian cycles and the progress of exogenous hormone treatment. In a previous study, basal body temperature (BBT), as measured telemetrically, was shown to be closely correlated with urinary pregnanediol glu-

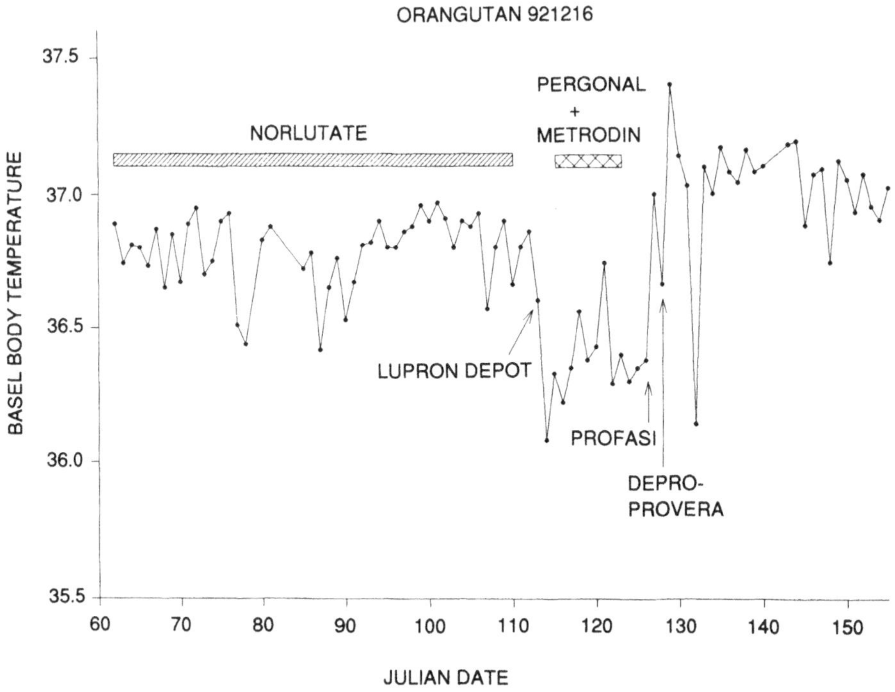

Figure 1. Daily basal body temperature of donor female orangutan #921216 during the regimen of cycle suppression and follicle stimulation.

curonide, a metabolite of progesterone [Asa et al., 1994]. Thus, at least in this species, temperature telemetry can serve as an indirect measure of circulating progesterone or progestins. Each female was implanted abdominally under general anesthesia with a radiotransmitter (Wildlife Materials, Inc., Carbondale, IL) to measure core body temperature. The mean of temperatures recorded every 2 min between 0300 h and 0400 h was used as the BBT for that day.

The hormone regimen for each female is summarized in Figures 1 and 2. Cycles in both females were suppressed with the synthetic progestin, norethindrone acetate (Norlutate, Parke-Davis, Morris Plains, NJ), 10 mg/day in tablet form, starting on the first day of menses, then by a single 7.5 mg intramuscular (IM) injection of luprolide acetate (Lupron Depot, TAP Pharmaceuticals, North Chicago, IL), a synthetic luteinizing hormone releasing hormone (LHRH) agonist. Two days following Lupron administration to the donor female, follicle stimulation began. This consisted of 225 International Units (IU) human follicle stimulating hormone (hFSH) and 75 IU human luteinizing hormone (hLH) (2 ampules Pergonal and 1 ampule Metrodin menotropins, Serono Laboratories, Norwell, MA) by daily IM injection for 9 days. The menotropins were dissolved in 2 ml diluent and injected by the keeper without restraint. The animal had been previously acclimated to the procedure by sham injections, followed by injections of saline. Human chorionic gonadotropin (hCG: Profasi, Serono Laboratories) was given IM by dart (10,000 IU in 3 ml diluent) 36 hours after the last menotropin injection.

Starting on the first day of donor treatment with menotropins, daily micronized estradiol tablets (Estrace, Mead Johnson Laboratories, Evansville, IN) were presented to the recipient in food treats (Days 1-3: 2 mg, Days 4-6: 4 mg, Days 7-11: 6 mg; Day 12: 8 mg). On Day 11, the recipient was injected IM with the synthetic progestin medroxy-progesterone acetate (Depo-Provera, Upjohn Co., Kalamazoo, MI; 10 mg/kg body weight).

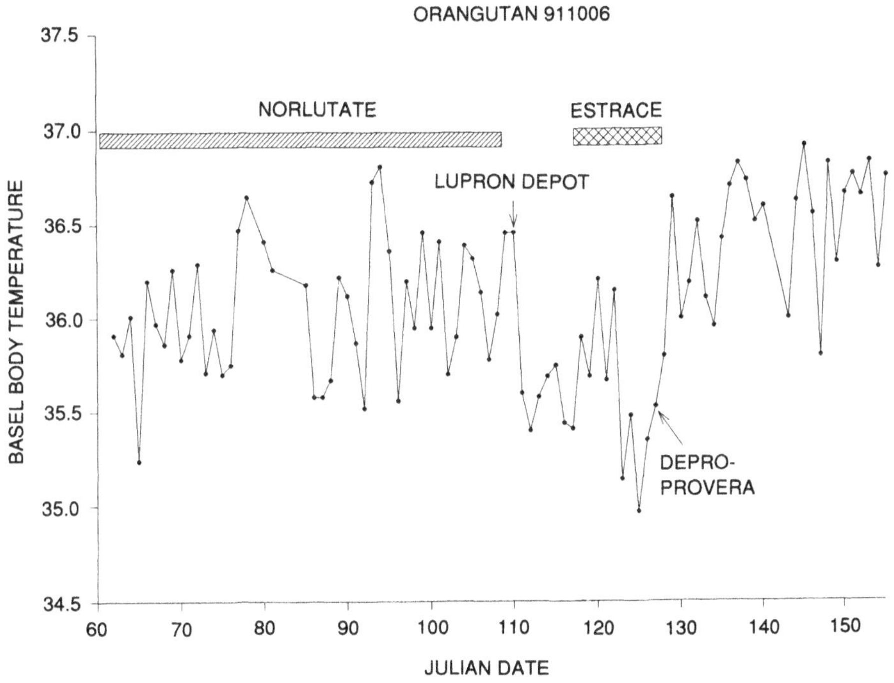

Figure 2. Daily basal body temperature of potential recipient female #911006 during the regimen of cycle suppression and uterine stimulation.

Progress of follicle stimulation was monitored by transrectal ultrasound and by assay of serum estradiol on Days 4 and 9 of treatment with Pergonal and Metrodin. The donor was anesthetized for sonography and blood collection using 250 mg Telazol (tiletamine and zolazepam, A.H. Robbins, Richmond, VA).

An Aloka 500V ultrasound unit with a 7.5 megaHertz linear array transducer was used to visualize follicular and uterine development via the rectum. Attempts to visualize ovaries with a 7.5 megaHertz linear-array vaginal transducer designed for human application were unsuccessful, apparently due to differences in uterine anatomy. The Fallopian tubes of the orangutan appear to lie more cephalad than do those of the human, putting them out of range of the 7.5 transducer. The ultrasound images produced were recorded on videotape for later review and reference. Serum was assayed for estradiol by radioimmunoassay by Smith Kline Laboratories, St. Louis, MO, using an assay developed for human serum.

Thirty-six hours after hCG administration, following immobilization with Telazol, the donor was intubated and maintained on 1.5% isoflurane (Aerrane, Anaquest, Madison, WI) for the ovum retrieval procedure. Sonography was again performed and follicles aspirated after exposing the ovaries through midline laparotomy, just above the pubis. Following aspiration, Depo-Provera (10 mg/kg) was given IM in anticipation of the female serving as her own recipient if more than three embryos resulted.

Semen was collected by electroejaculation (Stimulator: S.W.J. Seager, Washington, DC; rectal probe, #6: P-T Electronics, Boring, OR) from male orangutan #921507 following immobilization with Telazol on the morning of oocyte retrieval and again 24 hour later.

RESULTS AND DISCUSSION

Norlutate treatment successfully suppressed menstrual cycles, based on the absence of blood in urine as measured by Hemastix. In addition, the sustained basal body temperature (BBT) elevation during that period (Fig. 1) is consistent with this interpretation. Cessation of Norlutate administration and injection with Lupron Depot were rapidly followed by a drop in BBT, indicating continued cycle suppression without the progestin influence on temperature. During the period of daily injections of the menotropins, Pergonal and Metrodin (Days 1 through 9), BBT fluctuated, then increased dramatically following injections of Profasi (Day 10) and Depo-Provera (Day 12). Likewise for the potential recipient, BBT was consistent with the pattern expected for the treatment regimen. Although this female's mean BBTs were somewhat more variable than those of the donor, in general, BBT was higher during treatment with Norlutate and Depo-Provera than during treatment with Lupron Depot and Estrace.

The donor female's serum estradiol measured 712 pg/ml on Day 4 of treatment with menotropins and 2,641 pg/ml on Day 9. Transrectal ultrasound detected multiple follicles of 0.25 cm to 0.5 cm diameter on Day 4 and more than 15 follicles of 1 cm to 1.5 cm diameter on Day 9. By Day 10, only one follicle had reached 2 cm. That follicle and all follicles equal to or greater than 1.5 cm were aspirated. However, only the 2 cm follicle yielded an apparently mature ovum.

The semen sample collected on the morning of oocyte retrieval was inadequate, i.e., it contained only a few drops of fluid with occasional immotile sperm. Because adequate samples had been collected previously from this male, electroejaculation was performed again the next morning. The resulting semen sample was washed by centrifugation through a Percoll density gradient and then added to the single ovum that had been held in a CO_2 incubator overnight. Fertilization did not result.

In spite of elevated estradiol in the donor female on Day 9, failure to collect more than one apparently mature egg suggests that the stimulation regimen was inadequate. In women [Smitz, et al., 1992], multiple follicles of about 1.9 cm would be expected by Day 9 of Pergonal stimulation, whereas, in this orangutan, the largest follicles were only 1.5 cm by that time. In women, mature eggs typically are not obtained until follicles reach 2 cm or greater. An additional feature of the collection that differed from the human was the viscosity of the follicular fluid. In fact, in other species with which we have experience, follicular fluid is typically thin and non-viscous. However, in this orangutan, it was quite thick and difficult to aspirate. Whether this is typical of orangutans is unknown.

Subsequent trials will evaluate either a higher gonadotropin dose or a longer period of stimulation. Another possibility is to establish the methodology for *in vitro* oocyte maturation, so that even immature ova can be fertilized upon maturation.

The additional problem introduced by the inadequate semen sample may be addressed in two possible ways. One approach is to collect and freeze semen samples to be available on demand, although freezing diminishes semen quality. However, because the male used in this procedure, although genetically valuable, typically had semen samples with low sperm counts and high percentages of abnormal forms, freezing might result in his samples being unusable. The other approach is to use the newer technique of intracytoplasmic sperm injection [Silber et al., 1994], in which individual sperm cells are micro-injected into an ovum. This technique is especially appropriate for males whose sperm cannot accomplish ovum penetration unassisted, such as is likely for this orangutan.

This situation highlights what is often a problem with breeding programs for endangered species. Genetically valuable individuals that are not represented in the population, i.e., that have not reproduced, often have not done so precisely because they are either infertile or subfertile. Thus, attempts to develop assisted reproductive techniques for these animals are hampered by confounds related to the individuals' various abnormalities. However, assigning reproductively normal females to such projects takes them, at least temporarily, out of production. Using closely related, more abundant species as models is helpful, but not always possible, and there are differences, even among closely related species, that might preclude direct transfer of techniques.

The results from this study also demonstrate the usefulness of temperature telemetry for indirectly monitoring the endocrine status of animals that might be difficult or impossible to restrain for collection of more traditional samples, such as blood or for ultrasound examinations. Further trials are necessary, however, to refine the follicle stimulation protocol or to mature oocytes *in vitro*, and to increase the likelihood of fertilization with sperm from potentially subfertile males.

ACKNOWLEDGMENTS

The authors thank SmithKline Laboratory, St. Louis, MO, for serum estradiol analysis; TAP Pharmaceuticals for donation of Lupron Depot; Serono Laboratories for donation of Pergonal and Profasi; and BioGenex Laboratories for donation of OvuKit. We also thank veterinarians Eric Miller, Mike Talcott and Linda Wolf; veterinary technicians Barbara Jenness and Kelly Mooney; and animal keeper Meg White.

REFERENCES

Asa, C.S., Fischer, F., Carrasco, E., and Puricelli, C., 1994, Correlation between urinary pregnanediol glucuronide and basal body temperature in female orangutans, *Pongo pygmaeus*. *Amer. J. Primatol.* 34:275-281.

Silber, S.J., Nagy, Z.P., Liu, J., Godoy, H., Devroey, P., and Van Steirteghem, A.C., 1994, Conventional in-vitro fertilization versus intracytoplasmic sperm injection for patients requiring microsurgical sperm aspiration. *Hum. Reprod.* 9:1705-1709.

Silber, S.J., Ord, T., Balmaceda, J., Patrizio, P., and Ascho, R., 1990, Congenital absence of the vas deferens. The fertilizing capacity of human epididymal sperm. *New England J. Med.* 323:1788-1792.

Smitz, J. Van, Den Abbeel, E., Bollen, N., Camus, M., Devroey, P., Tournaye, H., and Van Steirteghem, A.C., 1992, The effect of gonadotrophin-releasing hormone (GnRH) agonist in the follicular phase on in-vitro fertilization outcome in normo-ovulatory women. *Hum. Reprod.* 8:1098-1102.

22

SEXUAL BEHAVIOR OF ORANGUTANS (*PONGO PYGMAEUS*)

Basic and Applied Implications

R. D. Nadler

Yerkes Regional Primate Research Center
Emory University
Atlanta, Georgia 30322

ABSTRACT

Early observations in zoological gardens suggested that orangutans mated relatively frequently, without relation to the phase of the female's menstrual cycle, and that copulation was often initiated forcefully by the male in disregard of the female's resistance. These observations were interesting because they suggested that orangutans more closely resembled humans in these respects than did other nonhuman primates, including the other great apes. Two types of laboratory pair-test were conducted to clarify the sexual proclivities of orangutans, 1) the conventional pair-test, in which the male and female are freely accessible to each other in a single cage and 2) a restricted-access test in which the female controls access. In the conventional tests, mating occurred on 85% of the test days due, primarily, to forcibly initiated copulation by the male and little indication of cyclicity in the sexual behavior of the female. When the males were prevented from pursuing the females in the female-controlled tests, however, they solicited the females nonaggressively and copulation initiated by the female occurred primarily during the midcycle phase in association with elevated levels of estrogen in the female. The experimental laboratory data have both scientific and practical significance. They suggest that orangutans resemble other nonhuman primates in terms of the cyclic enhancement of female sexual motivation at midcycle, i.e., female orangutans exhibit estrus, and they suggest that environmental influences have significance for captive management. An important determinant of species-typical sexual behavior in fully mature captive orangutans is an environment that permits the female to regulate its access to the male. A captive environment that affords the female such an option, simulates a crucial aspect of the natural habitat to which the species is biologically adapted.

The Neglected Ape, Edited by Ronald D. Nadler et al.
Plenum Press, New York, 1995

INTRODUCTION

Early observations on orangutans in zoological gardens suggested that this species of great ape mated frequently and without relation to the phase of the female's menstrual cycle (Fox, 1929; Asano, 1967; Heinrichs and Dillingham, 1970; Coffey, 1972). Such mating reportedly occurred as a result, primarily, of the forcibly initiated copulation by the male, irrespective of resistance by the female (Fox, 1929; Seitz, 1969; Coffey, 1972). These observations were interesting because they suggested that orangutans more closely resembled humans in their sexual proclivities than did the other great apes, that is, 1) female hormonal state appeared to have little influence on the sexual activity of orangutans and 2) the forcible initiation of copulation by a male resembled a form of behavior that was considered to be uniquely human. These observations on captive orangutans might have been dismissed as an aberration resulting from the artificial conditions of captivity were it not for comparable observations of indiscriminate, forcible copulation by male orangutans in the wild (MacKinnon, 1971, 1974; Rodman, 1973; Rijksen, 1978; Galdikas, 1981). Since the forcible initiation of copulation by the male was observed under natural conditions, it was considered to be an adaptive mating strategy, appropriate for a species in which males and females lived in a dispersed pattern of social organization (Rodman, 1973). The behavior of the orangutan was also of theoretical interest in terms of the issue of estrus, i.e., whether the female orangutan exhibited a special period of heightened sexual motivation during the menstrual cycle, similar to most other female mammals. This paper reports the results of laboratory research conducted with orangutans on these issues, much of which was reported earlier. The particular analyses and comparisons presented here, however, have not been presented before.

LABORATORY PAIR-TESTS OF SEXUAL BEHAVIOR

Conventional Laboratory Pair-Tests

In order to assess the sexual interactions of orangutans under more controlled conditions, research was initiated using the conventional laboratory pair-test in which the male and female are confined in a single cage in which they are freely accessible to each other throughout the entire test (Nadler, 1977). Tests of different duration were conducted daily throughout the cycle of the female in four oppositely sexed pairs of orangutans; 1) tests of 30 minutes duration with an observer recording the behavior and 2) tests of 5-6 hr duration in which the animals were monitored by a video camera after the initial 30 min test with an observer. Under these conditions, the males immediately and aggressively pursued the females as soon as the door separating their compartments was opened at the beginning of the tests. The males quickly caught the females, wrestled them to the floor (Fig. 1) and generally initiated copulation forcibly in a variation of the ventro-ventral position with the females reclining on their backs (Fig. 2). Copulation continued for prolonged periods of time, in comparison to the other great apes, approximately 14 minutes, on average. The males frequently interrupted their thrusting, sometimes repositioned the females at these times and frequently examined the females' genitals by smelling or licking. The forcible pattern of sexual initiation was pursued daily by three of the four males that were tested, while the remaining male displayed the same type of behavior, but did so less frequently.

It was clear from the results of the foregoing study that the experimenter's ability to detect cyclicity in the sexual behavior of female orangutans was compromised in the conventional laboratory pair-test by the assertiveness and aggressiveness of the males in

Figure 1. Male orangutan forcibly positions female for copulation during a conventional laboratory pair-test of sexual behavior

Figure 2. A form of ventro-ventral copulation in orangutans (male on top) during a conventional laboratory pair-test of sexual behavior.

initiating copulation. Despite the males' aggressiveness, however, there was some indication of increased sexual motivation by the females during the midcycle phase of the cycle. The only examples of female masturbatory behavior occurred at midcycle and the females less often avoided the males during midcycle by suspending themselves above the floor from the cage fencing. When the pairs were allowed to remain together for prolonged periods of time, moreover, i.e., for 5-6 hr, copulations beyond the initial one each day occurred more frequently at midcycle than at other phases of the cycle. These data, combined with a few reports on orangutans living in zoological gardens (Chaffee, 1967; Coffey, 1975; Maple, Zucker and Dennon, 1979; see also, Markham, 1990) suggested that cyclicity in female sexual motivation would be more apparent in laboratory tests if the female was given greater control over the initiation of copulation.

Restricted-Access Pair-Tests with Female Control of Access

In order to reduce the male orangutan's influence over the female and thereby, improve the opportunity for assessing fluctuations in female sexual behavior during the menstrual cycle, a new type of pair-test was devised, the restricted-access test with female control of access. In this test, the doorway separating the male's and female's compartments was covered with a sturdy metal plate which reduced the opening to a size through which only the smaller female could pass. A total of 11 oppositely sexed pairs of orangutans, consisting of 6 females and 4 males, participated in the study with the restricted-access test (Nadler, 1988; Nadler and Collins, 1991). Several dramatic differences occurred in the behavior of the animals under these new test conditions. Initially, some of the females entered the male's compartment as they had done in the earlier tests and were similarly pursued and subdued by the males which then initiated copulation forcibly. Most of the females, however, appeared to learn the option that was afforded them by the restricted doorway and either remained in their compartments during the early follicular phase of the cycle or escaped after entering the male's compartment and returned to their own compartments. Several of the males initially struggled mightily to remove the metal plate from the doorway or to squeeze through the opening, but lacking success, they all eventually desisted.

There followed what may be described as a standoff. In some pairs, the female occasionally ventured out into the male's compartment and rapidly returned to its own compartment at the slightest indication of pursuit by the male. In others, the female entered the male's compartment and was essentially ignored by the male. As midcycle approached, however, several of the males began displaying a form of behavior that had never before been reported for a male of this species; the male reclined on its back with a penile erection and directed a "penile display" toward the female on the other side of the doorway (Fig. 3). The penile display was sometimes maintained for prolonged periods of time, during which, the female entered the male's compartment. At such times and in a number of pairs, the male maintained its position, the female closely examined the male's penis and then mounted the male and effected intravaginal penile insertion (Fig. 4). In some cases, the female initiated thrusting which resulted in ejaculation by the male; in others, the male reversed positions with the female nonaggressively and thrusted to ejaculation in a more typical ventro-ventral position.

The result of these novel behavioral interactions was that copulation during the menstrual cycle was considerably reduced, in comparison to the conventional pair-tests, and there was a much clearer indication of cyclicity in sexual activity.

Figure 5 presents the comparison between the conventional pair-tests and restricted-access tests in the percentage of total copulations that occurred on each day of the cycle and the relation between these behavioral data and the serum levels of 17β-estradiol and progesterone during the menstrual cycle. The hormonal data were obtained in a separate

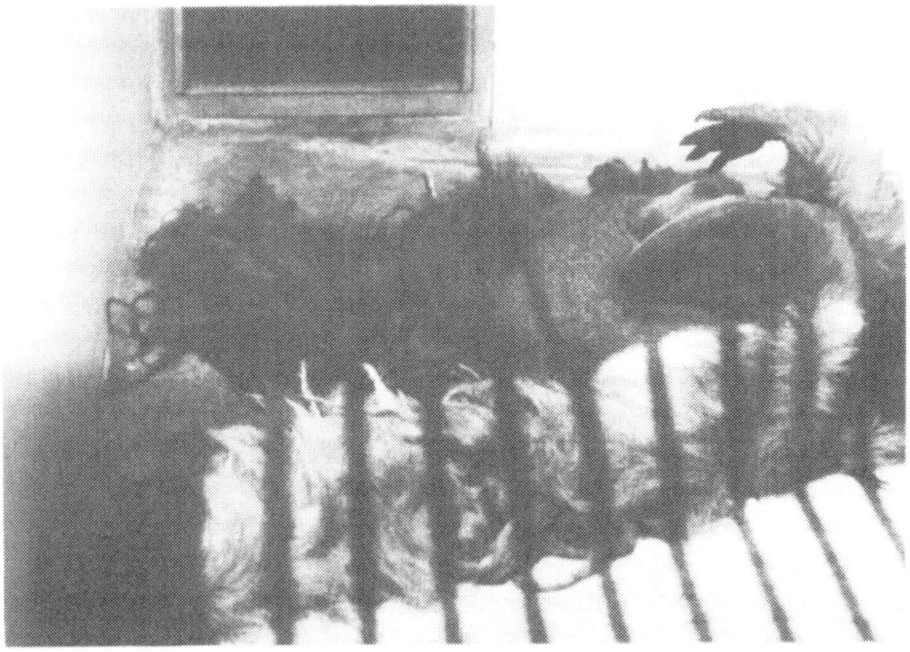

Figure 3. Penile display by male orangutan directed toward a female on the other side of a reduced doorway during a restricted-access test of sexual behavior with female control of access.

Figure 4. Ventro-ventral copulation by orangutans (female on top) during midcycle in a restricted-access test of sexual behavior with female control of access.

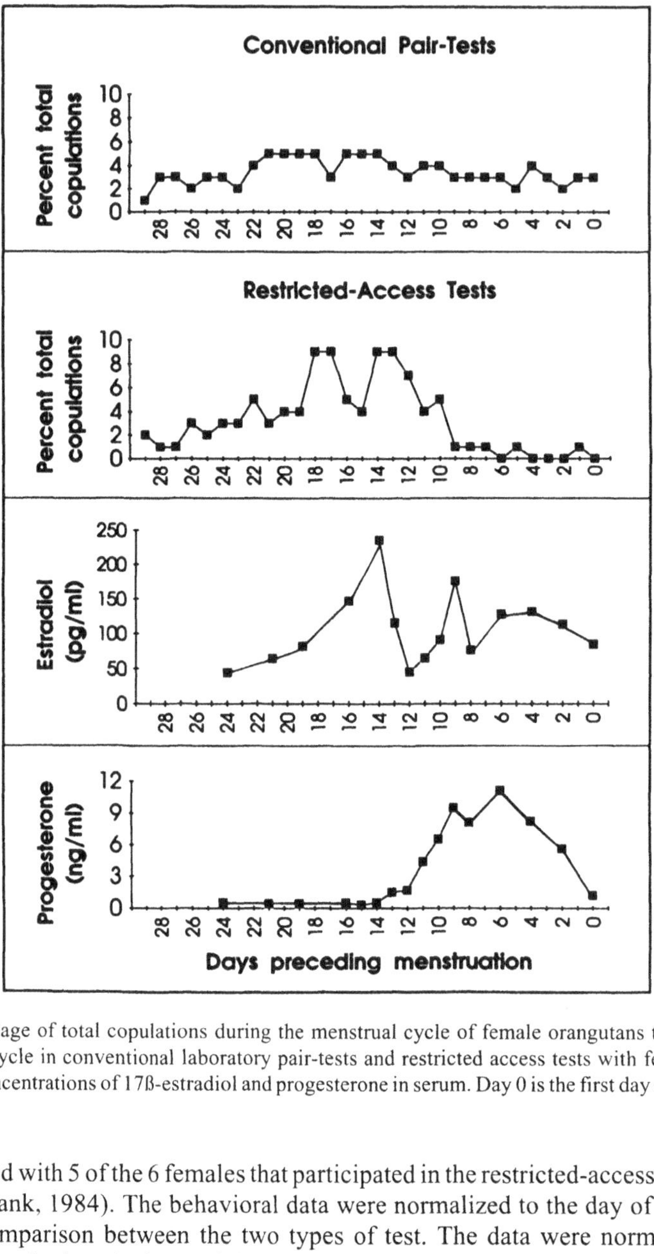

Figure 5. Percentage of total copulations during the menstrual cycle of female orangutans that occurred on each day of the cycle in conventional laboratory pair-tests and restricted access tests with female control of access and the concentrations of 17ß-estradiol and progesterone in serum. Day 0 is the first day of menstruation.

study conducted with 5 of the 6 females that participated in the restricted-access tests (Nadler, Collins and Blank, 1984). The behavioral data were normalized to the day of menstruation to permit a comparison between the two types of test. The data were normalized in this manner because the luteal phase of the cycle is less variable in length than the follicular phase. Counting days backward from menstruation, therefore, gives a more accurate approximation of the midcycle phase and the time of ovulation than counting forward from the previous menstruation. The hormonal data were normalized to the day of the midcycle peak in luteinizing hormone (LH) and then were aligned with the behavioral data on the basis that day 13 preceding menstruation was, on average, the day of the LH peak. When aligned in this way, the hormonal data provide an approximation of the hormonal conditions that were associated with the behavior during the menstrual cycle.

It can be seen in Figure 5 that copulations in the conventional pair-tests occurred fairly uniformly throughout the cycle, with little indication of an increased percentage of

Table 1. Sexual behavior of orangutans in conventional laboratory pair-tests and restricted-access tests with female control of access. Data are expressed in terms of days preceding menstruation

	Conventional pair-test	Restricted-access test	
Percentage of total copulations/day			
Days 28 - 19	3.3 ± 0.4	2.8 ± 0.4	t(10) = 1.2, ns
Days 18 - 13	4.5 ± -.3	7.5 ± 1.0	t(5) = 2.6, p < 0.05
Days 12 - 0	3.1 ± 0.2	1.6 ± 0.6	t(12) = 2.5, p < 0.05
Percentage of pairs copulating/day			
Days 28 - 19	85 ± 6	34 ± 3	t(10) = 6.5, p < 0.01
Days 18 - 13	100 ± 0	54 ± 7	t(5) = 7.1, p < 0.01
Days 12 - 0	82 ± 3	15 ± 5	t(12) = 9.1, p < 0.01

total copulations occurring during the midcycle phase when estradiol levels are elevated and progesterone levels are minimal. In the restricted-access tests, by contrast, a greater percentage of total copulations occurred during the midcycle phase. Although the midcycle phase cannot be specified with precision on the basis of average hormone levels, an estimate of the midcycle, preovulatory phase would include the 6-day period before ovulation, or, approximately days 18 - 13 preceding menstruation. Forty-five percent of all copulations in the restricted-access tests occurred during this 6-day midcycle period, in comparison to only 27% in the conventional tests. The difference between the two types of tests in the percentage of total copulations that occurred per cycle day was not significant for the early follicular phase (days 28 - 19), but it was significant for the midcycle (days 18 - 13) and luteal (days 12 - 0) phases of the cycle (Table 1).

There was a greater percentage of total copulations during midcycle and a smaller percentage during the luteal phase in the restricted-access tests with female control of access. A relatively small percentage of copulations in the restricted-access tests occurred during the luteal phase after day 10 preceding menstruation, i.e., few copulations occurred during the phase of the cycle when progesterone levels are elevated.

Figure 6 presents a comparable relationship between the two types of tests with respect to the percentage of test pairs that copulated each day throughout the cycle. In the conventional tests, all four of the pairs copulated on a majority of the days and 3 of the 4 pairs copulated on most of the days remaining; 87% of the pairs copulated each day, on average. In these tests, there was little difference in the percentage of pairs that copulated per day during the early follicular (85%), midcycle (100%) and luteal phase (82%) of the cycle and no significant correlation between the behavior and serum levels of estradiol during the follicular phase ($r(4) = 0.50$, ns). There was a significant negative correlation between the behavioral measure and serum levels of progesterone throughout the cycle ($r(13) = -0.61$, $p < 0.05$). The percentage of pairs that copulated on any given day was lower in the restricted-access tests than in the conventional tests for all phases of the cycle (Table 1) and a greater percentage of pairs tested per test day copulated during the early follicular (34%) and midcycle phases (54%) than in the luteal phase (15%). There was a significant positive correlation between the mean level of serum estradiol and the percentage of pairs that copulated per day during the follicular phase in the restricted-access tests ($r(4) = 0.93$, $p < 0.01$) and a significant negative correlation between the mean level of serum progesterone and the same behavioral measure throughout the cycle ($r(13) = -0.59$, $p < 0.05$).

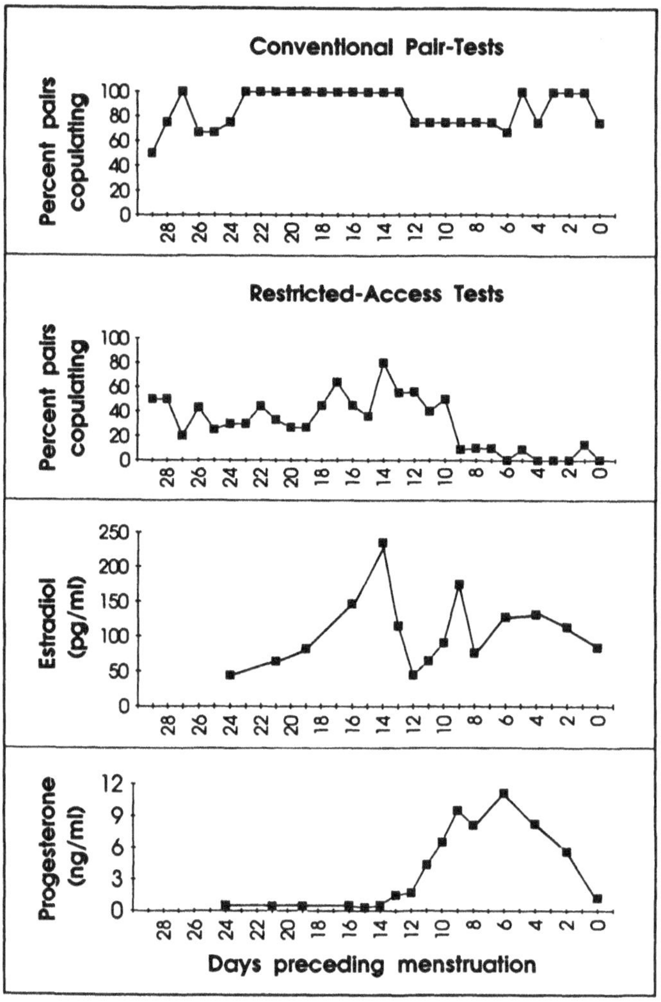

Figure 6. Percentage of pairs tested that copulated on each day during the menstrual cycle of female orangutans in conventional pair-tests and restricted-access tests with female control of access and the concentrations of 17ß-estradiol and progesterone in serum. Day 0 is the first day of menstruation.

PROXIMATE INFLUENCES ON SEXUAL BEHAVIOR OF ORANGUTANS

The laboratory research described above is compatible with the proposal that at least two classes of proximate variables exert a measurable influence on the sexual behavior of orangutans, one internal or organismic variable and the other external or environmental. The organismic variable was female hormonal state, implicated on the basis of the temporal relationship between the sex hormone profile of the females and the pattern of copulatory behavior during the menstrual cycle in the female-controlled restricted-access tests. This hormone-behavior relationship is comparable to that found in the other great apes (Nadler, et al., 1983, 1994; Nadler and Collins, 1991) and is consistent with the results of laboratory studies on other nonhuman primates, especially rhesus monkeys, in which hormone levels of the female were manipulated experimentally (Nadler, et al., 1986). These studies, overall, suggest that increasing levels of estradiol in the female facilitate sexual interactions, whereas, increasing levels of progesterone inhibit them.

The facilitation of sexual interactions by estradiol likely results from the induction of estrus in the female, a special period of heightened female sexual motivation, reflected, primarily, in the enhancement of sexual proceptivity (Heape, 1900, Beach, 1976, Nadler, 1992, 1994). Under natural conditions, fully mature orangutans consort with each other primarily as a result of a female seeking out a male for copulation. Such seeking out of the male is a form of sexually proceptive behavior. A comparable form of proceptive behavior in the restricted-access tests was shown to be related to, and presumably, facilitated by, elevated levels of estradiol in the female. These data, therefore, suggest that it is the hormonal induction of estrus in the female that determines whether and when copulation occurs among fully mature orangutans.

Since a proceptive female is more attractive to males than one that is not proceptive (Beach, 1976), the hormonal effect on the female's behavior also has an indirect effect on the male, namely, by enhancing female sexual attractivity. Female attractivity may also be enhanced by an hormonal effect on the female's genitals. The female orangutan's genitals do not undergo cyclical patterns of swelling like the genitals of the other female great apes (Schultz, 1938; Lippert, 1974), but they are apparently of some interest to the male. The male orangutans in the laboratory tests frequently examined the genitals of the females during copulation, in a manner suggesting that olfactory or gustatory stimuli may contribute to female attractivity (Nadler, 1977). The hormonal facilitation of sexual interactions in orangutans, therefore, likely results from the combined enhancement of sexual proceptivity and attractivity, much as it does in most other nonhuman primate females.

The second proximate variable that influenced the sexual behavior of the orangutans was the type of pair-test in which the animals were tested. The conventional pair-test, which favored the male's regulation of sexual interactions, resulted in a relatively acyclic pattern of copulation during the menstrual cycle. The factors that likely contributed to the male's preeminence in these tests were the male's dominance over the female and the limited spatial conditions of the test cage; under such conditions the female's options for refusing, avoiding or escaping from the male were limited. In the restricted-access test with female control of access, there was a difference in both the form and the frequency of sexual interactions. The main factor that contributed to these differences were clearly the female's ability to regulate the sexual interactions. Since the pattern of interactions in the restricted-access tests closely resembled the sexual relations that are considered to be characteristic of fully mature orangutans in the wild (see below), they support the interpretation that the restricted-access test simulates the conditions of female control that are typical of the species in its natural habitat, i.e., the conditions for which the species is behaviorally adapted.

ULTIMATE INFLUENCES ON SEXUAL BEHAVIOR OF ORANGUTANS

At about the same time that the restricted-access tests of sexual behavior were being conducted in the laboratory, comparable observations of penile displays and female proceptivity were made on wild orangutans in Sumatra (Schürmann, 1981; 1982) and proceptive behavior was reported for female orangutans in Borneo (Galdikas, 1981). These data, in conjunction with other observations on wild orangutans (MacKinnon, 1971, 1974; Rodman, 1973, Rijksen, 1978; Mitani, 1985), suggest that the species-typical pattern of sexual interactions in orangutans varies with the stage of male maturation. The forcible initiation of copulation by males is most common in the subadult stage of maturation, prior to the time when males establish themselves in a home-range. The forcible initiation of copulation by fully adult males in the conventional pair-test appears to represent a natural form of behavior, therefore, but one which is less common in fully mature males under natural conditions. Although it is likely that such behavior was made possible by male dominance over the female and the limited spatial conditions of a laboratory cage, the specific stimulus conditions that account for the frequency of this behavior remain hypothetical (Nadler, 1989). Reunion with a familiar female after a period of separation and meeting an unfamiliar female for the first time were reported to be the most common conditions under which forcible copulation occurred in the natural environment (Rijksen, 1978). If such conditions represent a "natural" stimulus for the forcible initiation of copulation by male orangutans, then the daily protocols for conducting the conventional laboratory pair-test may be sufficient to account for the results which were obtained. In the conventional pair-test, the animals are reunited each day after a period of separation, under conditions that favor the prerogatives of the male. Under such conditions, the fully adult male responds to the stimuli in the manner which is characteristic of a subadult, i.e., the male attempts to initiate copulation with the female. The less frequent pursuit of this strategy by fully adult males in the natural habitat may merely reflect their diminished ability to pursue and to catch smaller and more agile females under those more spacious and complex conditions.

INTERMALE COMPETITION FOR ESTROUS FEMALES

It was proposed more than a decade ago that various aspects of reproductive behavior, physiology and anatomy of the great apes are adapted to the species' mating systems (Short, 1979; Harcourt, 1981; Nadler, 1982). Table 2 illustrates these species-typical relationships for fully adult individuals, in light of the most recent data. The reproductive characteristics of the orangutan, overall, more closely resemble the characteristics of the gorilla (unimale mating system) than the chimpanzee (multimale mating system). Male courtship and sexual initiative are less pronounced in the former two species, consistent with the low degree of intermale competition for estrous females, characteristic of unimale mating systems. Although subadult male orangutans attempt to copulate opportunistically with females, they are not effective competitors of fully adult males for estrous females. When they are in estrus, female orangutans appear to choose and to establish proximity with fully adult males. Subadult males do not intrude into these relationships. The adult male orangutan's courtship display, therefore, is the sexual solicitation of a male that has already been chosen by an estrous female, not a competitive act of one male among many for the attention of a female. As such, there is no selective advantage for the adult male orangutan to develop as elaborate a display as the male of a multimale mating system.

The external genitals of male and female orangutans represent a mixed picture. The penis of the male orangutan is larger than originally thought and closer in size to the penis of the chimpanzee than the gorilla (Nadler, 1982, 1988; Dahl, 1988). Penile visibility, therefore, is more directly adapted for a role in courtship, rather than to a mating system *per se*. The female orangutan lacks a cyclic genital swelling altogether; it advertises estrus behaviorally with assertive proceptive behavior directed toward a given male. The duration of estrus and mating during the female cycle is intermediate in the orangutan to that of the chimpanzee and gorilla, as is the testis weight-body weight ratio. The longer duration of estrus in the orangutan, in comparison to the gorilla, may be an adaptation to the dispersed nature of social organization in this species; the female orangutan likely requires more time than the female gorilla to establish proximity and sociosexual relations with the male at the time of estrus. The difference in the testis weight-body weight ratio between the chimpanzee

Table 2. Reproductive physiology, behavior and anatomy of the great apes in relation to mating system

Genus	Mating system	Sexual initiative	Male courtship display	Penile visibility	Testis weight:body weight ratio	Duration of estrus (days)	Cyclic perineal swelling
Chimpanzee	Multimale	Male/female	Conspicuous, active	High	0.269	10 - 14	Large, conspicuous
Orangutan	Unimale	Female/male	Conspicuous, passive	Moderate	0.035	4 - 6	Absent
Gorilla	Unimale	Female/male	Inconspicuous, passive	Low	0.015	1 - 3	Small, inconspicuous

and the other great apes is typically attributed to differences in sperm competition between males of multimale and unimale mating systems (Harcourt, Harvey, Larson and Short, 1981; Dahl, Gould and Nadler, 1993), but this does not account for the difference between the orangutan and gorilla. This latter difference may also reflect a difference in sperm competition, but it could also reflect the difference in semen/sperm requirements for fertilizing a female after a greater (orangutan) and lesser (gorilla) number of days of estrus and ejaculation prior to ovulation. Since ovulation occurs at the end of estrus, a male must possess testes of sufficient functional capacity to fertilize a female after ejaculating for a number of days equal to the duration of estrus. The intermediate position of the male orangutan's testis weight-body weight ratio corresponds to the intermediate position of the duration of estrus of the female.

IMPLICATIONS OF THE RESEARCH FOR CAPTIVE MANAGEMENT OF ORANGUTANS

Field research on orangutans revealed that this species lives a semi-solitary life-style, with adult males generally living alone and females living alone or with their dependent offspring in home-ranges subsumed within the larger home-range of one or more males (MacKinnon, 1971, 1974; Rodman, 1973, Rijksen, 1978; Galdikas, 1981; Schürmann, 1981, 1982). Field research suggested that male orangutans pursue at least two different strategies in their sexual relations with females. When males are subadult they actively pursue females and frequently initiate copulation forcibly, irrespective of female hormonal state and fertility. When males are fully adult and established in a home-range, they pursue females less frequently and instead, solicit copulation nonaggressively from females that are likely in estrus. The behavior of orangutans in zoological gardens and in conventional laboratory pair-tests suggested that disregard of the species' life-style and spatial prerogatives in the natural habitat resulted in the arrest of the male's behavioral maturation with respect to sexual and reproductive behavior, i.e., the behavioral maturation of the male was halted in, or regressed to, the subadult stage. The female-controlled laboratory tests revealed that fully adult males that had never displayed the nonaggressive form of sexual solicitation, quickly began to display this behavior when they were prevented from pursuing the females as they had always done. These results suggest that the nonaggressive form of sexual solicitation, including the penile display, is a species-typical form of behavior, but one that is highly responsive to environmental conditions. The data from zoological gardens, laboratory and field settings demonstrate the complementary nature of the contributions to our understanding of orangutans that have been obtained by investigators under widely different conditions of observation. They suggest, moreover, that an enlightened approach to the captive management of orangutans requires consideration of these complementary data.

It is generally accepted nowadays that the most appropriate conditions in which to maintain a species, from the perspective of reproduction, maturation of young and esthetics, are those that simulate critical aspects of the species' natural habitat. *Critical* aspects of the natural habitat may be interpreted as those conditions, social and physical, that facilitate the expression of species-typical behavior, behavior for which the species is biologically adapted. The function of *simulation* is to produce an effect on the animal; simulation does not require transfer of the forest into the cage. A few branches or shrubs may serve and even polyvinyl chloride or steel pipes may be adequate, from the animal's perspective. As is well known, this philosophical position was not always appreciated, and once appreciated, was not always realized for both theoretical and practical reasons. There was frequently inadequate information available on the natural lives of animals, inadequate imagination and understanding of the concept of simulation and inadequate funds for the renovation or construction of facilities.

Research indicates that orangutans differ from many other primates with respect to sociality; orangutans are biologically adapted to a semi-solitary life. Adult females rarely socialize with adult males, and when they do, it is primarily in the context of relatively brief periods of consortship and mating. Since males are generally dominant over females, species-appropriate management of orangutans requires that the captive environment provide the female with options for avoiding and escaping from a male that simulate conditions that are normally available to female orangutans in the wild. Clearly, the provision of an environment with sufficient space to enable adequate separation of the animals, combined with objects or structures that permit the female to avoid visual contact with the male, is one approach that has been used effectively in some zoological gardens. When circumstances do not permit this approach, however, the provision of opportunities for avoiding the male, similar to that used in the restricted-access tests, may find useful application.

CONCLUSIONS

1. The sexual behavior of orangutans in two different pair-test paradigms approximated the species-typical behavior of these animals at two different stages of male maturation, the subadult stage, when the animals were tested in conventional laboratory pair-tests and the adult stage, in tests where females controlled sexual access.
2. Sexual behavior occurred frequently and irrespective of menstrual cycle phase in the conventional tests as a result of forcibly initiated copulation by the males, whereas, sexual behavior occurred cyclically in relation to female hormonal state in the female-controlled tests.
3. Female hormonal state exerts a measurable influence on the sexual behavior of orangutans, including the induction of estrus at midcycle; this influence is most apparent when the female is given the option of regulating sexual interactions in a manner that is comparable to that which exists in the wild.
4. An important consideration in the captive maintenance of orangutans, which is consistent with their behavioral adaptations in the natural habitat, is the provision of conditions that permit the female to avoid or escape the sexual initiative of the male.

ACKNOWLEDGMENTS

Preparation of this paper was supported by the National Science Foundation with Grant No. IBN91-09441 to the author and by the National Institutes of Health with Grant No. RR-00165 from the National Center for Research Resources to the Yerkes Regional Primate Research Center. I thank Ms. B. McElduff for secretarial assistance. The Yerkes Center is fully accredited by the American Association for Accreditation of Laboratory Animal Care.

REFERENCES

Asano, M., 1967, A note on the birth and rearing of an orang-utan *Pongo pygmaeus* at Tama Zoo, Tokyo, *Int. Zoo Yearb.* 7:95-96.

Beach, F. A., 1976, Sexual attractivity, proceptivity, and receptivity in female mammals, *Horm. Behav.* 7: 105-138.

Chaffee, P.S., 1967, A note on the breeding of orang-utans at Fresno Zoo, *Int. Zoo Yearb.* 7:94-95.

Coffey, P.F., 1972, Breeding Sumatran orang utan *Pongo pygmaeus abelii* Lesson 1827, *Ann. Rep. Jersey Wildl. Preserv. Trust*, pp. 15-17.

Coffey. P.F., 1975, Sexual cyclicity in captive orang utans *Pongo pygmaeus* with some notes on sexual behaviour, *Ann. Rep. Jersey Wildl. Preserv. Trust*, pp. 54-55.

Dahl, J.F., 1988, The external genitalia. Pp. 133-144 in *Orang-utan Biology*, (Ed. J.H. Schwartz), New York, Oxford University Press.

Dahl, J.F., Gould, K.G. and Nadler, R.D., 1993, Testicle size of orang-utans in relation to body size, *Am. J. Phys. Anthropol.* 90:229-236.

Fox, H., 1929, The birth of two anthropoid apes. *J. Mammal.* 10:37-51.

Galdikas, B.M.F., 1981, Orangutan sexuality in the wild. Pp. 281-300 in: *Reproductive Biology of the Great Apes: Comparative and Biomedical Perspectives*, (Ed. C.E. Graham), New York, Academic Press.

Harcourt, A.H., 1981, Intermale competition and the reproductive behavior of the great apes. Pp. 301-318 in: *Reproductive Biology of the Great Apes: Comparative and Biomedical Perspectives*, (Ed. C.E. Graham), New York, Academic Press.

Harcourt, A.H., Harvey, P.H., Larson, S.G. and Short, R.V., 1981, Testis weight, body weight and breeding system in primates, *Nature* 293:55-57.

Heape, W., 1900, The "sexual season" on mammals and the relation of the "pro-estrum" to menstruation, *Quart. J. Microscop. Sci.*, 44:1-70.

Heinrichs, W.L., and Dillingham, L.A., 1970, Bornean orang-utan twins born in captivity, *Folia Primatol.* 13:150-154.

Lippert, W., 1974, Beobachtungen zum Schwangerschafts- und geburtsverhalten beim Orang-Utan (*Pongo pygmaeus*) im Tierpark Berlin, *Folia Primatol.* 21:108-134.

MacKinnon, J.R., 1971, The orang-utan in Sabah today. *Oryx* 11:141-191.

MacKinnon, J.R., 1974, The behaviour and ecology of wild orang-utans (*Pongo pygmaeus*), *Anim. Behav.* 22:3-74.

Maple, T.L., Zucker, E.L., and Dennon, M.B., 1979, Cyclic proceptivity in a captive female orang-utan (*Pongo pygmaeus abelii*), *Behav. Proc.* 4:53-59.

Markham, R.J., 1990, Breeding orang-utans at Perth Zoo: Twenty years of appropriate husbandry, *Zoo Biol.* 9:171-182.

Mitani, J.C., 1985, Mating behaviour of male orangutans in the Kutai Game Reserve, Indonesia, *Anim. Behav.* 33:392-402.

Nadler, R.D., 1977, Sexual behavior of captive orang-utans, *Arch. Sex. Behav.* 6:457-475.

Nadler, R.D., 1982, Reproductive behavior and endocrinology of orang utans, Pp. 231-248 in: *The Orang Utan: Its Biology and Conservation*, (Ed. L.E.M. de Boer), The Hague, Dr. W. Junk Publ.

Nadler, R.D., 1988, Sexual and reproductive behavior, Pp. 105-116 in: *Orang-Utan Biology*, (Ed. J.H. Schwartz), New York, Oxford University Press.

Nadler, R.D., 1989, Sexual aggression in common chimpanzee, gorilla and orang-utan, *Prim. Rep.* 23:5-11.

Nadler, R.D., 1992, Sexual behavior and the concept of estrus in the great apes, Pp. 191-206 in: *Topics in Primatology, Vol II, Behavior, Ecology, And Conservation*, (Eds. N. Itoigawa, Y. Sugiyama, G.P. Sackett, R.K.R. Thompson), Tokyo, University of Tokyo Press.

Nadler, R.D., 1994, Walter Heape and the issue of estrus in primates, *Am. J. Primatol.* 33:83-87.

Nadler, R.D., and Collins, D.C., 1991, Copulatory frequency, urinary pregnanediol and fertility in great apes, *Am. J. Primatol.* 24:167-179.

Nadler, R.D., Collins, D.C., and Blank, M.S., 1984, Luteinizing hormone and gonadal steroid levels during the menstrual cycle of orang-utans, *J. Med. Primatol.* 13:305-314.

Nadler, R.D., Herndon, J.G., and Wallis, J., 1986, Adult sexual behavior: hormones and reproduction, Pp. 363-407 in: *Comparative Primate Biology, (Series Ed. J. Erwin) Vol. IIA - Behavior, Conservation, And Ecology*, (Eds. G. Mitchell and J. Erwin), New York, Alan R. Liss.

Nadler, R.D., Collins, D.C., Miller, L.C., and Graham, C.E., 1983, Menstrual cycle patterns of hormones and sexual behavior in gorillas, *Horm. Behav.* 17:1-17.

Nadler, R.D., Dahl, J.F., Collins, D.C., and Gould, K.G., 1994, Sexual behavior of chimpanzees (*Pan troglodytes*): Male vs female regulation, *J. Comp. Psychol.* 108:58-67.

Rijksen, H.D., 1978, *A Field Study on Sumatran Orang Utans (Pongo Pygmaeus Abelii* Lesson 1827), *Ecology, Behaviour and Conservation*. Wageningen, H. Veenman and Zonen B.V.

Rodman, P.S., 1973, Population composition and adaptive organization among orang-utans of the Kutai Reserve, Pp. 171-209 in: *Comparative Ecology and Behaviour of Primates*, (Eds. R.P. Michael and J.H. Crook), London, Academic Press.

Schultz, A.H., 1938, Genital swelling in the female orang-utan, *J. Mammal.* 19:363-366.

Schürmann, C.L., 1981, Courtship and mating behavior of wild orang-utans in Sumatra, Pp.130-135 in: *Primate Behavior and Sociobiology*, (Eds. A.B. Chiarelli and R.S. Corruccini), Berlin, Springer Verlag.

Schürmann, C.L., 1982, Mating behavior of wild orang-utans, Pp. 269-284 in: *Biology and Conservation of the Orang-Utan*, (Ed. L.E.M. de Boer), The Hague, W. Junk B.V.

Seitz, A., 1969, Einige Feststellungen zur Pflege und Aufzucht von Orang-Utans, *Pongo pygmaeus* Hoppius 1763, *Zool. Gart. (N.F.)* 36:225-245.

Short, R.V., 1981, Sexual selection in man and the great apes, Pp. 319-341 in: *Reproductive Biology of the Great Apes: Comparative and Biomedical Perspectives*, (Ed. C.E. Graham), New York, Academic Press.

23

PROXIMITY, CONTACT, AND PLAY INTERACTIONS OF ZOO-LIVING JUVENILE AND ADULT ORANGUTANS, WITH FOCUS ON THE ADULT MALE

E. L. Zucker and S. C. Thibaut

Department of Psychology
Loyola University
New Orleans, Louisiana 70118

INTRODUCTION

The behavioral development of young apes is influenced by their physical and social environments. Deficient or limited environments can result in reproductive insufficiencies (see Beck and Power, 1988) and behavioral deficits, such as stereotypic movements (see Puleo, et al., 1983; Meder, 1989). The role of the environment in behavioral deficits is evident from studies where stereotypies were reduced or eliminated when the physical or social environments were enhanced (Clarke, et al., 1982; Puleo, et al., 1983; Meder, 1989). The mere presence of conspecifics, however, does not guarantee the development of species-typical behavior, even in complex social groups. For example, a young orangutan reared in a social group containing only adults is limited in its opportunities for social play (Zucker, et al., 1986; see Baldwin, 1986, for more about social and ecological constraints on play). Although adult orangutans of both sexes do play, the rate of adult play is much lower than that of immature orangutans (Zucker, et al., 1986), and in playing with immature orangutans, the style of play of adult females differs from that of adult males (Zucker, et al., 1978). Nursery-reared apes are at risk for behavioral deficits as their social and physical environments usually are less complex than their respective natural environments, although nursery environments and rearing protocols can be made relatively complex and stimulating.

In zoos, the reintroduction of nursery-reared apes into groups of conspecifics has been a successful management strategy (Hamburger, 1988; Keiter, et al., 1983; Meder, 1985, 1989, 1990). The social environment into which the reintroduced animals are placed further shapes the development of human-reared apes; the interactions of reintroduced orangutans has been the focus of study during the past 10 years at the Audubon Park and Zoological Garden. Here, we describe the parameters of proximity, contact, and social play of five reintroduced orangutans, with emphasis on quantitative differences in the immatures'

The Neglected Ape, Edited by Ronald D. Nadler et al.
Plenum Press, New York, 1995

interactions with the adult male orangutan, in comparison with their interactions with the adult females.

The interest in the interactions, in captivity, of immatures with adult males (and adult males with immatures) stems from the reported differences in sociality of adult males in captivity and in the wild. In the wild, adult male orangutans tend to be solitary, and interact little with conspecifics in nonsexual contexts (Davenport, 1967; Galdikas, 1985a, b; Horr, 1975; 1977; MacKinnon, 1974; Rijksen, 1978; Rodman, 1973; Rodman and Mitani, 1987). Interactions between adult males and immature orangutans do occur on a limited basis and to a limited extent (MacKinnon, 1974; Rijksen, 1978; Galdikas, 1985a, b), with the probability of interactions increasing at times of more abundant food availability (Sugardjito, et al., 1987) or in the context of competition for receptive females (Mitani, et al., 1991). Adult female, subadult, and immature orangutans are the age/sex classes likely to be found in social groupings (Horr, 1977; Galdikas, 1985b).

For captive adult males, however, there are reports of relatively extensive interactions with adult female and immature conspecifics. Several authors have emphasized that the behaviors expressed by adult males in captivity reflect the behavioral potential of adult male orangutans (Edwards and Snowdon, 1980; Zucker, et al., 1978, 1986), the expression of which is facilitated by concentrated resources and spatially limited environments. A summary of the interactions reported in the published literature for captive male orangutans is presented in Table 1.

The interactions reported in the literature range in intensity from proximity and contact to social grooming to vigorous social play. Proximity, of measurable duration, has been studied in orangutan groups in five different settings, and social grooming has been reported for four different groups of captive orangutans. In addition to the references cited in Table 1, Maple (1980) provided a description of grooming for orangutans at the Atlanta Zoo (now known as Zoo Atlanta) and the Yerkes Regional Primate Research Center; the results of this study of grooming were reported only in Maple's (1980) volume, and did not appear in a separate publication.

Table 1. Summary of interactions and behaviors of adult males in captivity

	Source									
Behavior	ES	K	P	TGRM	WMK	Z	ZDPM	ZF	ZMM	ZTW
Proximity	X		X	X		X		X	X	X
Contact	X	X		X						X
Male-male aggression			X							
Other agonism	X				X					
Grooming	X	X	X		X					
Social play					X		X		X	X
Agonistic interventions	X					X				

Sources:
ES = Edwards and Snowdon (1989); Vilas Park Zoo, Madison, WI (USA).
K = Keiter (1983); Jersey Wildlife Preservation Trust, Channel Islands.
P = Poole (1987); Singapore Zoological Gardens.
TGRM = Tobach, Greenberg, Radell, and McCarthy (1989); Sedgwick County Zoo, Wichita, KS (USA).
WMK = Williams, Mellen, and King (1984); Metro Washington Park Zoo, Portland, OR (USA).
Z = Zucker (1987); Audubon Zoological Park, New Orleans, LA (USA).
ZDPM = Zucker, Dennon, Puleo, and Maple (1986); Atlanta Zoo and Yerkes Regional Primate Research Center, Atlanta, GA (USA).
ZF = Zucker and Ferrera (1990); Audubon Zoological Park, New Orleans, LA (USA).
ZMM = Zucker, Mitchell, and Maple (1978); Atlanta Zoo, Atlanta, GA (USA).
ZTW = Zucker, Thibaut, and Watts (1991); Audubon Zoological Park, New Orleans, LA (USA).

An important conclusion to be drawn from the information in this table is that the behaviors listed have occurred in more than one social group, in groups containing different age and sex class members, and in different captive habitats. Thus, it is unlikely that the interactions of these various adult males were merely idiosyncratic behaviors.

The data presented in this chapter come from observations of interactions involving an adult male, two adult females, and five immature orangutans at the Audubon Park and Zoological Garden in New Orleans, Louisiana. The adults of this group have been housed together continuously since 1982. The immature orangutans have been hand-reared, and reintroduced to the adults as two cohorts (see Zucker, et al., 1991). As the immature orangutans were more active and socially involved, the data for their interactions with each other and the adults are more extensive. However, similarities and differences in the interaction patterns of the adult male and adult females with the immatures, and with each other, were assessed for the behaviors of proximity, contact, and social play.

METHOD

Subjects

Subject information is presented in Table 2. The adult male and one adult female were wild born, whereas the second adult female was zoo-born. The immature orangutans, when reintroduced, were functionally independent, and thus are referred to as "juveniles" in the remainder of this chapter. The first set of three juveniles, reintroduced when approximately 1.5 years of age in June, 1986, consisted of one male and a set of female twins. The second set of juveniles, also reintroduced when approximately 1.5 years of age in August 1989, consisted of two females. Soon after the introduction of these latter two females, the twins were transferred to another zoo; the juvenile male from the first cohort remained in the social group.

Exhibit

This group is exhibited in a naturalistic, outdoor area approximately 19.5 m X 13.5 m (see diagram in Zucker and Ferrera, 1990). The exhibit contained numerous tree trunks, branches, ropes, and elevated platforms, which enabled brachiation and other arboreal activities.

Table 2. Subject information

Age class/name	ISIS no.	Sex	Date of birth	Parentage/rearing	When studied	Hours of focal data
Adults						
Frankie (FR)	031	M	c. 1962	Wild-born	1989	5.67
Mama (MA)	032	F	c. 1963	Wild-born	1989	5.67
Sara (SA)	372	F	8/17/70	Philadelphia Zoo	1989	6.00
Juveniles						
Jambu (JM)	707	M	4/28/85	MA X FR	1989	6.00
					1991	33.60
Lagniappe (LN)	674	F	1/26/85	SA X FR	1989	5.67
BonTemps (BT)	673	F	1/26/85	SA X FR	1989	6.00
Siabu (SI)	998	F	3/25/89	MA X FR	1991	33.80
Feliz (FZ)	1005	F	12/23/88	Gladys Porter Zoo	1991	33.60

Data Collection

The data included in this report were obtained via focal animal sampling (Altmann, 1974), collected January through April of 1989 and 1991. In 1989, each of the three adults (FR, MA, SA) and three juveniles (BT, LN, JM) were sampled for 20 minutes every 2 days. The observation order over the two days for the six orangutans was determined randomly. Data were collected only on days when all animals were outdoors. One of the adult females (MA) was pregnant during this time, and frequently chose to remain indoors, limiting this phase of the study to 35 hours of focal observations (see Table 2).

In 1991, each of the three juveniles (JM, SI, FZ) was observed via focal animal sampling for a 15-minute period, with the order of observations for each set of three focal samples determined randomly. During this phase of the study, 101 hours of focal animal data were obtained (see Table 2).

Data Analysis

Proximity was defined as being stationary and within 1 m of another orangutan, and contact was recorded only when no other behavior was ongoing. The social play of orangutans has been described in structural terms elsewhere (Maple and Zucker, 1978; Zucker, et al., 1978, 1986), and these interactions are readily discernible by observers as play. The frequencies of proximity, contact, and social play were tabulated and analyzed, as were the durations of proximity, contact, and social play. As focal animal sampling was used, rates of each of the three types of interactions were determined, as were the mean durations. For interactions between juveniles and adults, only those interactions initiated during an individual's focal sample were included in the analyses. For interactions between adults, those interactions initiated and received during an individual's focal sample were included in the analyses. The durations of the three types of interactions by each juvenile with each possible partner were tested for differences among partners, using either analyses of variance or Kruskal-Wallis tests, with the former used when the juvenile orangutan interacted more than twice with every possible partner.

The number of interactions between juveniles and adults of each sex were tested for differences using chi-square tests. The null hypothesis for these tests was that the interactions would be distributed equally without regard to the sex of the adult. Expected values were generated based on there being two adult females and one adult male, so two-thirds of the total number of interactions should involve an adult female and one-third should involve the adult male.

Mean durations were calculated for each of the three types of interactions, and durations of the interactions involving juveniles and the adults were tested with t-tests for differences as a function of the sex of the adult. For the purpose of these analyses, the data for all juveniles were combined, as were the data for the two adult females.

Duration data for proximity between adult orangutans similarly were tested with t-tests to assess differences between female-female and male-female dyads. Contact and play between two adults were too infrequent to test in a similar manner.

RESULTS

Interactions of Juveniles: Individual Data

Rates and mean durations for all three types of interactions by the juveniles are presented in Table 3, for each juvenile and for the three combined. The mean rates and mean

Table 3. Rate (per hour) and mean duration (min) of proximity, contact, and play for each juvenile orangutan

	Behavior					
	Proximity		Contact		Play	
	Rate	Duration	Rate	Duration	Rate	Duration
1989						
BT	13.33	0.69	4.67	1.86	2.50	1.19
LN	8.11	0.47	2.47	0.99	1.06	0.84
JM	11.67	0.38	5.83	0.67	5.50	1.90
Combined	11.09	0.53	4.36	1.16	3.06	1.34
1991						
FZ	7.35	1.79	3.39	1.77	2.92	1.95
SI	6.09	1.67	5.92	1.96	3.37	2.08
JM	5.80	1.85	6.61	2.07	3.57	1.79
Combined	6.41	1.77	5.31	1.96	3.29	1.94
Overall	7.11	1.48	5.16	1.87	3.34	1.84

durations also are presented for each of the two observation periods, as well as for the two study periods combined.

Whereas the total time (in minutes) that juveniles spent interacting with other juveniles and adults differed as a function of the recipient or co-actor (see Zucker, et al., 1991), there were few differences when the data were analyzed with regard to mean durations. Each of the juveniles interacted fairly equally, in terms of mean durations, with each of the partners with whom it interacted. In several cases, interactions did not occur between a pair of individuals, so in these cases, there were obvious differences in mean durations between possible partners. The frequencies and mean durations of interactions for each juvenile with each possible partner are presented in Table 4 for proximity, Table 5 for contact, and Table 6 for social play. Pairs not interacting are evident from the data in these tables.

Considering only those pairs where there were interactions, only three deviated from the overall pattern of relatively equal interaction. In 1989, BT spent significantly more time per instance of proximity with the adults MA and FR than any other orangutan ($F = 2.51$, $df = 4, 65$, $p = .05$), and JM spent significantly more time per episode of proximity with the adult female MA, his biological mother ($F = 3.39$, $df = 4, 65$, $p < .02$). The only significant difference in 1991 was that SI spent more time in contact with adult female SA, per instance, than with any other orangutan ($F = 5.27$, $df = 4, 210$, $p = .0005$).

Table 4. Mean duration (and frequency) of proximity for each juvenile with each possible partner

		Recipient						
Year	Subject	MA	SA	FR	BN	LN	JM	Total
1989	BN	1.57 (11)	0.52 (19)	1.37 (6)	—	0.39 (22)	0.44 (22)	80
	LN	0.32 (10)	0.75 (7)	0.35 (6)	0.55 (14)	—	0.57 (9)	46
	JM	0.69 (16)	0.29 (10)	0.42 (9)	0.24 (21)	0.24 (14)	—	70
1991		MA	SA	FR	FZ	SI	JM	Total
	FZ	2.02 (57)	1.71 (45)	2.03 (39)	—	1.76 (42)	1.52 (64)	247
	SI	1.79 (52)	1.71 (38)	1.50 (16)	1.60 (48)	—	1.63 (52)	206
	JM	1.52 (31)	1.77 (56)	1.67 (27)	1.53 (30)	2.41 (51)	—	195

Table 5. Mean duration (and frequency) of contact for each juvenile with each possible partner

		Recipient						
Year	Subject	MA	SA	FR	BN	LN	JM	Total
1989	BN	— (0)	2.52 (15)	2.14 (4)	—	0.36 (3)	0.79 (6)	28
	LN	0.08 (1)	1.32 (10)	— (0)	0.33 (1)	—	0.18 (2)	14
	JM	0.43 (5)	0.42 (10)	0.68 (7)	0.93 (10)	0.74 (3)	—	35
1991		MA	SA	FR	FZ	SI	JM	Total
	FZ	1.80 (20	1.82 (11)	2.09 (23)	—	1.58 (26)	1.68 (34)	114
	SI	1.67 (24)	2.27 (115)	2.50 (4)	1.53 (19)	—	1.39 (38)	200
	JM	1.92 (25)	2.17 (95)	2.02 (43)	1.73 (34)	2.36 (25)	—	222

Juvenile-Adult Interactions: Overall Patterns

Proximity. Juvenile orangutans initiated more proximity with adults than they did with other juveniles, and the mean duration of proximity with adults was significantly greater than the mean duration with other juveniles (see Table 7; $t = 3.81$, $df = 839$, $p = .0001$).

Juvenile orangutans initiated significantly more instances of proximity with adult females than they did with the adult male ($X^2 = 23.37$, $df = 1$, $p < .0001$), although there was no difference in the mean duration of the interaction as a function of the sex of the adult ($t = 0.10$, $df = 451$, $p > .05$).

Contact. The parameters for contact were similar to those of proximity. Juvenile orangutans were in contact more often with adults than they were in contact with other juveniles, and were in contact with adults for significantly longer periods of time than they were in contact with juveniles (Table 8; $t = 4.57$, $df = 624$, $p < .0001$). Adult females were more frequently contacted by juveniles than was the adult male ($X^2 = 38.66$, $df = 1$, $p < .0001$), but the mean durations did not differ significantly as a function of sex of the adult ($t = 0.37$, $df = 424$, $p > .05$).

Play. Juvenile orangutans played more frequently with other juveniles than they played with adults, although there was no significant difference in the mean duration of play bouts with adults or other juveniles (Table 9; $t = 1.46$, $df = 382$, $p > .05$). Juveniles played significantly more with the adult male than was expected by chance ($X^2 = 4.48$, $df = 1$, $p < .03$). However, the mean durations of juveniles' play bouts with adults did not differ as a function of the sex of the adult ($t = 0.39$, $df = 83$, $p > .05$).

Table 6. Mean duration (and frequency) of social play for each juvenile with each possible partner

		Recipient						
Year	Subject	MA	SA	FR	BN	LN	JM	Total
1989	BN	— (0)	—(0)	0.22 (2)	—	0.49 (4)	1.72 (9)	15
	LN	0.12 (1)	—(0)	2.05 (2)	0.20 (2)	—	0.43 (1)	6
	JM	0.43 (2)	—(0)	1.48 (8)	1.89 (7)	3.01 (6)	—	33
1991		MA	SA	FR	FZ	SI	JM	Total
	FZ	1.71 (7)	2.17 (6)	1.00 (1)	—	1.88 (42)	2.05 (42)	98
	SI	2.00 (1)	1.81 (16)	1.33 (3)	2.53 (47)	—	1.77 (47)	114
	JM	1.75 (4)	1.40 (10)	1.77 (22)	1.58 (45)	2.15 (39)	—	120

Table 7. Frequency and mean duration (min) of interactions of juvenile and adult orangutans: Proximity

Dyad	Frequency	Mean duration (SD)
Juvenile → Juvenile	390	1.29 (0.93) -
		|[a]
Juvenile → Adult	453	1.57 (1.14) -
Juvenile → Adult Female	351 -	1.57 (1.14) -
	|[b]	| n.s.
Juvenile → Adult Male	102 -	1.58 (1.16) -

[a] $t = 3.81$, $p = .0001$.
[b] $X^2 = 23.37$, $p < .0001$

Table 8. Frequency and mean duration (min) of interactions of juvenile and adult orangutans: Contact

Dyad	Frequency	Mean duration (SD)
Juvenile → Juvenile	200	1.52 (0.98) -
		|[a]
Juvenile → Adult	426	2.01 (1.36) -
Juvenile → Adult Female	345 -	2.02 (1.34) -
	|[b]	| n.s.
Juvenile → Adult Male	81 -	1.95 (1.47) -

[a] $t = 4.57$, $p = .0001$.
[b] $X^2 = 38.66$, $p < .0001$

Table 9. Frequency and mean duration (min) of interactions of juvenile and adult orangutans: Play

Dyad	Frequency	Mean duration (SD)
Juvenile → Juvenile	299	1.89 (1.38) -
		| n.s.
Juvenile → Adult	85	1.65 (1.27) -
Juvenile → Adult Female	47 -	1.70 (1.07) -
	|[a]	| n.s.
Juvenile → Adult Male	38 -	1.59 (1.48) -

[a] $X^2 = 4.48$, $p < .03$

Interactions of Adults

During focal samples of adult orangutans, proximity was the only one of the three types of interactions observed. The frequency and duration data for these spatial interactions are presented in Table 10. Adults, as an age class, were in proximity to other adults more frequently than they were in proximity to juvenile orangutans, but the mean duration of this proximity did not differ as a function of the age class of the partner ($t = 1.36$, $df = 41$, $p > .05$). Proximity between an adult female and a juvenile was more common that proximity between the adult male and a juvenile, although the mean durations of these interactions did not differ significantly ($t = 1.10$, $df = 14$, $p > .05$). Proximity between two adults was more likely to involve a heterosexual dyad than a same-sex dyad, although the mean durations of these two types of associations did not differ significantly ($t = 0.92$, $df = 25$, $p > .05$).

Table 10. Frequency and mean duration (min) of proximity between adult orangutans

Dyad	Frequency	Mean duration (SD)
Adult → Adult	27	1.82 (2.56) -
		| n.s.
Adult → Juvenile	16	0.90 (1.06) -
Adult Female → Juvenile	14	1.01 (1.09) -
		| n.s.
Adult Male → Juvenile	2	0.13 (0.14) -
Adult Female → Adult Male	24	1.98 (2.68) -
		| n.s.
Adult Female → Adult Female	3	0.52 (0.18) -

Note: These data are for adults approaching and being approached (within 1 m) during focal animal samples of adults.

DISCUSSION

Upon reintroduction, all juveniles interacted in some way with all available animals (peers and adults), and the few behavioral preferences expressed were independent of parent-offspring and sibling relationships. That each juvenile interacted with a variety of conspecifics suggests the need to reintroduce human-reared immature orangutans to socially complex environments. The presence of only one adult female might be too limiting socially, whereas the presence of two (or more) females and an adult male provides social diversity and the opportunity for more varied interactions.

With respect to the overall pattern of interactions between juvenile and adult orangutans, adult females, in the wild, are considered to be more social than adult males (see Galdikas, 1985b; Mitani, et al., 1991; Rodman and Mitani, 1987), and a similar conclusion derives from studies of captive orangutans (e.g., Poole, 1987). The results presented here show that in a zoo-living group of orangutans, adult females did interact more frequently with immature orangtuans than did an adult male when the behaviors of proximity and contact were considered; the adult male, however, engaged in play behavior with young orangutans more frequently than did the adult females. For all three types of behaviors, however, the adult male and the adult female orangutans did not differ in the mean durations of their interactions with these juveniles. While adult males generally do interact less frequently with conspecifics than do adult females, when adult males did interact, they did so to the same extent as did adult females.

The interaction patterns of adults at the Audubon Zoo were consistent with those for other captive groups (see Table 1), and the duration data presented provide additional detail to descriptions of orangutan social behavior. Interactions involving two adults were limited to proximity, and proximity between an adult female and male was more common than was proximity between the two adult females, a result consistent with previous findings (Edwards and Snowdon, 1980). In these interactions, however, there were no statistically significant differences in the mean durations as a function of the composition of the dyads. Similarly, for play between juveniles and adults, adult males have been described as playing more frequently and more intensely with juveniles than adult females (Zucker, et al., 1978, 1986). While the current data continue to support this conclusion, these data also indicate that the duration of play bouts did not differ as a function of the adult's sex. For that matter, juveniles engage in much more play behavior than do adults (see Zucker, et al., 1986), but the mean durations of play bouts between two juvenile orangtuans did not differ from those between a juvenile and an adult of either sex.

There are several plausible reasons, both proximate and ultimate, for the apparently greater involvement in social activities by captive adult males in comparison to wild adult males. First, and most obviously, is that the foraging demand is reduced in captivity, allowing more time for social interaction (see Erwin and Deni, 1979). Secondly, in captivity, and even when housed in large, naturalistic habitats, spatial and social densities are relatively higher than they are in the wild, which should increase the likelihood that individuals encounter each other and interact with each other (see Williams, et al., 1984; Zucker, et al., 1986). Thirdly, and related to increased social and spatial densities in captivity, is that animals are highly visible to each other, likely resulting in greater familiarity and knowledge of behavioral tendencies (see Zucker and Kaplan [1981] regarding interactions between a male patas monkey and juveniles). A fourth factor possibly contributing to increased interactions by males with juveniles in captivity is that males in captivity have greater confidence in paternity (see Galdikas, 1989; Mitani, 1985; Schürmann, 1982; Schürmann and van Hooff, 1986). The adult male engaged in more play with juveniles than did the females, and if play is beneficial to animals as it is hypothesized to be (see Fagen, 1974; Baldwin and Baldwin, 1977), then play can be seen as a form of parental investment. In captivity, with increased confidence in paternity, extensive interactions with immature animals could contribute to behaviorally competent immatures, and eventually, to increased reproductive success.

The proximate factors of available time, higher densities, and high visibility, as applied here to an adult male, also apply to adult females in captivity. While conditions exist in captivity that promote interactions by both sexes, adult females still interacted more frequently than did adult males, paralleling the patterns seen in the wild (e.g., Galdikas, 1985b). These proximate factors, then, might influence the quality of interactions more than they influence the quantity (frequency) of interactions. Interaction quality, measured by mean durations, was the same for both sexes of adults.

It is recognized that the data presented here are limited in that they come from only one group, containing only one adult male, and some types of interactions occurred infrequently (e.g., adult-adult interactions). Data from other social groups are needed, and comparisons with the findings presented here are encouraged. Despite these limitations, as the types of social interactions of adult males have been found to be fairly consistent across a number of social groups and housing situations, the finding of equally extensive interactions by adult males and females likely will also be found when data from additional groups become available.

ACKNOWLEDGMENTS

We gratefully acknowledge the assistance and cooperation provided by the administration and staff of the Audubon Zoo, particularly the staff of the Primate Department during the time these data were collected (E. Watts, M. Deitchman-Fernandez, C. Kennedy-Munn, K. Manceaux, and D. McCarthy). We thank Y. Tineo and A. Canapary for assistance with tabulating the 1991 data, S. Stacks and R. Fleming for assistance with literature searches, and for assistance with data collection in 1991, we thank, in alphabetical order: J. Bertsch, M. Blankenship, M. Bordlee, D. Cager, K. Cassell, A. Cigali, A. Coco, R. Derrick, K. Dietschy, M. Glorioso, W. Herbert, K. McGrath, N. Nelson, M. Neumann, J. Powers, S. Spahr, E. Stanley, and D. Walker. The Audubon Zoo is fully accredited by the American Zoo and Aquarium Association (AZA).

REFERENCES

Altmann, J., 1974, The observational study of behavior: Sampling techniques. *Behaviour* 49: 227-267.

Baldwin, J. D., 1986, Behavior in infancy: Exploration and play. In G. Mitchell and J. Erwin (eds.), *Comparative Primate Biology, Vol. 2A: Behavior, Conservation, and Ecology*. Alan R. Liss, Inc., New York, p. 295-326.

Baldwin, J. D., and Baldwin, J. I., 1977, The role of learning phenomena in the ontogeny of exploration and play. In S. Chevalier-Skolnikoff and F. E. Poirier (eds.), *Primate Bio-social Development: Biological, Social, and Ecological Determinants*. Garland Publishing, New York, p. 343-406.

Beck, B. B., and Power, M. L., 1988, Correlates of sexual and maternal competence in captive gorillas. *Zoo Biol.* 7: 339-350.

Clarke, A. S., Juno, C. J., and Maple, T. L., 1982, Behavioral effects of a change in the physical environment: A pilot study of captive chimpanzees. *Zoo Biol.* 1: 371-380.

Davenport, R. K., Jr., 1967, The orang-utan in Sabah. *Folia Primatol.* 5: 247-263.

Edwards, S. D., and Snowdon, C. T., 1980, Social behavior of captive, group-living orang-utans. *Int. J. Primatol.* 1: 39-62.

Erwin, J., and Deni, R., 1979, Stangers in a strange land: Abnormal behaviors or abnormal environments? In J. Erwin, T. L. Maple, and G. Mitchell (eds.), *Captivity and Behavior: Primates in Laboratories, Breeding Colonies, and Zoos*. Van Nostrand Reinhold, New York, p. 1-28.

Fagen, R., 1974, Selective and evolutionary aspects of animal play. *Am. Naturalist* 106: 850-858.

Galdikas, B. M. F., 1979, Orangutan adaptation at Tanjung Puting Reserve: Mating and ecology. In D. Hamburg and E. McCown (eds.), *The Great Apes*. Benjamin Cummings, Menlo Park, p. 194-233.

Galdikas, B. M. F., 1985a, Subadult male orangutan sociality and reproductive behavior at Tanjung Puting. *Am. J. Primatol.* 8: 87-99.

Galdikas, B. M. F., 1985b, Orangutan sociality at Tanjung Puting. *Am. J. Primatol.* 9: 101-119.

Hamburger, L., 1988, Introduction of two young orang-utans *Pongo pygmaeus* into an established family group. *Int. Zoo Yrbk.* 27: 273-278.

Horr, D. A., 1975, The Borneo orang-utan: Population structure and dynamics in relationship to ecology and reproductive behavior. In L. A. Rosenblum (ed.), *Primate Behavior: Recent Developments in Field and Laboratory Research, Vol 4*. Academic Press, New York, p. 307-323.

Horr, D. A., 1977, Orang-utan maturation: Growing up in a female world. In S. Chevalier-Skolnikoff and F. E. Poirier (eds.), *Primate Bio-social Development: Biological, Social, and Ecological Determinants*. Garland Publishing, New York, p. 289-321.

Keiter, M. D., 1983, A study of the integration of an adult Sumatran orang-utan female *Pongo pygmaeus abelii* to an existing adult pair at the Jersey Wildlife Preservation Trust. *Dodo, J. Jersey Wildl. Pres. Trust* 20: 53-65.

Keiter, M. D., Reichard, T., and Simmons, J., 1983, Removal, early hand rearing, and successful reintroduction of an orangutan (*Pongo pygmaeus abelii*) to her mother. *Zoo Biol.* 2: 55-59.

MacKinnon, J., 1974, The behaviour and ecology of wild orang-utans (*Pongo pygmaeus*). *Anim. Behav.* 22: 3-74.

Maple, T. L., 1980, *Orang-utan Behavior*. Von Nostrand Reinhold, New York.

Maple, T. L., and Zucker, E. L., 1978, Ethological studies of play behavior in captive great apes. In E. O. Smith (ed.), *Social Play in Primates*. Academic Press, New York, p. 113-142.

Meder, A., 1985, Integration of hand-reared gorilla infants in a group. *Zoo Biol.* 4: 1-12.

Meder, A., 1989, Effects of hand-rearing on the behavioral development of infant and juvenile gorillas (*Gorilla g. gorilla*). *Dev. Psychobiol.* 22: 357-376.

Meder, A., 1990, Integration of hand-reared gorillas into breeding groups. *Zoo Biol.* 9: 157-164.

Mitani, J. C., 1985, Mating behaviour of male orangutans in the Kutai Game Reserve, Indonesia. *Anim. Behav.* 33: 392-402.

Mitani, J. C., Grether, G. E., Rodman, P. S., and Priatna, D., 1991, Associations among wild orang-utans: Sociality, passive aggregations, or chance? *Anim. Behav.* 42: 33-46.

Poole, T. B., 1987, Social behavior of a group of orangutans (*Pongo pygmaeus*) on a artificial island in Signapore Zoological Gardens. *Zoo Biol.* 6: 315-330.

Puleo, S. G., Zucker, E. L., and Maple, T. L., 1983, Social rehabilitation and foster mothering by captive female orangutans. *Zool. Gart.* 53: 196-202.

Rijksen, H. D., 1978, *A Field Study on Sumatran Orang Utans* (*Pongo pygmaeus abelii* Lesson 1827). Wageningen, H. Veenman and Zonen B.V.

Rodman, P. S., 1973, Population composition and adaptive organization among orang-utans at the Kutai Reserve. In J. H. Crook and R. P. Michael (eds.), *Comparative Ecology and Behaviour of Primates.* Academic Press, London, p. 171-209.

Rodman, P. S., and Mitani, J. C., 1987, Orangutans: Sexual dimorphism in a solitary species. In B. B. Smuts, D. L. Cheney, R. M. Seyfarth, R. W. Wrangham, and T. T. Struhsaker (eds.), *Primate societies.* University of Chicago Press, Chicago, p. 146-154.

Schurmann, C., 1982, Mating behaviour of wild orangutans. In L. de Boer (ed.), *The Orangutan: Its Biology and Conservation.* W. Junk, The Hague, p. 269-284.

Schurmann, C., and van Hooff, J. A. R. A. M., 1986, Reproductive strategies of the orang-utan: New data and reconsideration of existing sociosexual models. *Int. J. Primatol.* 8: 265-287.

Sugardjito, J., te Brockhorst, I. J. A., and van Hooff, J. A. R. A. M., 1987, Ecological constraints on the grouping of wild orang-utans (*Pongo pygmaeus*) in the Gunung Leuser National Park, Sumatra, Indonesia. *Int. J. Primatol.* 8: 17-41.

Tobach, E., Greenberg, G., Radell, P., and McCarthy, T., 1989, Social behavior in a group of orang-utans (*Pongo pygmaeus abelii*) in a zoo setting. *Appl. Anim. Behav. Sci.* 23: 141-154.

Williams, C. J., Mellen, J. D., and King, N. E., 1984, Behavior of a group of captive orangutans (*Pongo pygmaeus abelii*) in two different enclosures. *AAZPA Reg. Proc.*: 186-201.

Zucker, E. L., 1987, Control of intragroup aggression by a captive male orangutan. *Zoo Biol.* 6: 219-223.

Zucker, E. L., Dennon, M. B., Puleo, S. G., and Maple, T. L., 1986, Play profiles of captive adult orangutans: A developmental perspective. *Dev. Psychobiol.* 19: 315-326.

Zucker, E. L., and Ferrera, P. S., 1990, Sociospatial relationships during gestation in captive lowland gorilla and orangutan groups. *Zool. Gart.* 60: 253-262.

Zucker, E. L., and Kaplan, J. R., 1981, A reinterpretation of 'sexual harassment' in patas monkeys. *Anim. Behav.* 29: 957-958.

Zucker, E. L., Mitchell, G., and Maple, T., 1978, Adult male—offspring paly interactions within a captive group of orangutans (*Pongo pygmaeus*). *Primates* 19: 379-384.

Zucker, E. L, Thibaut, S. C., and Watts, E., 1991, Social environments and the behavior of juvenile orangutans. *AAZPA Ann. Conf. Proc.*: 421-427.

24

A BRIEF REVIEW OF STUDIES OF ORANGUTAN MORPHOLOGY AND DEVELOPMENT WITH A DISCUSSION OF THEIR RELEVANCY TO PHYSICAL ANTHROPOLOGY

L. A. Winkler

Department of Anthropology
University of Pittsburgh
Titusville, Pennsylvania 16354

ABSTRACT

Studies of comparative anatomy and development have frequently been used to debate issues of hominoid phylogeny and taxonomy and to explain fossil ape and human morphologies. However, although a large corpus of information exists on human morphological development, information on the extant great apes, particularly the orangutan, remains deficient in areas critical for comparison with relevant fossil material, including the patterns, rates and variability of dental and skeletal development. Such data are particularly applicable in evaluating the life histories and phylogenetic relationships of recently discovered immature fossil apes. Recent studies have described (Winkler, 1987 et seq.) various aspects of orangutan anatomy and dental and somatic development documenting similarities between the orangutan and the other great apes as well as several differences in anatomical structure and developmental pattern. Many great ape similarities may be shared generalized anthropoid traits, whereas the differences can be explained, in many cases, by morphologies or developmental patterns unique to the orangutan. Because of extreme variability in many aspects of orangutan development and morphology, a great deal more information is needed to clarify taxonomic issues.

INTRODUCTION

> "The attention of the medical profession to comparative anatomy, and the interest which the naturalist feels in prosecuting this interesting study, are my inducements for offering the following account of an animal which forms, in the chain of created beings, the connecting link of brutes to man." (Jeffries, 1826).

By the beginning of the nineteenth century, the Western world was aware of the existence of the great apes and was able to differentiate the African apes from the orangutan

The Neglected Ape, Edited by Ronald D. Nadler et al.
Plenum Press, New York, 1995

(see Rohrer-Ertl (1988) for historical review). Although Darwin's revolutionary theory of natural selection (Darwin, 1871), which suggested an apelike ancestry for humans, was not published until the second half of the nineteenth century, earlier scientists also viewed the great apes (especially the orangutan) as a link between humans and other life forms in the animal kingdom (Jeffries, 1826; Linnaeus, 1767). The work of Huxley (1863) and the publication of Darwin's theories reiterated this relationship between apes and humans, albeit apes were now viewed as the source of human origins (Darwin looked to Africa for the origin of humanity, thereby suggesting the African apes were closer to humans than the orangutan).

In order to explore this relationship between the great apes and humans and the relationships among the great apes themselves, numerous studies were completed in the nineteenth and earliest part of the twentieth centuries including studies on orangutan anatomy (Beddard, 1893; Bluntschli, 1929; Boyer, 1935, 39; Brandes, 1929; Delisle, 1895; Deniker and Boulart, 1895; Fick, 1895a, b; Hepburn, 1892; Hrdlicka, 1907; Jeffries, 1826; Krogman, 1931; Lightoller, 1928; Owen, 1830/31, 1843; Primrose, 1900; Schultz, 1933, 1936, 1941; Selenka, 1898; Sonntag, 1924; Straus, 1942; Sullivan and Osgood, 1924/25; Traill, 1818; Trinchese, 1870). Although many of these studies were basic descriptions of anatomical structure, others explored the development of morphological features. Studies on the orangutan became much rarer after the mid-twentieth century due to the widespread opinion that the African apes, specifically the chimpanzee, were humans' closest living relatives (see discussion below) and, therefore, more important to study than the other apes. The discovery of new fossil hominoids and reassessments of taxonomic relationships among Hominoidea have in the last decade regenerated interest in the orangutan (Andrews, 1982; Andrews and Cronin, 1982; Andrews and Tekkaya, 1980; Pilbeam, 1982; Schwartz, 1984, 1988, 1990; Ward and Brown, 1988; Wu, et al., 1982).

These studies of great ape (hominoid) morphology and its development have served a variety of functions including: 1) assessing taxonomic relationships to reconstruct phylogeny, 2) describing morphological structure and investigating its ontogeny, 3) establishing age standards and norms for extant apes, 4) establishing ranges of variation in growth and development, 5) investigating factors affecting growth and development, and 6) producing information on primate life history which by analogy can be used to hypothesize about extinct hominoids. This paper briefly reviews studies of orangutan morphology and development and their relevance to the six functions mentioned above.

TAXONOMIC AND PHYLOGENETIC ASSESSMENTS

Using data from studies of comparative morphology, a number of hypotheses have been suggested regarding the specific relationships between the extant apes and humans over the last century and one half (see Fig. 1). Although all of these hypotheses agreed that the great apes are closer relatives of man than the lesser apes (*Hylobates*), they differ in their conclusions regarding the closeness of relationship among the great apes themselves and between one or more of the great apes and humans. Reconstructing this hominoid phylogeny is heavily dependent on comparisons of morphological character states or anatomy as well as molecular evidence of genetic similarity and developmental similarities between these extant hominoids.

The conclusions of Schultz (1936) (Fig. 1A) reflect traditional opinion which viewed the great apes as composing a monophyletic group (sharing common ancestry) more closely related to each other than any are to humans. Although most scientists today no longer view the great apes as a monophyletic group due to evidence from studies of molecular evolution and genetic distance (Sarich and Cronin, 1976), Kluge (1983) has suggested the need for additional research in resolving hominoid relationships before dismissing this phylogeny.

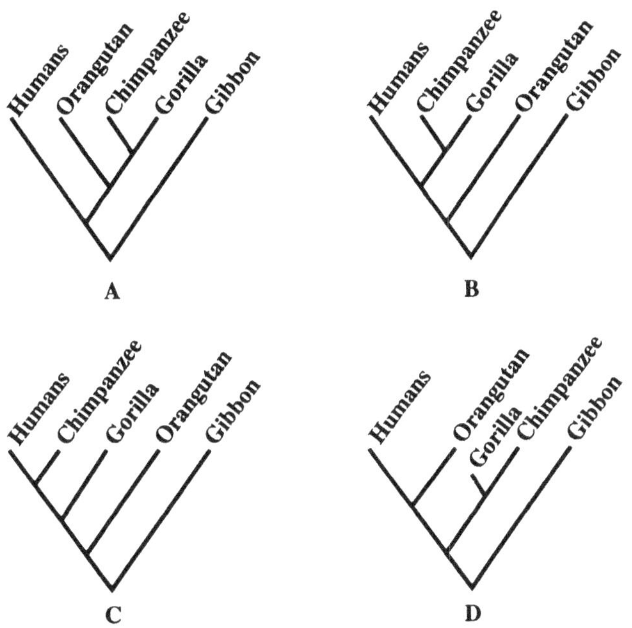

Figure 1. Theories of relationships of extant hominoids.

In contrast, phylogenies which separate the orangutan from the other great apes have become widely accepted in the last twenty years. Szalay and Delson (1979) and several scientists working with molecular evidence of evolution (Goodman, 1976; Sarich and Cronin, 1976) have suggested that the orangutan is more distantly related to humans than the African apes, which are more closely related to each other than either is to humans (Fig. 1B). On the other hand, Weinert (1932), Sibley and Ahlquist (1984) and Yunis and Prakash (1982) (working with chromosomal evidence of evolution) espouse the opinion that the chimpanzee is closest to human, with the gorilla and the orangutan being more distant relatives (Fig. 1C). Groves (1986) provides a thorough review of the evidence for these two alternative phylogenies noting that the evidence supports a chimpanzee-human grouping but also suggests trichotomous ancestral branching (where three share a common ancestor) among humans, chimpanzees, and gorillas. Contrary to the previous two hypotheses, which place the African apes closer to humans than the orangutan, Schwartz (1984) suggests a fourth hypothesis of hominoid relationships based on his reanalysis of comparative morphological data (Fig. 1D) which places the orangutan as human's closest extant relative.

DESCRIBING MORPHOLOGY AND ASSESSING ONTOGENY

As mentioned above, reconstructing evolutionary history or phylogeny is heavily dependent on knowledge of comparative anatomy. Although there have been several excellent descriptions of the skeletal system (see discussion below), most of what is known about orangutan anatomy has been derived from studies of regional anatomy of one or two morphological systems (musculature, teeth, etc.) in a limited number of specimens where descriptions have been, in some cases, somewhat cursory.

	Dental Status[1]	Sex[2]	Distal epiphysis of ulna	Distal epiphysis of radius[3]	Capitulum of radius	Distal epiphysis of tibia	Proximal epiphysis of tibia	Distal epiphysis of fibula	Number of carpels	Number of tarsals
Orangutans										
Newborn	N	M	-	P	?	-	?	-	2	2
18 day old	N	F	-	-	-	-	-	-	2	4
2 month old	N	F	-	-	?	-	-	-	2	4
9 month old	P	F	-	P	?	P	?	P	3	6
12 month old	D	F	P	P	P	P	P	P	5	7
16 month old	D	F	P	P	?	P	P	P	7	7
17 month old	D	F	P	P	P	P	P	P	8	7
6 year old	M	F	P	P	P	P	P	P	9	7
4 year old	M	M	P	P	P	P	P	P	7	7

[1] - Dental Status = N: no teeth in occlusion, P: partial deciduous, D: full deciduous,
M: M1 emerging or in occlusion
[2] - M = male, F = female
[3] - P = present

(Revised from Winkler, in press)

Figure 2. The appearance of the secondary epiphyses in immature orangutans (*Pongo pygmaeus*).

Descriptions of the entire skeleton have been provided by Owen (1836), Trinchese (1870), Delisle (1895), and Schultz (1941). Schultz' (1941) research is particularly noteworthy in that it includes a large cross-sectional sample of all age groups (although the sample is limited for early immature age categories). He provides excellent descriptions of variation and pathology, changes related to aging (including changes in relative proportions of various sections of the body and skull), and the only information other than that of a recent study (Winkler, in press) available on the appearance and fusion of secondary epiphyses. In addition, Schultz (1936, 1941) compares the orangutan with the other great apes, noting that the presence of an os centrale in the orangutan reflects a developmental difference by which this bone fuses with the scaphoid prenatally or neonatally in the other hominoids but remains separate in the orangutan (although it occasionally fuses with the scaphoid in old age in the orangutan). Winkler's manuscript documents the appearance and sequence of the secondary epiphyseal ossification centers of the wrist and ankle (see Fig. 2).

Further information on the skeleton is provided by Hrdlicka (1907), Selenka (1898), and Krogman (1930, 1931), who produced in-depth studies of cranial morphology. Hrdlicka (1907) and Selenka (1898) describe and measure a series of skulls and both note a high degree of variation. In addition, Selenka as well as Krogman (1931) discuss growth changes in cranial dimensions.

Other systems of the body have been less thoroughly studied and described than the skeleton. Initial knowledge of orangutan anatomy comes from the autopsy results and descriptions of the superficial anatomy of three juvenile orangutans by Jeffries (1826) and Owen (1830/31, 1843). Further details of the appendicular musculature of the orangutan have been provided by Primrose (1900), Sigmon (1974), Sigmon and Farslow (1986) and Anderton (1988). Sigmon's (1974, Sigmon and Farslow, 1986) descriptions, particularly that of the hindlimb, are excellent.

The craniofacial muscles have been described by Huber (1931), Lightoller (1928), and Sullivan and Osgood (1924/25), with the studies of Huber and Lightoller being particularly noteworthy in their attempts to compare specimens and account for variation. Huber (1931) also provides descriptions of the male fatty cheek flanges (pads) and their relationship to the craniofacial musculature. Strauss (1942) and Winkler (1989) likewise describe the structure of these fatty cheek pads, with Winkler describing their ontogeny in a sample which included both immature and adult animals. Detailed descriptions of the muscles of mastication are provided by Bluntschli (1929), Boyer (1939), and Winkler (1991). Bluntschli produces excellent illustrations and both he and Winkler note possible age-related changes between immature specimens and the adults. In addition, both he and Winkler discuss relationships between the muscles and bony anatomy and furnish muscle weights for several of the animals.

A few studies have described two or more anatomical systems (circulatory, musculature, skeletal, etc.) and the relationships between them including that of Sonntag (1924), Hill (1939), and Winkler (1987). Sonntag provides a general account of the anatomy of a single young female. Hill describes the external, osteological, and muscular features of an adult male orangutan. Winkler provides a description of craniofacial and cervical anatomy including musculature, nerve supply, circulation, and other soft tissues of this region including a discussion of age-related variation.

The development and morphology of the teeth of the orangutan are final areas which have merited several studies. Knowledge of tooth structure and ontogeny is particularly important for comparison with extinct hominoids since teeth make up a significant portion of the fossil record. In addition, variation in the patterns of primate tooth development (as indicated by the emergence sequences seen in Fig. 3) may be important in assessing the evolutionary patterns of the group.

Descriptions of orangutan tooth morphology have been provided by Swindler and Olshan (1988) and Swarts (1988). Brandes (1928, 1939), Beynon, et al. (1991), Fooden and Izor (1983), Krogman (1930, 1931), Lippert (1977), Seitz (1969), Selenka (1898), Schultz (1941), Singleton (1992) and Winkler, et al. (1991, 1994, in press) discuss various aspects of orangutan tooth development including tooth emergence. Fooden and Izor (1983), Lippert (1977), Schultz (1941), Selenka (1898), Singleton (1992) and Winkler, et al. (1991) provide information on deciduous tooth emergence. The study of Fooden and Izor (1983) is particularly noteworthy in that it documents mean age and range of eruption for each of the deciduous teeth and investigates other variables which might affect tooth development (i.e.,

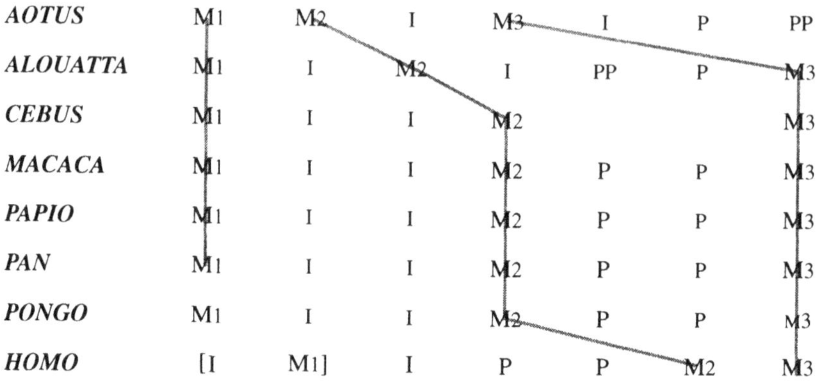

After Schultz (1956) and Smith (1992)

Figure 3. Sequence of emergence of the permanent teeth in primates.

sex of the individual, rearing patterns). Of these studies, Brandes'(1928, 1939) are the only ones to report on the timing of the emergence of the permanent teeth based on longitudinal data, in this case a single animal. (See comments below regarding variability in orangutan tooth development.)

Although not based solely on the orangutan, Dean and Wood (1981) provided a comprehensive study of relative permanent tooth growth among all extant great apes in which they summarize mean eruption times for the three genera and also estimate chronological stages of calcification and eruption. These tabulations have been revised by Beynon, et al. (1991) and Winkler, et al. (1991, 1994, in press) for the orangutan. The latter two groups of authors document earlier calcification commencement times for the permanent dentition than indicated by the work of Dean and Wood (1981), differences between the upper and lower jaws in calcification commencement sequences, the presence of eruption sequence polymorphisms (I1-I2/M2, M3/C), significant variability in states of dental development in specimens of approximately the same age or similar tooth emergence status, and greater variability in the state of root development at emergence than previously indicated. Since evidence of the patterns of permanent tooth development and emergence demonstrates considerable variability in the orangutan (Bond, personal communication, Winkler, et al., (in press), a great deal more research needs to be done in this area.

Although, as indicated above, there have been many descriptive studies of orangutan morphology and, to a lesser degree, the ontogeny of orangutan morphology, much more work remains to be done. The production of a comprehensive description of orangutan anatomy, integrating the anatomical systems in each of the regions of the body (similar to Raven's (1950) *The Anatomy of the Gorilla* or Swindler and Wood's (1973) *An Atlas of Primate Gross Anatomy* would be useful. Longitudinal studies of tooth and skeletal development and general morphological changes throughout development, particularly beyond infancy, are also needed in order to have data for comparison with other hominoids and for assigning age to wild orangutans and orangutans in rehabilitative programs destined for return to the wild.

In addition, studies integrating developmental data from two or more anatomical regions are also needed. This sort of study has been attempted by Winkler (in press), comparing orangutan dental development with carpal and tarsal development (see Fig. 4), and by Smith (1992), who examined correlations between the timing of developmental events such as the emergence of M1 with those of other morphological systems in primates.

	NO TEETH	PARTIAL DECIDUOUS	ALL DECIDUOUS	M_1^1 ERUPTED
Carpals	1-2 / 2-3	? / 3-5	3-7 / 5-8	8 / 7-9
Tarsals	2-5 / 2-4	5 / 6	6-7 / 6-7	7 / 7

Figure 4. Associations between the appearance of chimpanzee/orangutan ossification centers and dental eruption (after Winkler, in press).

The studies of anatomy mentioned above indicate that the orangutan differs from the other great apes and from humans in a number of ways. Some of these differences are specialized structures such as the male fatty cheek flanges, whereas others are differences in anatomical features, such as the lack of an anterior belly of the digastric muscle, differences in facial sinus morphology and facial morphology, and the presence of a scansorius muscle. Studies of ontogeny can often clarify whether such differences are constant throughout the life of an individual of that species or whether they reflect ontogenetic changes. Schultzs (1936) explanation of the presence of an unfused os centrale in the orangutan (as mentioned above) demonstrates the merits of such studies.

More studies are also needed to investigate variability. Many of the previously discussed studies suggest that the orangutan is variable in its morphological patterns. This variation includes functional changes related to age and the growth of the skeleton (e.g., changes in proportional muscle size, expansion of muscle attachment area, tooth eruption, etc.), sexual differences (e.g., cheek flanges, laryngeal air sacs, and other dimorphic differences such as larger muscles with larger attachment points which reflect the larger size of adult males), and interindividual variability. Interindividual variability is substantial in some muscle groups such as the muscles of facial expression (see Winkler, 1987). Additional anatomical studies may clarify where variation exists and the ranges of this variation.

ESTABLISHING AGE STANDARDS AND RANGES OF VARIABILITY

Knowledge of what constitutes a normal growth pattern is crucial to assessing the health of living animals. Since "normal" development frequently refers not only to the most common pattern (the mean) or growth trajectory of any population, but also to variants which constitute the range of what is considered normal, determining the ranges of variation in morphologies and their development (including both the patterns and rates of development) becomes extremely important. Although there have been studies examining morphological variation throughout the developmental period (see above), many of these were based on cross-sectional samples of animals of unknown ages. In instances where age is unknown, dental emergence status has been used to sort specimens into categories which are taken as reflective of age (Schultz, 1941; Winkler, 1987; Winkler, et al., 1991). These studies have demonstrated fairly extensive variation in development within any given dental category.

Figures 5-10 demonstrate this variation in cranial dimensions and dental development relative to dental emergence or dental development categories. Figures 5-8 are based on a cross-sectional sample of 182 orangutans (see Winkler, 1987 for details), ten of which are of known age and are plotted on the bar graphs for the sake of reference. However, since dental eruption itself varies by age, these studies using the dentition as an age indicator do not establish age norms and variation around these norms, but rather variability of one developmental system relative to the developmental status of another

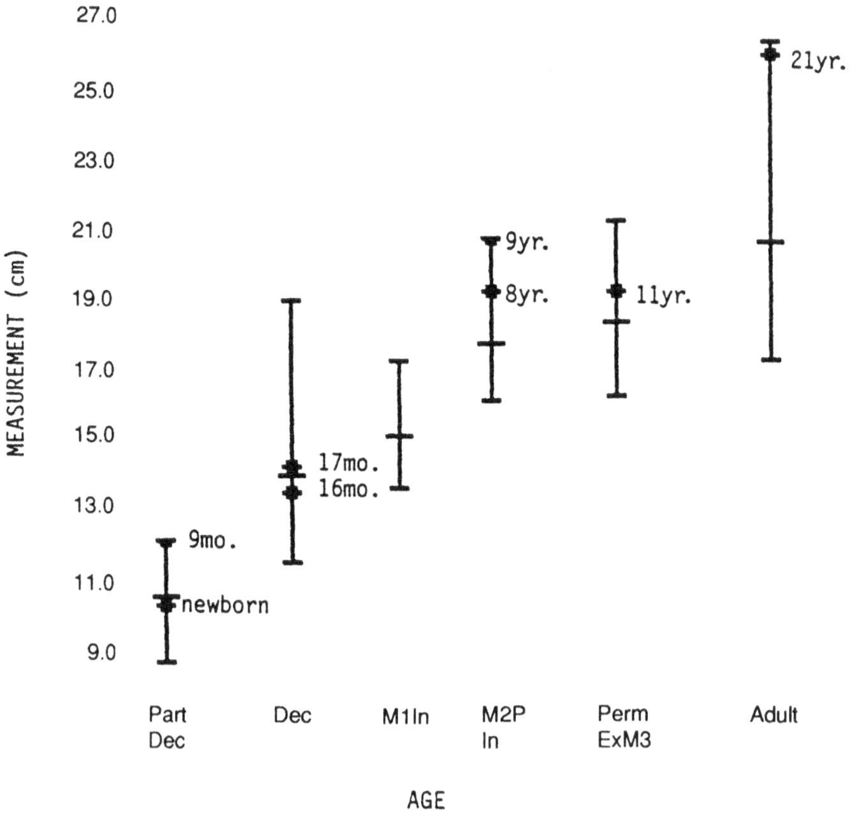

Figure 5. Skull length.

(see Figs. 9, 10). One of the few studies available which does establish age norms and variation around these norms is that of Fooden and Izor (1983) mentioned above. They provide developmental information for deciduous tooth emergence, weight gain, and linear dimensions from a relatively large sample of orangutans from the Brookfield Zoo and Lincoln Park Zoological Gardens.

In addition to studies of anatomical morphology as discussed above, several studies document changes in behavior and indicators of sexual maturity throughout orangutan development (Fooden and Izor, 1983; Graham, 1988; Kingsley, 1988; MacKinnon, 1974; Maple, 1980; Rijksen, 1978; Rodman, 1988). Graham (1988) discusses reproductive anatomy and physiology and some aspects of its development (including hormonal relationships, menstrual cycle length, etc.). Kingsley (1988) reviews the onset of sexual maturity in male orangutans and hormonal factors influencing this onset. Both Graham (1988) and Kingsley (1988) note that the timing of reproductive maturity and the development of secondary sexual characteristics may differ in animals raised in captivity compared to those in the wild.

Another intriguing aspect of orangutan sexual development which merits further study is the timing of testicular descent. Schultz (1941) and Fooden and Izor (1983) have both suggested that in male orangutans, the testes may descend into the scrotum during infancy to then be subsequently retracted until adolescence. If true, this may have implications for male fertility and a wide range of secondary sexual characteristics.

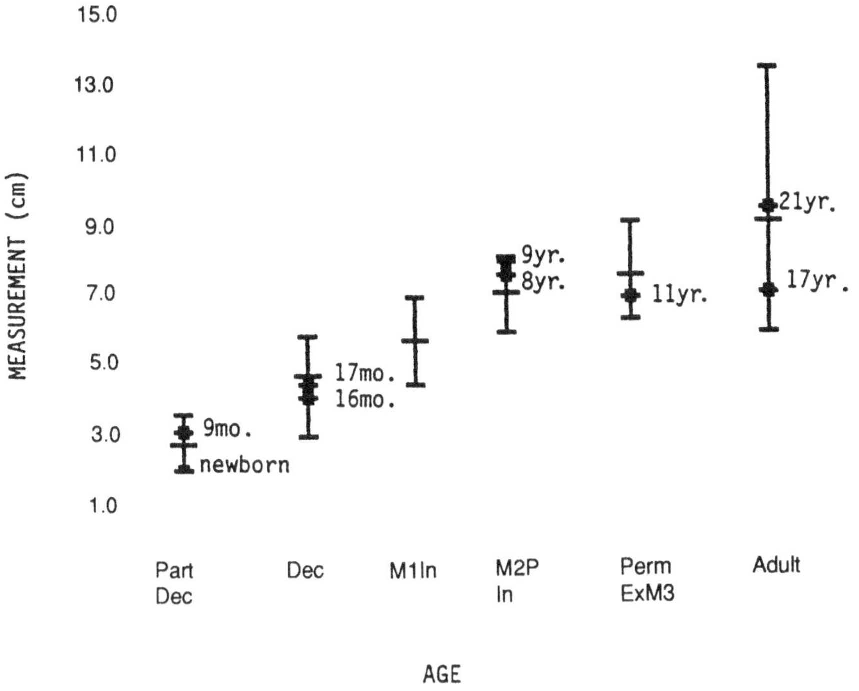

Figure 6. Upper facial height.

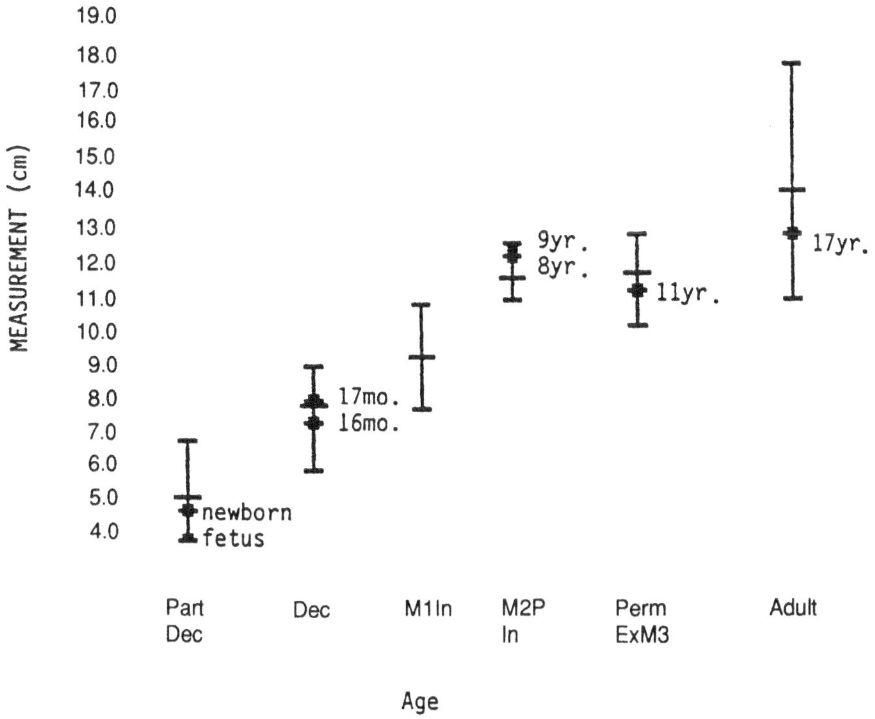

Figure 7. Bizygomatic breadth.

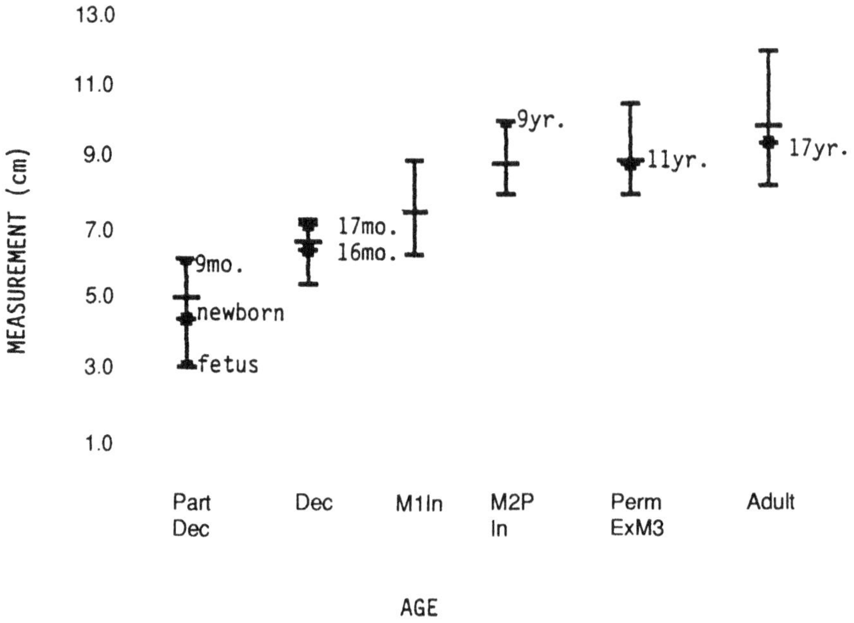

Figure 8. Biorbital breadth.

FACTORS AFFECTING DEVELOPMENT

In addition to the studies of Kingsley (1988) and Graham (1988) mentioned above, several other recent studies of great ape development have suggested possible differences in development between mother-reared and bottle-reared animals (Fooden and Izor, 1983; Marzke, et al., 1993, 1995; Winkler, et al., 1991; Winkler, in press) and between animals raised in captivity and those raised in the wild (Maple, 1980). Much more research is needed to determine the scope of these differences and to seek explanations for their occurrence. For example, although Graham (1988), Kingsley (1988), and Nadler (1988) have explored

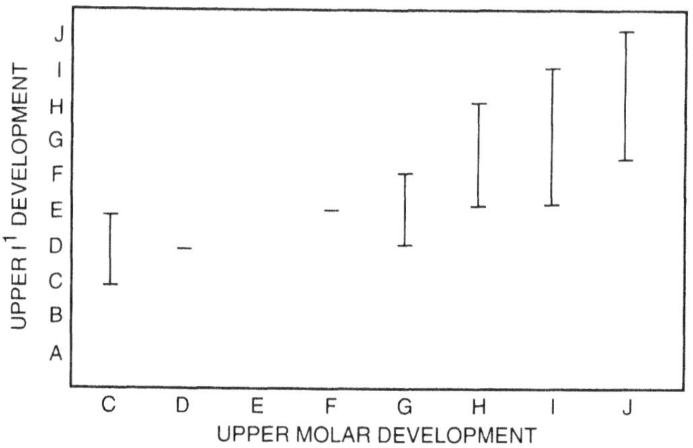

Figure 9. Upper I^1 development relative to M^1 (after Winkler, et al. in press).

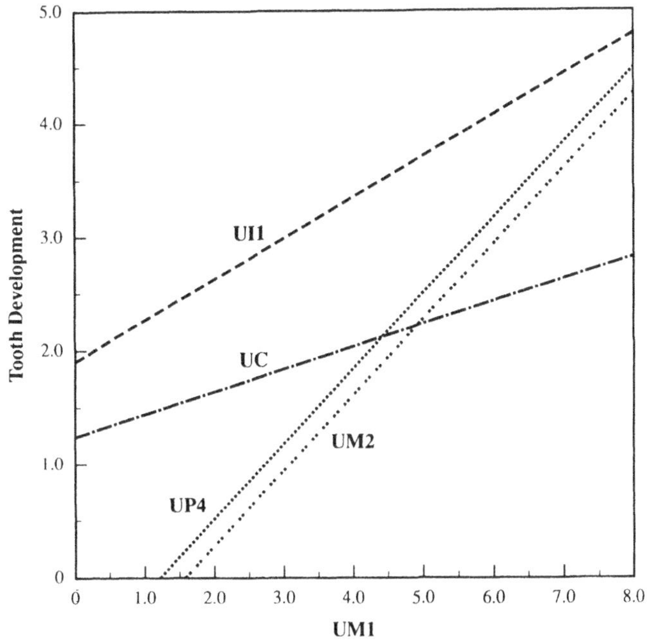

Figure 10. Upper tooth development relative to M^1 (after Winkler, et al. in press).

the hormones associated with sexual development in captivity, such information is unavailable for wild orangutans. This information may help explain why puberty and the first pregnancy in females is reported to occur slightly later in the wild than in captivity (Galdikas, 1981; Graham, 1988). The suggestions of developmental differences between animals which are bottle or mother-reared and between those raised in captivity and those raised in the wild certainly indicate that caution is needed when hypothesizing about behavior and development of animals in the wild when using evidence obtained from captive animals.

LIFE HISTORY AND THE FOSSILS

The discoveries of new hominoid fossils and reclassification of others have regenerated interest in orangutan life history in recent years (Kelley, 1993). Many scientists (Andrews and Cronin, 1982; Ward and Brown, 1988) have proposed a close phylogenetic link between the fossil ape *Sivapithecus* and the orangutan. The discovery of the Asian fossil *Lufengpithecus*, which resembles *Sivapithecus*, and the recent suggestion that *Dryopithecus*, a European fossil ape, is also closely related to the orangutan (Martin and Andrews, 1993; Sola and Kohler, 1993) have raised further questions regarding the taxonomic relationships between the fossil and extant Asian apes.

Attempts to sort out these relationships are heavily dependent on comparative anatomy and also on comparisons of the biological history of a taxon (its life history)(Harvey and Clutton-Brock, 1985). The work of Schultz (1936) has been one of the first to compare life history data among primates, demonstrating the trend throughout the group toward expansion of the child dependency period (and, hence, increasing and enhancing time spent in the development of learned behavior). As indicated above, knowledge of many aspects of orangutan life history (e.g., skeletal development, dental

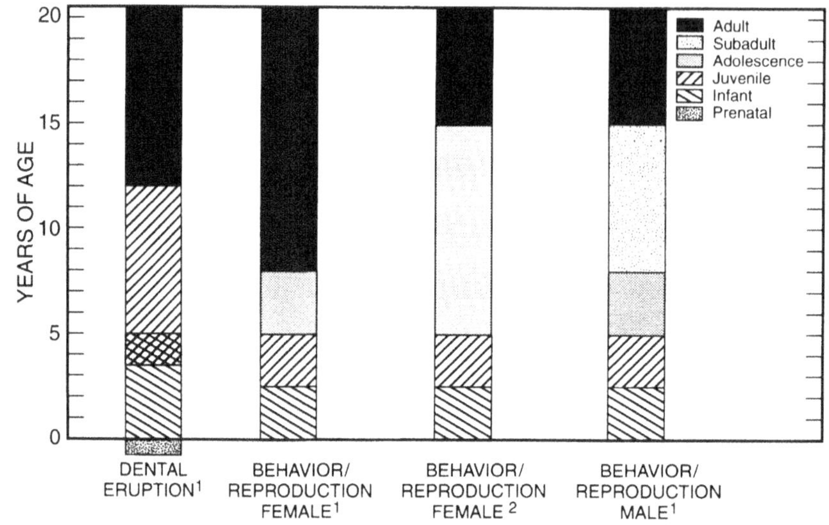

[1]Based in part on Brandes (1939), Fooden and Izor (1983), MacKinnon (1974), Maple (1980), Rijksen (1978), and Schultz (1941).
[2]Based on new field data from Leighton (1994) and the Orangutan Population and Habitat Viability Analysis Workshop (1993).

Figure 11. A comparison of different types of life history data from the orangutan.

development), particularly that of animals in the wild, is very limited. Although the life history sequences presented in Figure 11 are primarily based on captive animals, new information on female reproductive variables is included (Leighton, et al., 1994; Markham, 1994; Orangutan Population and Habitat Viability Analysis Workshop, 1993, this volume). As indicated in this figure, recent assessments of orangutan development in the wild suggest different development parameters than those of animals raised in captivity. More information is needed on the life history variables of wild animals since they remain the ideal source for comparison with the fossil evidence.

In addition, some caution is needed to ensure that life history data comparisons across taxa are done with homologous data. Major life history events (M1 eruption, first breeding age, epiphyseal appearance and fusion, completion of the permanent dentition) follow different trajectories. And, as demonstrated in Figure 11, growth and development cease at different times in different morphological systems and may even be different between the sexes. What precisely is being compared needs to be clarified before using life history data in discussing phylogeny and in hypothesizing about the life history of extinct fossils.

CONCLUSIONS

Although much has been done to describe the anatomy and the development of the morphological systems of the orangutan in the last two centuries, more work remains. Longitudinal information is needed on the growth and development of the skeleton, teeth and other morphological systems. Attempts should be made to clarify the factors affecting this growth. And, although several studies have provided life history data on the orangutan, much more information is needed, particularly for orangutans in the wild.

ACKNOWLEDGMENTS

I wish to thank the sponsors of the International Conference on the Orangutan (California State University/Fullerton, the Zoological Society of San Diego, and the Atlanta/Fulton County Zoo, Inc.) for inviting me to present a paper and be part of the conference and for financial support. I am also grateful to Lori Perkins, Lori Sheeran and Jackie Ogden for their assistance in making arrangements, etc. and to Melanie Bond for her knowledge of orangutans and support of my research. Many thanks to Mrs. Cynthia Strickland, Mr. Dan Smith, Dr. Philip Atteberry, and Ms. Penny Horn for a variety of assistance in manuscript presentation. I also wish to thank Drs. Mary Marzke and Jo Fritz and the Primate Foundation of Arizona for allowing the use of unpublished figures in my oral presentation. Figures 1, 3, 4, 9 - 11 were prepared by Mr. Bill Johnston and Mr. Bill Brent of the University of Pittsburgh. Funding was provided by the University of Pittsburgh and the Conference Committee. Much of my own research cited in this paper was based on material obtained from Yerkes Regional Primate Center, Emory University (NIH RR-00165). In addition, I very gratefully acknowledge information on known-age orangutans provided by Utah's Hogle Zoo (courtesy of Kim Davison) and by The Zoo of Gainesville, Florida (courtesy of Linda Pastorello).

REFERENCES

Anderton, J., 1988, Anomalies and atavisms in appendicular myology. In J. Schwartz (ed.): *Orang-utan Biology*, New York: Oxford University Press, pp. 331-345.

Andrews, P., 1982, Hominoid evolution. *Nature* 295:185-86.

Andrews, P. and Cronin, J., 1982, The relationship of *Sivapithecus* and *Ramapithecus* and the evolution of the orangutan. *Nature* 297:541-546.

Andrews, P. and Tekkaya, I., 1980, A revision of the Turkish Miocene hominoid *Sivapithecus meteai. Palaeontology* 9:85-95.

Beddard, F., 1893, Contributions to the anatomy of the anthropoid apes. *Trans. Zool. Soc. Lond.* 13:177-218.

Beynon, A. D., Dean, M. C., and Reid, D. J., 1991, Histological study on the chronology of the developing dentition in gorilla and orangutan. *Am. J. Phys. Anthropol.* 86:189-204.

Bluntschli, H., 1929, Die Kaumuskulaturdes Orang-Utan und Ihre Bedeutung fur die Formung des Schadels I. Teil: Das morphologische Verhalten. *Gehen. Morphol. Jahrb.* 63: 531-606.

Boyer, E., 1935, The musculature of the inferior extremity of the orangutan *Simia satyrus. Am. J. Anat.* 56:193-256.

Boyer, E., 1939, The cranio-mandibular musculature of the orang-utan *Simia satyrus. Am. J. Phys. Anthropol.* 24:417-427.

Brandes, G., 1928, Der Durchbruch der Zahne beim orang-utan. *Zool. Gart.* 1:25-28.

Brandes, G., 1929, Die Backenwulste des Orang-Mannes. *Zool. Gart.* 1:365-368.

Brandes, G., 1939, *Buschi: Vom Orang-Saugling zum Backenwulster*, Leipzig: Verlagsbuchhandlung Quelle and Meyer.

Darwin, C., 1871, *The Descent of Man, Vol. 1*, London: John Murray.

Dean, M., and Wood, B., 1981, Developing pongid dentition and its use for ageing individual crania in comparative cross-sectional growth studies. *Folia Primatol.* 36:111-127.

Delisle, F., 1895, Notes sur l'osteometrie et la Craniologie des Orang-Outans. *Nouv. Arch. Mus. d'Hist. nat.* 7, 3 ser:83-118.

Deniker, J. and Boulart, R., 1895, Notes Anatomiques sur les saces laryngiens, les excroissances adipeuses, les poumonts, les cerveaux, etc. des Orang-Outans. *Nouv. Arch. Mus. d'Hist. nat.* 7:35-56.

Fick, R., 1895a, Beobachtungen an einan zweiten erwachsenen Orang-Utang und einenem Schimpansen. *Arch. Anat. Physiol. Anatomische Abtheilung.* 289-318.

Fick, R., 1895b, Vergleichend anatomische studien an einen erwachsenen Orang-Utang. *Arch. Anat. Physiol. Anatomie Abtheilung.* 1-100.

Fooden, J. and Izor, R., 1983, Growth curves, dental emergence norms, and supplementary morphological observation in known-age captive orangutans. *Am. J. Primatol.* 5:285-301.

Galdikas, B., 1981, Orangutan reproduction in the wild. In C. E. Graham (ed.): *Reproductive Biology of the Great Apes*, New York: Academic Press, pp. 281-300.

Goodman, M., 1976, Protein sequences in phylogeny. In F. Ayala (ed.): *Molecular Evolution*, Sunderland, MA: Sinauer Assoc., Inc., pp. 141-159.

Graham, C.E., 1988, Reproductive Physiology. In J. Schwartz (ed.): *Orang-utan Biology*, New York: Oxford University Press, pp. 91-104.

Groves, C., 1986, Systematics of the great apes. In D. Swindler and J. Erwin (eds.): *Comparative Primate Biology, Vol. 1*, New York: Alan R. Liss, pp. 187-217. @RefAuSty = Harvey, P. and Clutton-Brock, T., 1985, Life history variation in primates. *Evolution* 39:559-81.

Hepburn, D., 1892, The comparative anatomy of the muscles and nerves of the superior and inferior extremities of the anthropoid apes. *J. Anat. Physiol.*, London 26:149-186, 324-356.

Hill, W., 1939, Observations on a giant Sumatran orang. *Am. J. Phys. Anthropol.* 24:449-505.

Hrdlicka, A., 1907, Anatomical observations on a collection of orang skulls from western Borneo. *Proceedings of the United States National Museum*. 31:539-568.

Huber, E., 1931, *Evolution of Facial Musculature and Facial Expression*, Baltimore: John Hopkins Press.

Huxley, T., 1863, *Man's Place in Nature*, New York: D. Appleton and Co.

Jeffries, J., 1826, Some accounts of the dissection of *Simia satyrus* orang-outang, or wildman of the woods. *Philosophical Magazine and Journal* 67:182-190.

Kelley, J., 1993, Life history profile of *Sivapithecus*. *Am. J. Phys. Anthropol.* Supplement 16:123.

Kingsley, S., 1988, Physiological development of male orang-utans and gorillas. In J. Schwartz (ed.): *Orang-utan Biology*, New York: Oxford University Press, pp. 123-132.

Kluge, A., 1983, Cladistics and the classification of the great apes. In R. Ciochon and R. Corruccini (eds.): *New Interpretations of Ape and Human Ancestry*, New York: Plenum Press, pp. 151-177.

Krogman, W., 1930, Studies in growth changes in the skull and face of anthropoids. I. The eruption of the teeth in anthropoids and old world apes. *Am. J. Anat.* 46:303-313.

Krogman, W., 1931, Studies in the growth changes in the skull and face of the orangutan. *Am. J. Anat.* 47:343-365.

Leighton, M., Seal, U., and Tilson, R., 1994, Population viability analysis of orangutans in Gunung Leuser National Park. *Paper presented at Int. Orang utan Conf., The Neglected Ape.*

Lightoller, G., 1928, The facial muscles of three orang-utans and two cercopithecidae. *J. Anat.* 63:19-81.

Linnaeus, C., 1767, *Systema Naturae, per Regna etc., Tomus I, Editio Decima Tertia ed Editionem Duodecimam Reformatum Holmiensem*, Vindobona: Thoma de Trattnern.

Lippert, W., 1977, Erfahrungen bei der Aufzucht von Orang-Utans (*Pongo pygmaeus*, im Tierpark Berlin. *Zool. Gart.* 47:209-225.

MacKinnon, J., 1974, The behavior and ecology of wild orang-utans (*Pongo pygmaeus*). *Anim. Behav.* 22:3-74.

Maple, T., 1980, *Orangutan Behavior*. New York: Van Nostrand Reinhold.

Markham, Rosemary, 1995, Doing it naturally: reproduction in captive orangutans (*Pongo pygmaeus*), this volume.

Martin, L. and Andrews, P., 1993, Renaissance of Europe's ape. *Nature* 365:494.

Marzke, M., Hawkey, D., Young, D., and Fritz, J., 1993, Comparative analysis of weight gain, hand/wrist maturation, and dental emergence rates in chimpanzees in varying captive environments. *Am. J. Phys. Anthropol.* Suppl. 16:139.

Nadler, R.D., 1988, Sexual and reproductive behavior. In J. Schwartz (ed.): *Orang-utan Biology*, New York: Oxford University Press, pp. 105-116.

Orangutan Population and Habitat Viability Analysis Workshop., 1993, Medan, North Sumatra.

Owen, R., 1830/1831, On the anatomy of the orang utan (*Simia satyrus*, L.). *Proc. Zool. Soc. Lond.* 1930/31:9-10.

Owen, R., 1836, On the specific distinctions of the orangs. *London and Edinburgh Philosophical Magazine* 10:295-301.

Owen, R., 1843, Notes on the dissection of a female orang-utan (*Simia satyrus*, Linn.). *Proc. Zool. Soc. Lond.* 1843:123-124.

Pilbeam, D., 1982, New hominoid skull material from the Miocene of Pakistan. *Nature* 295:232-34.

Primrose, A., 1900, *The Anatomy of the Orang Outan, Toronto (University of Toronto Studies, Anatomical Series 1.*

Raven, H., 1950, *The Anatomy of the Gorilla*. New York: Columbia University Press.

Rijksen, H., 1978, *A Field Study on Sumatran Orang Utans (Pongo pygmaeus abelii* Lesson 1827), Wageningen, the Netherlands: H. Veenman and Zonen B.V.

Rohrer-Ertl, O., 1988, Research history, nomenclature, and taxonomy of the Orang-utan. In J. Schwartz (ed.): *Orang-utan Behavior*, New York: Oxford University Press.

Rodman, P., 1988, Ecology and behavior of orangutans. In J. Schwartz (ed.): *Orang-utan Behavior*, New York: Oxford University Press.

Sarich, V. and Cronin, J., 1976, *Molecular systematics of the primates*. In M. Goodman and R. Tashian (eds.): Molecular Anthropology, New York: Plenum Press, pp. 141-170.

Schultz, A., 1933, Notes of the growth of anthropoid apes. Report of the Laboratory and Museum of Comparative Pathology. *Zoo. Soc. Philadelphia* 34-45.

Schultz, A., 1936, Characters common to higher primates and characters specific to man. *Quart. Rev. Biol.* 11:259-283; 425-455.

Schultz, A., 1941, Growth and development of the orangutan. *Contrib. Embryol.*, Carneg. Inst. 29:57-110.

Schwartz, J., 1984, The evolutionary relationships of man and orang-utans. *Nature* 308:501-505.

Schwartz, J., 1988, *Orang-utan Biology*. New York: Oxford University Press.

Schwartz, J., 1990, *Lufengpithecus* and its potential relationship to an orang-utan clade. *J. Hum. Evol.* 19:591-605.

Seitz, A., 1969, Einige Feststellungen zur Pflege und Aufzucht von Orang-Utans, *Pongo pygmaeus* Hoppius 1763. *Zool. Gart.* 36:225-245.

Selenka, E., 1898, Rassen, Schadel und Bezahnung des Orangutan. *Studien uber Entwickelung und Schadelbau* 1:1-99.

Sibley, C. and Ahlquist, J., 1984, The phylogeny of the hominoid primates, as indicated by DNA-DNA hybridization. *J. Molec. Evol.* 20:2-15.

Sigmon, B., 1974, A functional analysis of pongid hip and thigh musculature. *J. Hum. Evol.* 3:161-185.

Sigmon, B. and Farslow, D., 1986, The primate hindlimb. In D. Swindler and J. Erwin (eds.): *Comparative Primate Biology Vol. 1: Systematics, Evolution, and Anatomy*, New York: Alan R. Liss, Inc., pp. 671-718.

Singleton, I., 1992, A brief account of the hand rearing and development of a female Bornean orangutan *Pongo pygmaeus pygmaeus* at the Jersey Wildlife Preservation Trust. *Dodo. J. Wildl. Preserv. Trust* 28:78-84.

Smith, B., 1992, Life history and the evolution of human maturation. *Evol. Anthropol.* 1:134-142.

Sola, S. M. and Kohler, M., 1993, Recent discoveries of *Dryopithecus* shed new light on evolution of great apes. *Nature* 365:543-545.

Sonntag, C., 1924, On the anatomy, physiology,and pathology of the chimpanzee. *Proc. of the Zool. Soc. Lond.* 1:323-429.

Strauss, W., 1942, The structure of the crown-pad of the gorilla and of the cheek-pad of the orang-utan. *J. Mammal.* 0:276-281.

Sullivan, W. and Osgood, C. (1924/25, The facialis musculature of the orang, *Simia Satyrus*. *Anat. Rec.* 29:195-243.

Swarts, D., 1988, Deciduous dentition: implications for hominoid phylogeny. In J. Schwartz (ed.): *Orang-utan Behavior*, New York: Oxford University Press, pp. 263-270.

Swindler, D. and Olshan, A., 1988, Comparative and evolutionary aspects of permanent dentition. In J. Schwartz (ed.): *Orang-utan Behavior*, New York: Oxford University Press, pp. 271-282.

Swindler, D. and Wood, C., 1973, *An Atlas of Primate Gross Anatomy*. Seattle: University of Washington Press.

Szalay, F. and Delson, E., 1979, *Evolutionary History of the Primates*, New York: Plenum Press.

Traill, T., 1818, Observations on the anatomy of the orang-outang. *Mem. Mernerian Nat. Hist. Soc. Edinburgh* 3:1-49.

Trinchese, S., 1870, Descrizione di un feto di Orang-Utan. *Annali del Museo Civico di Storia Naturale*. December, 1870.

Ward, S. and Brown, B., 1988, Basicranial and facial topography in *Pongo* and *Sivapithecus*. In J. Schwartz (ed.): *Orang-utan Biology*, New York: Oxford University Press, pp. 247-260.

Weinert, H., 1932, *"Ursprung der Menschheit." Ueber den engeren Anschluss des Menschen-geschlechtes an die Menschenaffen*. Stuttgart.

Winkler, L., 1987, *Age Related Changes in the Craniofacial Anatomy of the Orangutan*. Ph.D. Dissertation: University of Pittsburgh.

Winkler, L., 1989, The fatty cheek pads of the orangutan and their relationship to facial musculature. *Am. J. Primatol.* 17:305-319.

Winkler, L., 1991, The morphology and variability of the masticatory structures of the orangutan. *Int. J. Primatol.* 12:45-65.

Winkler, L., 1991, Aspects of dental development in the orangutan prior to the eruption of the permanent dentition. *Am. J. Phys. Anthropol.* 86:255-272., with J.H. Schwartz and D.R. Swindler)

Winkler, L., Schwartz, J., and Swindler, D., 1994, Variation in patterns of orangutan tooth formation: implications for comparing hominoids., abstract, *Am. J. Phys. Anthropol.* Supplement 18.

Winkler, L., Schwartz, J., and Swindler, D., in press, The development of the orangutan permanent dentition: assessing patterns and variation in tooth development. *Am. J. Phys. Anthropol.*

Winkler, L., in press, Relationships between the appearance of ossification centers of the wrist and ankle and dental development in immature orangutans and chimpanzees. *Am. J. Phys. Anthropol.*

Wu, R., Han, D., Xu, Q., Qi, G., Lu, Q., Pan, Y., and Chen, W., 1982, More *Ramapithecus* skulls found from Lufeng, Yunnan-report on the excavation of the site in 1981. *Acta Anthrop. Sinica* 5:26-30.

Yunis, J. and Prakash, O., 1982, The origin of man: A chromosomal pictoral legacy. *Science* 215:1525-1530.

25

GENETIC VARIABILITY IN ORANGUTANS

C. Muir,[1] B. M. F. Galdikas,[2] and A. T. Beckenbach[1]

[1] Department of Biological Sciences
Institute of Molecular Biology and Biochemistry and
[2] Department of Archaeology
Simon Fraser University
Burnaby, B.C. V5A 1S6, Canada

INTRODUCTION

Population Subdivision

The orangutan (*Pongo pygmaeus*) once ranged widely over Southeast Asia and Indonesia. The species is now reduced to the islands of Sumatra and Borneo. Within Borneo, continued habitat destruction is fragmenting the remaining wild populations into isolated subpopulations or demes. These two processes, overall reduction in population numbers and fragmentation of surviving populations, pose the greatest threat to the ultimate survival of endangered species in the wild.

Part of that threat lies in the loss of genetic variation. Severe reduction in population numbers results in the immediate loss of rare alleles (genetic variants) at most loci. If population numbers remain low through a sequence of generations, genetic drift becomes important. Genetic drift is the random, rather than adaptive, change in allele frequencies and ultimately results in the loss of variation. Eventually all members carry the same allele, which is then said to be "fixed" in the population. If populations are very small, drift may even fix mildly deleterious alleles, reducing the absolute fitness of the populations. In any event, loss of genetic variation reduces the ability of the species to adapt to changing environmental conditions.

Population subdivision accelerates the loss of variation through genetic drift within subpopulations. At the species level it has the opposite effect; allelic variation lost in one subpopulation may be fixed in others and ultimately restored through migration. It is essential, therefore, to obtain an accurate picture, not only of the total population numbers, but also of the degree of subdivision and levels of migration between subpopulations.

Population subdivision may have important ecological effects as well. W.C. Allee noted that for many animal species there is a minimum sustainable population density (see Allee, et al., 1949, pp. 395-403). If a population is reduced below that number, social or other interactions necessary for successful reproduction are lacking and the population will continue to decline to extinction. This observation is sometimes called the "Allee effect".

The Neglected Ape, Edited by Ronald D. Nadler et al.
Plenum Press, New York, 1995

The Allee effect may be exacerbated by population subdivision. Even if the total population numbers are adequate, local demes may be too small to reproduce. Thus the species as a whole may continue to decline. It is essential, therefore, to understand the sizes of local breeding populations (N_e - the effective population size) to evaluate the risk that subdivision has reduced local breeding populations below minimal sustainable size.

Heirarchical Statistics

Assessment of population subdivision is more than a simple count or mapping of individuals. It is essential to analyse the breeding structure and extent of migration using genetic techniques. Wright's F statistics were developed specifically for this purpose (see Hartl and Clark, 1989, pp. 293-300). These statistics allow for the partitioning of genetic variation into inbreeding at the individual level and at the levels of the deme and subpopulation, relative to the species as a whole. The analysis can also be used to interpret subdivision within Borneo in relation to the divergence between remnant populations on Borneo and Sumatra.

Genetic Loci

Genetic variation can be analysed from either the nuclear genome or the mitochondrial genome. The nuclear genome is biparentally inherited, with equal contributions from each parent. The mitochondrial genome, while present in both sexes, is inherited (almost) exclusively from the mother (reviewed in Wolstenholme, 1992). Analysis of mitochondrial genes is simpler, for a variety of reasons, and can provide valuable information regarding genetic subdivision of populations. It cannot, however, provide information about migratory exchange of genetic information if migrants are predominantly males. Because of the social structure of orangutans, much of the migration may indeed be males: non-resident males may travel in loosely organized sib-structured packs (Galdikas, 1985). In this study we examined mitochondrial variation. We intend to expand the work to include representative nuclear genetic loci, as well.

Sampling DNA from Endangered Species

One of the major problems faced when attempting a genetic study of endangered and reclusive species is the source of DNA samples. A relatively new technique, the Polymerase Chain Reaction (PCR) (Mullis and Faloona, 1987), allows the production of highly purified DNA of selected genes from extremely small samples. The samples may even be partially degraded or contaminated with bacterial or other DNA sources. The technique relies on short pieces of DNA (oligonucleotide primers) that are designed for specific orangutan sequences, to amplify selectively the chosen sequences. Still, some form of tissue sample is necessary, often requiring direct invasive contact with the animal. A single drop of blood can provide sufficient DNA for genetic characterization of an individual. For wild individuals, sampling requires the use of a tranquilizer dart to immobilize the animal. A less invasive option involves the use of a plug biopsy dart, which can be used to tear off a piece of tissue from a distance. Even this technique, however, is somewhat invasive.

We have been using two other DNA sources which do not require direct contact with the animal: hair follicles which cling to individual hairs and may be collected from nests (Vigilant et al., 1989; Morin, 1993) and fecal samples which can simply be collected from the ground. From the standpoint of conservation biology and the goal of minimizing the impact on endangered species, feces are an ideal source of DNA. From a technical standpoint feces are perhaps less ideal. The DNA is highly contaminated with bacterial and dietary items. Feces, however, also contain cells from the orangutan's body: sloughed epithelial cells from the colon, white blood cells and other damaged cells that are being expelled from the body. PCR amplification can be extremely selective. With carefully chosen primers and optimized conditions, the technique ignores the contaminating DNA and amplifies only the selected orangutan sequences.

METHODS

Blood samples were obtained from 24 wild born orangutans from a number of different locations. Samples were stored at -70°C. DNA extractions were performed with IsoQuick® extraction kit (MicroProbe Corporation) following the manufacturer's suggested protocol.

Hair samples were obtained from a number of wild born and captive orangutans. DNA was extracted from hair follicles by first suspending the follicle and ~1 cm of shaft in 200 μl of a 5% Chelex 100® (BioRad) mixture followed by a 20 min incubation at 56°C. The samples were then vortexed vigorously for 10 sec, followed by boiling for 8 min. The samples were then centrifuged for 3 min at top speed (15,000 g) in a microfuge. Twenty μl of the supernatant was isolated from the top of the suspension.

Feces were obtained from 12 wild born orangutans. About 10 gm of feces from each individual were suspended in 70% EtOH and stored at -20°C. One hundred μl of the liquid from the bottom (with sediment) of the tube was used as source for DNA extraction with the IsoQuick® kit according to manufacturer's suggested protocol.

PCR

Primers for amplifying and sequencing three regions of the mitochondrial genome were based on the human mtDNA sequence (Anderson, et al., 1981). Primer positions are indicated according to the 3′ nucleotide position (H = heavy strand; L = light strand): D-loop region, L15996, H00607; ND3 gene, L10026, H10433; COII-ATPase 8 junction, L08124, H08366. These regions were chosen based on the likelihood of intraspecific sequence variation.

Amplifications were run for 32 cycles at 94°C (1.5 min), 52°C (1 min) and 72°C (1.5 min) using a Precision Scientific GTC-2 Genetic Thermal Cycler. These cycles were followed by a single cycle with a 5 min extention at 72°C.

Sequencing

Gel-isolated PCR products were used as template for double strand sequencing reactions, using Sequenase® PCR Product sequencing kits (US Biochemical), following the manufacturer's suggested protocol. The PCR primers were used to prime the sequencing reactions.

RESULTS

Blood, hair and feces all proved to be reliable sources of DNA for PCR analysis. Figure 1 shows amplified PCR products from DNA extracted from each of these sources. Amplification with the ND3 primers yielded fragments of the expected size (about 450 base pairs) from each extraction. Preliminary sequence analysis verified that the amplified products were from the animal sampled, rather than any contaminating source.

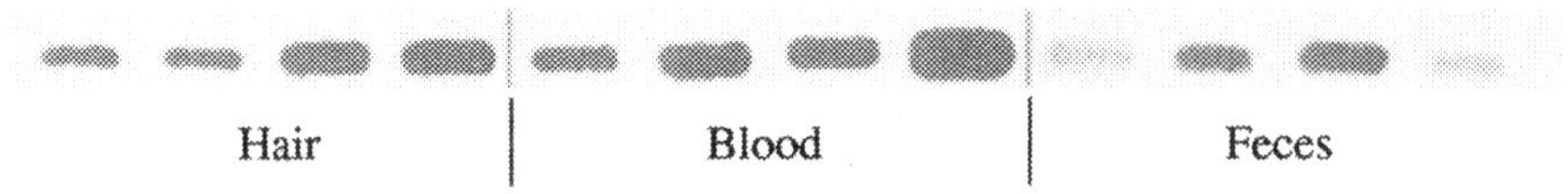

Figure 1. Amplified PCR products from orangutan DNA samples. Primers L10026 and H10433 were used for amplification of the entire ND3 gene.

Sequence comparisons of portions of the two protein coding genes, COII and ND3, showed that there is a substantial amount of variation in orangutan populations. Figure 2 gives an alignment of the 3′ end of the mitochondrial COII gene. We have included a human sequence ("Cam") and the published sequence from a gorilla (Horai, et al., 1992) for comparison. Most of the observed changes are silent, third codon position variants; they do not alter the protein product. Such changes are expected to have little or no adaptive effect. Nonetheless, since they are inherited they serve as excellent markers of population structure.

DISCUSSION

The definition of what constitutes a species has long been a source of debate (Avise, 1994, pp. 252-261). One of the most popular concepts is the biological species concept (Dobzhansky, 1937). Mayr (1940) defined species as "groups of actually or potentially interbreeding natural populations which are reproductively isolated from other such groups." Other species definitions exist, both for philosophical reasons and because the biological species definition cannot be applied in all cases. Cracraft (1983) introduced the phylogenetic species concept, placing the emphasis on monophyly; common ancestry and descent.

```
Roger       GTA TAT TAT GGC CAA TGC TCA GAG ATC TGT GGA GCC AAA CAC AGC TTT ATG CCT ATC   57
Abigail     ... ... ... ..A ... ... ... ..A ... ... ... ... ... ... ... ... ... ... ...
Siswi       ... ... ... ... ... ..T ... ... ... ... ... ..T ..C ... ... ... ... ... ...
Mark        ... ... ... ... ... ..T ... ... ... ... ... ..T ..C ... ... ... ..A ... ...
Hoblerlily  ... ... ... ... ... ..T ... ... ... ... ... ..T ..T ... ... ... ..A ... ...
Brook       ... ... ... ... ... ..T ... ... ... ... ... ..T ..T ..T ... ... ..A ... ...
Aye         ... ... ... ... ... ... ... ..A ... ... ... ..T ..C ... ... ... ..A ... ...
Stan        ... ... ... ... ... ... ... ... ... ... ... ..T ..C ... ... ... ..A ... ...
Bella (Sum) ... ... ... ..N ... ..N ... ..N ... ... ... ... ..G ... ..T ... ... ... ...
Gorilla     ... ..C ..C ... ..G ... ... ..A ... ... ... ... ..C ... ..T ... ... ..C ..T
Cam         ... ..C ..C ..T ... ... ..T ..A ... ... ... ..A ... ... ..T ... ... ... ...

Roger       GTT CTA GAA CTA ATC CCC CTA AAA ATC TTC GAA ATA GGA CCC GTA TTC ACT TTA TAG  114
Abigail     ... ... ... ... ..T ... ... ... ..T ... ... ... ... ..T ... ... ... ... ...
Siswi       ... ... ... ... ..T ... ... ... ... ... ... ... ... ..T ... ... ... ... ...
Mark        ... ... ... ... ..T ... ... ... ... ... ... ... ... ..T ... ... ... ... ...
Hoblerl     ... ... ... ... ..T ... ... ... ... ... ... ... ... ..T ... ... ... ... ...
Brook       ... ... ... ... ..T ... ... ... ... ... ... ... ... ..T ... ... ... ... ...
Aye         ... ... ... ... ..T ... ... ... ... ... ... ... ... ... ... ... G.. ... ..A
Stan        ... ... ... ... ..T ... ... ... ... ... ... ... ... ..T ... ..G G.. ... ...
Bella (Sum) ... ... ... ... ... ... ... ... ... ... ... ... ... ... ... ... ... ... ...
Gorilla     ... ... ..G ... ... ... ... ... ... ..T ... ... ... ... ... ... G.C C.. ..A
Cam         ... ... ... T.. ..T ... ... ... ... ..T ... ... ... ... ... ..T ..C C.. ...
```

Figure 2. Aligned sequence from the 3′ end of COII. The sequences are arranged in triplets corresponding to codons. All sequences are compared to a Bornean orangutan, "Roger", with a dot indicating identical nucleotides and a letter indicating an altered nucleotide. The bottom two sequences are from a gorilla (Horai, et al., 1992) and human (the senior author). Bella is Sumatran; the remainder are from a single population on Borneo. The sample labelled "Aye" is from feces; other orangutan sequences are taken from blood or hair samples.

Orangutans from Borneo and Sumatra, when brought together in captivity, are known to freely interbreed, producing fertile offspring. By the biological species definition, these two subspecies cannot be considered separate species. Janczewski et al. (1990) recommended raising the subspecies to species status based on the phylogenetic species definition. Their study, however, was based on only three individuals from Borneo and five from Sumatra. Ryder and Chemnick (1993) supported this recommendation in studies of chromosome structure and restriction analysis of mitochondrial DNA. The DNA analysis included six individuals ascribed to Borneo, six ascribed to Sumatra, and two others. Although differences were observed between the two subspecies, even the phylogenetic species definition requires that the differences observed between groups be placed in the context of within group variation.

Our pilot study had as its main goals, (1) to establish a reliable, non-invasive source of DNA for genetic characterization of endangered species and (2) to identify genomic regions that may be informative for the analysis of population substructure. We have found that both hair folicles, available from nests, and fecal samples can provide adequate DNA samples.

Our preliminary results, although they show population level variation in these genes, do not provide evidence for separation of Sumatran and Bornean orangutans. They are based, however, on a single individual of Sumatran origin and representatives from only one population from Borneo. A much more extensive sampling of populations from Borneo and Sumatra is required for definitive conclusions to be drawn.

ACKNOWLEDGMENTS

Hair samples from captive orangutans were made available through the kind donation from the Metropolitan Calgary and Toronto Zoos. This association was facilitated by Lori Perkins of the Species Survival Plan centered in Zoo Atlanta. We want to acknowledge the invaluable technical and editorial help received from Karen Beckenbach, Allan Arndt and Charlene Mayes. We also want to acknowledge the help and support received from the International and Canadian Chapters of the Orangutan Foundation. This work was supported in part by a grant from NSERC of Canada to ATB.

REFERENCES

Allee, W.C., Emerson, A.E., Park, O., Park, T., and Schmidt, K.P., 1949, *Principles of Animal Ecology*, W.B. Saunders Co., Philadelphia.

Anderson, S., Bankier, A.T., Barrell, B.G., de Bruijn, M.H.L., Coulson, A.R., Drouin, J., Eperon, I.C., Nierlich, D.P., Roe, B.A., Sanger, F., Schreier, P.H., Smith, A.J.H., Staden, R., and Young, I.G., 1981, Sequence and organization of the human mitochondrial genome, *Nature* 290:457-465.

Avise, J.C., 1994, *Molecular Markers, Natural History and Evolution*, Chapman and Hall, New York.

Cracraft, J., 1983, Species concepts and speciation analysis, Pp. 159-187 in *Current Ornithology*, R.F. Johnston (ed.), Plenum Press, New York.

Dobzhansky, Th., 1937, Genetic nature of species differences, *Am. Nat.* 71:404-420.

Galdikas, B.M.F., 1985, Adult male sociality and reproductive tactics among orangutans at Tanjung Puting, *Folia Primitol.*, 45:9-24.

Hartl, D.L., and Clark, A.G., 1989, *Principles of Population Genetics*, 2nd Ed., Sinauer Associates, Sunderland, Massachusetts.

Horai, S., Satta, Y., Hayasaka, K., Kondo, R., Inoue, T., Ishida, T., Hayashi, S., and Takahata, N., 1992, Man's place in Hominoidea revealed by mitochondrial DNA genealogy, *J. Molec. Evol.*35:32-43.

Janczewski, D.N., Goldman, D., and O'Brien, S.J., 1990, Molecular genetic divergence of orang utan (*Pongo pygmaeus*) subspecies based on isozyme and two-dimensional gel electrophoresis, *J. Hered.* 81:375-387.

Mayr, E., 1940, Speciation phenomena in birds, *Am.. Natur.* 74:249-278.

Morin, P.A., Wallis, J., Moore, J.J., Chakraborty, R., and Woodruff, D.S., 1993, Non-invasive sampling and DNA amplification for paternity exclusion, community structure, and phylogeography in wild chimpanzees, *Primates* 34:347-356.

Mullis, T., and Faloona, F.A., 1987, Specific synthesis of DNA *in vitro* via a polymerase-catalyzed chain reaction, *Meth. Enzymol.* 155:335-350.

Ryder, O.A., and Chemnick, L.G., 1993, Chromosomal and mitochondrial DNA variation in orang utans, *J. Hered.* 84:405-409.

Vigilant, L, Pennington, R., Harpending, H., Kocher, T.D., and Wilson, A.C., 1989, Mitochondrial DNA sequences in single hairs from a southern African population, *Science* 253:1503-1507.

Wolstenholme, D.R., 1992, Animal mitochondrial DNA: Structure and evolution, *Int. Rev. Cytol.* 141:173-216.

26

DOING IT NATURALLY

Reproduction in Captive Orangutans (*Pongo pygmaeus*)

R. Markham

Perth Zoo
20 Labouchere Road, South Perth, Western Australia 6151

SOCIAL ORGANIZATION IN THE NATURAL HABITAT

Field studies over the past 30 years have shown that orangutans lead predominately solitary lives within a loosely structured, highly dispersed, social system (Davenport, 1967; Galdikas, 1978, 1982, 1985; Harrisson, 1962; Horr, 1977; MacKinnon, 1974; Rijksen, 1978; Rodman, 1973; Schaller, 1961; Schürmann, 1981). The first basic population unit of this system is the adult female with a single infant, often accompanied by a juvenile/adolescent daughter. Each of these adult female units occupies a home range which overlaps on the periphery with the home ranges of other adult females, but with a core range used only by that female unit. The second basic population unit, the solitary adult male, is either resident in a particular area, occupying a very large range which cuts through the home ranges of several adult females , or is nomadic. The nomadic adult male is likely to be a younger male lacking the opportunity to establish a home range. Independent immature orangutans are relatively sociable, forming temporary associations and establishing dominance hierarchies before becoming solitary adults.

The most significant feature of the orangutan social system is the lack of physical contact and the dispersed nature of almost all social encounters. While all the orangutans in an area are familiar with one another, encounters between population units are generally infrequent, casual, and at a distance. Eye contact is almost always the only social behavior involved in such encounters between adults. Physical contact between adults occurs in only two circumstances - mating and fighting - and given the dispersed lifestyle and pattern of reproduction in this species, both of these are rare, brief occurrences in the lifetime of an orangutan.

REPRODUCTION AND DEVELOPMENT IN THE NATURAL HABITAT

It has been estimated that female orangutans in the wild reach menarche at between 9 and 11 years, and after an adolescent sterile period of some years, give birth for the first

The Neglected Ape, Edited by Ronald D. Nadler et al.
Plenum Press, New York, 1995

time at between 12 and 15 years. Only single infants have been observed in the wild, but as twinning occurs in 1.2% of births in captivity, it is likely that wild adult females cannot successfully care for two infants. The interbirth interval has been estimated to average 7 to 8 years, which is the longest of any mammal (Galdikas and Wood, 1990). Infants suckle for between 5 and 7 years, during which time the mother appears to be anovulatory. Infant mortality is believed to be low but has not been measured. A female orangutan in the wild may produce only four surviving offspring during her 25 years of reproductive life.

Young females have been observed to remain in their mother's home ranges for a total of 9 years, during which time they learn maternal behavior by observing their mothers with new infants. They then gradually move, or are driven away from their mothers' home ranges and set up one of their own in the maternal area. The females in any area of rain forest are therefore related to one another, and agonistic encounters are reduced to a minimum.

Young males leave the maternal area at around 7 years of age and wander large distances into new areas. Although capable of reproducing from a relatively early age, males probably do not get a chance to breed until they are as old as 18 to 20 years. Receptive females are scarce and almost invariably select resident adult males with fully developed secondary sexual characteristics to father their infants. These secondary sexual characteristics generally do not develop until males are between 15 and 20 years of age. Subadult male sexual aggression towards adult females is believed to be dominance-related rather than procreation-related behavior (Rijksen, 1978), but may be opportunist reproductive behavior related to male - male competition for females (Galdikas, 1985).

Orangutans in the wild are believed to have a potential lifespan of 50 to 60 years, with most adults surviving beyond 40 years (B. Galdikas, personal communication). It is probable that the highest mortality rate in orangutans occurs during the nomadic social period of their lives after they leave the maternal home range. The young males are especially inquisitive and adventurous at this time and are likely to fall victim to predators, falls, and encounters with adult males when wandering in unfamiliar territories.

REPRODUCTION AND DEVELOPMENT IN CAPTIVITY

In captivity, husbandry practices and diets have produced very different breeding and mortality patterns for this species than those observed in the wild (Perkins, 1991; 1993). In captive female orangutans, menarche tends to occur at 7 to 8 years of age, a full 2 to 3 years earlier than in the wild. Until recently, the majority of captive-born orangutans were hand-raised, and the levels of nutrition, especially the higher body fat levels, are likely to have resulted in early menarche (Zacharias and Wurtman, 1969). In Perth Zoo, it was found that 4 of the 6 hand-raised females reached menarche between 5 years, 10 months of age to 7 years, 10 months of age (Markham, 1990). The remaining 2 females were lactose intolerant and had poor appetites. They reached menarche at 9 years 2 months of age and 11 years 1 month of age. The only mother-raised female to have matured so far reached menarche at 8 years, 9 months of age.

The age of first parturition also is correspondingly early, often taking place long before the females are fully grown, i.e., before 12 years of age. Data from the International Studbook for the past 50 years shows that the mean age of first parturition in 372 captive females was 11 years, 3 months (SD = 3 years, 3 months), with a range of 5 years, 3 months to 32 years, 8 months. The mode, perhaps more indicative of the general situation , is the period between the ages of 9 years and 9 years, 11 months and represents 20% of first parturitions. Over 71% of females gave birth before the age of 12 years. There has been no significant increase in the mean over the 50-year period.

Interbirth intervals in captivity are perhaps the clearest indicator of differences in reproductive patterns between wild and captive orangutans. The mean value for all recorded interbirth intervals in captivity (n = 803) is 3 years (SD = 2 years), with a range of 7 months to 15 years, 3 months. The mode, again perhaps the best indicator, is the period between 1 year and 1 year, 11 months and represents 36 % of interbirth intervals. Only 16 % of interbirth intervals have been between 7 and 8 years, 76% have been less than 4 years, and 91% have been less than 6 years. There has been no significant increase in the mean over the 50-year period.

Infant mortality in captive-born purebred orangutans average 30% of births for 776 births, with an overall higher mortality rate in males (34%) than in females (26%), and is roughly the same for both subspecies. The death of captive-born orangutans in the period between infancy and maturity is 4% for males and 2% for females. So 62% of captive-born males and 72% of captive-born females survive to 11 years of age. Over the past 50 years, however, 90% of captive orangutans have died before the age of 31 years. Maternal mortality is undoubtedly one important cause of female deaths. Although most maternal deaths are not recorded as such, about 25% of adult female deaths have occurred within 1 month of parturition.

Fertility is high in captive orangutans, especially if only the animals over 15 years of age are considered. In the dead population, 75% of Bornean males, 78% of Bornean females, 87% of Sumatran males and 89% of Sumatran females have bred. In the living population 68% of Bornean males, 83% of Bornean females, 80% of Sumatran males, and 87% of Sumatran females have bred. The apparently lower rate of fertility in the males is likely to be due to a lack of opportunity to breed. The purebred orangutans in one-third of all collections are single specimens or are housed as single sex pairs or triads.

The exceptionally high level of premature adult mortality along with a relatively high level of infant mortality in this species in captivity has resulted in the number of surviving offspring left by each breeding female being low, despite the high level of fertility and the practice of breeding early and frequently from each female. The breeding females of the dead population averaged 1.5 surviving offspring for Borneans and 2.5 for Sumatrans. The living founder population of breeding females averages 3.1 surviving offspring for Borneans and 3.9 for Sumatrans, but this population is aging. The average age of the founder females is 22 years for Borneans and 30 years for Sumatrans. The living captive-born population averages 1.5 surviving offspring for Borneans and 1.8 for Sumatrans as of this writing. None of these figures take into account the number of subspecific hybrids produced by these females.

It should be noted that the above statistics for the captive population are largely a reflection of what has happened in the past with global population. In all the aspects mentioned, the managed populations of orangutan are beginning to show gradual progress towards improved husbandry practices and planned breeding progress that will hopefully reduce the mortality rates and slow down the loss of genetic diversity.

One of the most important contributions the captive population can make to conservation is to measure and record data that cannot be readily collected in the wild. The difficulties involved in following individual orangutans in the rain forest for any length of time means that reproductive data are almost impossible to collect. Unfortunately, the husbandry practices in most orangutan colonies also make it difficult, if not impossible, to record these data, and very little has been published.

ORANGUTANS AT PERTH ZOO

The orangutans at Perth Zoo are displayed in natural population units within group structure that mimics that of the wild (Markham, 1990). All except infants still suckling from their mothers are housed in separate dens at night and fed there. This has made it possible not only to gain control over breeding and individual diets, but it also has been possible to collect extensive data on most aspects of reproduction, some of which has been published already. For example, gestation periods have been calculated for 23 of the 25 births in the colony, based on menstrual cycles, males being with females for a short time only, mating records, and pregnancy tests (Table 1). There appears to be only 12 other published values for the orangutan gestation period based on accurate observations (Asano, 1967; Davis, 1977; Graham Jones and Hill, 1962; Hobson, 1976; Hodgen, *et al.*, 1977; Kingsley, 1981; Lippert, 1974). The most detailed of these, covering 6 pregnancies, were reported by Lippert, but as he counted gestation from the first day of the last menstruation, rather than midcycle, 14 days have been deducted from each of these 6 gestation periods to obtain the correct figures. If these 12 published values are added to the 23 from Perth Zoo, a mean value of 244 days (SD = 11 days, range = 223-267 days) is obtained. The problem with all the Perth Zoo data is that, with the exception of the 54-year-old Bornean female, it all comes from closely related females, and there is not sufficient data from elsewhere with which to compare these records.

Table 1. Gestation periods of captive orangutans

Female ISBN	Sub species	Collection	Gestation period	Reference
176	B	Honolulu	263	Graham-Jones & Hill (1962)
319	B	Tokyo-Tama	244	Asano (1967)
457	S	Berlin Tier.	248	Lippert (1974)
457	S	Berlin Tier.	240	Lippert (1974)
457	S	Berlin Tier.	251	Lippert (1974)
457	S	Berlin Tier.	254	Lippert (1974)
680	S	Berlin Tier.	263	Lippert (1974)
680	S	Berlin Tier	242	Lippert (1974
959	S	Jersey	228	Hobson (1976)
909	S	Yerkes RPRC	250	Hodgen, *et al.* (1977)
433	B	Portland	260	Davis (1977)
Unk	S	London	254	Kingsley (1981)
977	S	Perth	229	Perth Zoo Records
978	S	Perth	242	Perth Zoo Records
978	S	Perth	255	Perth Zoo Records
978	S	Perth	249	Perth Zoo Records
978	S	Perth	246	Perth Zoo Records
978	S	Perth	230	Perth Zoo Records
978	S	Perth	261	Perth Zoo Records
978	S	Perth	252	Perth Zoo Records
978	S	Perth	247	Perth Zoo Records
978	S	Perth	267	Perth Zoo Records
979	B	Perth	247	Perth Zoo Records
979	B	Perth	234	Perth Zoo Records
979	B	Perth	248	Perth Zoo Records
979	B	Perth	241	Perth Zoo Records
979	B	Perth	223	Perth Zoo Records
1105	X	Perth	224	Perth Zoo Records
1090	S	Perth	244	Perth Zoo Records
1090	S	Perth	238	Perth Zoo Records
1090	S	Perth	234	Perth Zoo Records
1090	S	Perth	225	Perth Zoo Records
1090	S	Perth	236	Perth Zoo Records
1427	S	Perth	239	Perth Zoo Records
1836	S	Perth	245	Perth Zoo Records

Aside from being unable to collect valuable reproductive data from the captive population, present husbandry practices are generally not designed to allow for the natural behavior of this predominantly solitary arboreal species. The mortality and breeding problems that have been encountered in captive orangutans are likely to have resulted from ignoring the natural history of the species. The simplest way to control breeding, after all, is to keep males separate from females. The simplest way to prevent stress and obesity in a solitary animal is to allow it to be solitary and feed it separately. The simplest way to avoid maternal mortality is to use the natural breeding pattern of the species. Captive management is not just concerned with breeding and conserving genetic diversity, it also must be concerned with conserving the whole animal over its natural lifespan so that at the end of 200 years, the species is genetically and behaviorally intact.

The orangutan is indeed the neglected ape - not because we have neglected to gather information about it, or because we have neglected to include it in our global management plans, but because we have neglected to use information wisely to provide it with an appropriate captive environment that allows it full expression of its natural physical and social behavior.

REFERENCES

Asano, M. A., 1967, Note on the birth and rearing of an orangutan *(Pongo pygmaeus)* at Tame Zoo, Tokyo, *Int. Zoo Yearb.* 5 7:95-96.

Davenport, R. K., 1967, The orangutan in Sabah, *Folia Primatol.* 5:247-263.

Davis, R. R., 1977, Pregnancy diagnosis in an orangutan using two pregnancy test kits, *J. Med. Primatol.* 6:315-318.

Galdikas, B.M.F., 1978, Orangutan Adaptation At Tanjung Puting Reserve: Mating And Ecology, *Doctoral Dissertation*, University of California.

Galdikas, B.M.F., 1982, Wild orangutan birth at Tanjung Puting Reserve, *Primates* 23:500-510, .

Galdikas, B.M.F., 1985, Orangutan sociability at Tanjung Puting, *Am. J. Primatol.* 9:101-119.

Galdikas, B.M.F., Woods, J.W., 1990, Birth spacing patterns in humans and apes, *Am. J. Phys. Anthropol.* 83:185-191.

Graham-Jones, O., Hill, W.C.O., 1962, Pregnancy and parturition in a Bornean orang, *Proc. Zool. Soc. Lond.* 139:503-510.

Harrisson, B., 1962, *Orangutan*, London, Collins.

Hobson, B.M., 1975, The diagnosis of pregnancy in the lowland gorilla and the Sumatran orangutan, *Dodo* 12:71-75.

Hodgen, G.D., Turner, C.K., Smith, E.E., Bush, R.M., 1977, Pregnancy diagnosis in the orangutan using the Sub-Human Primate Pregnancy Test Kit, *Lab. Anim. Sci.* 27:99-101.

Horr, D.A., 1977, Orangutan maturation: Growing up in a female world, Pp. 289-322 in *Primate Biosocial Development: Biological, Social And Ecological Determinants*, (Eds. Chevalier-Skolnikoff, F.E. Poirier), Garland Publishing Inc.

Kingsley, S. R., 1981, Reproductive Physiology And Behaviour In Captive Orangutans, *Doctoral Dissertation*, University of London.

Lippert, W., 1974, Beobachtungen zum Schwangerschufts und Geburtsverhalten beim Orangutan *(Pongo pygmaeus)* im Tiegarten Berlin, *Folia Primatol.* 21:108-134.

MacKinnon, J.R., 1974, The behaviour and ecology of wild orangutans, *Anim. Behav.* 22:3-74.

Markham, R.J., 1990, Breeding orangutans at Perth Zoo: Twenty years of appropriate husbandry. *Zoo Biol.* 9:171-182.

Perkins, L., 1991, *The 1990 International Studbook of the Orangutan (Pongo pygmaeus spp.)*. Zoo Atlanta.

Perkins, L., 1993, *The 1992 International Studbook of the Orangutan (Pongo pygmaeus spp.)*. Zoo Atlanta/Fulton County Zoo, Inc. (Zoo Atlanta).

Rijksen, H.D., 1978, *A Field Study on Sumatran Orangutans (Pongo pygmaeus abelii* Lesson 1827), *Ecology, Behaviour and Conservation*, 78-2, Wageningen, H., Veenman and Zonen B.V.

Rodman, P.S., 1973, Population, composition and adaptive organisation among orangutans of the Kutai Reserve, Pp.171-209 in *Comparative Behaviour And Ecology of Primates*, Michael and Crook. Eds., London, Academic Press.

Schaller, G.D., 1961, The Orangutan in Sarawak, *Zoologica* 46:73-82.
Schürmann, C.L., 1981, Courtship and mating behaviour of wild orangutans in Sumatra, Pp. 130-135 in *Primate Behaviour And Sociobiology*, (Eds. A.B. Chiarelli, R.S. Corruccini), Berlin, Springer-Verlag.
Zacharias, L.; Wurtman, R.J., 1969, Age of menarche: Genetic and environmental influences. *New England J.Med.* 280:868-875.

27

INTEGRATING NEEDS IN GREAT APE ACCOMMODATION

Sumatran Orangutan (*Pongo pygmaeus abelii*) "Home-Habitat" at the Jersey Wildlife Preservation Trust

J. J. C. Mallinson and J. B. Carroll

Jersey Wildlife Preservation Trust
Les Augrès Manor, Trinity, Jersey JE3 5BP Channel Islands

ABSTRACT

Changing fashions in ape accommodation have resulted in a move away from rocks and concrete grottos, and towards more naturalistic and stimulating environments. The design of the new accommodation at the Jersey Wildlife Preservation Trust (JWPT) undoubtedly improved the "quality of life" of the Trust's family group of Sumatran orangutans, takes account of their psychological and physical needs, and helps to maximize behavioral opportunities. At the same time, considerable effort has been made to inform and educate the public about both the captive breeding program and the wider efforts for orangutan conservation in Southeast Asia.

INTRODUCTION

As Gerald Durrell recorded: "I like orang-utans. I think we are privileged to share this planet with them. Whether orang-utans feel the same about us is another matter, because these enchanting apes, who are so closely related to you and me, have a problem. Their problem is survival. And being unable to address the problem themselves, they look to us, as it were, to resolve it for them" (Durrell, 1992, p.2).

It is often not fully appreciated that captive environments in *the modern zoo* have several functions to fulfil. First, they must provide acceptable conditions within which animals can breed, raise young, and perform as much of their natural behavioral repertoire as possible. Second, they must provide the visiting public with a positive image of both the animal and its captive environment. Third, they must inform the visitors of the conservation problems faced by the species and of strategies being used to overcome those problems. Fourth, they must provide an environment within which the animal caretakers can work effectively and, of particular relevance to any great ape species, safely. The weight given to

The Neglected Ape, Edited by Ronald D. Nadler et al.
Plenum Press, New York, 1995

each of these considerations will vary depending on the species, the mission of the zoo and on financial or space limitations, but all public exhibits of animals must integrate these functions successfully or they will be judged inadequate.

Captive environments for primates have improved dramatically over the last 20 years, although, regrettably, there are still many that take little account of the needs of the caged inhabitants and are remarkably dull for the animals and offensive to a well-informed visiting public. Concrete or tiled boxes with a few fixed metal structures relate poorly to what is known of the behavior, social relations and complex knowledge of the environment that orangutans use in the wild (Redshaw and Mallinson, 1991).

Improvements in captive environments are aimed at maximizing behavioral opportunities and increasing the chances of captive primates retaining and practicing as many of their natural manipulatory and social skills as possible (Mallinson and Redshaw, 1992). Naturalistic habitats and more innovative exhibits have also been found to be conducive to the expression of species-typical behavior patterns, such as suspensory quadrupedal and terrestrial bipedal locomotion in orangutans (Maple and Finlay, 1986).

During the last 15 years, it has been the policy of the JWPT either to redevelop or replace the majority of its primate cages with far more spacious and stimulating environments. The concept behind the design of the Trust's new orangutan accommodation was presented by JJCM at the international conference "The Neglected Ape" held at California State University, Fullerton (Mallinson, 1994). This paper expands on the latter presentation providing details of the design of the inside accommodation, the landscaping of the water-moated island habitats, and the provision of educational material to the visitors. Further details on the design and specifications are provided by Mallinson *et al.* (1994).

JWPT ORANGUTANS

Since the mid-1970s, the JWPT has been directly involved with the maintenance and breeding of both the Bornean (*Pongo pygmaeus pygmaeus)* and the Sumatran orangutan (*Pongo pygmaeus abelii*). Through multiple-generation births recorded from Jersey-born orangutans, the Trust has contributed materially to the goal of establishing viable, self-sustaining captive populations. These breedings are the direct result of cooperative loan agreements between zoos nationally, regionally and internationally.

In March 1993, the Trust's remaining Bornean orangutans (one male and two females) were sent on breeding loan to Zoo Parc de Beauval, France, leaving the Trust with two male and two female Sumatran orangutans in its collection.

Due to the considerable success of the Trust's accommodation for lowland gorillas (*Gorilla gorilla gorilla*), which includes a large, landscaped outside area of approximately 1950 m^2 (Mallinson, 1980), a feasibility study for the development of a similar new orangutan environment was undertaken. This involved visits to a number of zoos with good ape accommodations, and a review of the design criteria published by other zoos in Continental Europe (Mager and Griede 1986), Australia (Embury, 1992) and the USA (Bruner and Meller, 1992). Plans of various completed ape accommodations that were not visited and designs of developments that had yet to be constructed were also studied.

The previous accommodation for the Trust's family group of Sumatran orangutans was built in 1971. It covered a relatively small area containing three inside cages (each measuring 3 x 3.6 x 4 m), which also functioned as sleeping dens. The public were able to look into the dens, as well as into the outside area (measuring 15 x 10 m), through laminated glass windows. The whole structure was of concrete, metal and glass, with a barred weld-mesh roofing over the exterior.

As much as possible had been done to help the orangutans maximize the use of all available space by the provision of a series of metal climbing apparatus, platforms and ropes arranged vertically, horizontally and obliquely (Mallinson, 1982). It was recognized, however, that due to the considerable limitations of both the building and its site, any redevelopment would fail to provide the type of environment that would satisfactorily facilitate activity levels that are considered essential to the long-term welfare of this highly intelligent anthropoid.

Basic Concepts and Requirements

Although management and facilities for apes vary considerably between collections, in the majority of cases, orangutans are confined to indoor accommodation for over two-thirds of the 24 hours. Often in connection with this factor, insufficient thought has been given either to the inside space available or to the design of the cage furnishings (Mallinson, 1986). Professional zoo staff are of paramount importance, not only in providing satisfactory management but also, as Bruner and Meller (1992) noted, their involvement is a key element for success in exhibit design. Mammal Curator (Bryan Carroll), Section Head (Richard Johnstone-Scott), Staff Member (Ian Singleton) and the Zoological Director (Jeremy Mallinson) established the fundamental accommodation requirements for the JWPT's family group of Sumatran orangutans. This took account of two of the fundamental functions of the building: provision of appropriate caging for the animals, and the safety of the caretakers. Establishing these requirements used knowledge gained both from studies of orangutans in the wild (e.g. Mackinnon, 1974; Rijksen, 1978; Galdikas, 1988), and from the experience gained through establishing demographically and genetically viable captive populations.

In August 1990 the Trust commissioned its Architect (J. Douglas Smith) to design a new indoor habitat of "gymnasium type" proportions which would enable the orangutans to enjoy conditions reflecting their arboreal lifestyle and which also took into account their considerable strength, ingenuity and inherent ability to demolish all but the strongest of facilities. The building was to enable the animals to move arboreally, should be capable of maintaining a group of 5-6 individuals with separation facilities and, if necessary, be capable of holding two social groups.

In December 1993 the Trust invited the Landscape Consultants, Colvin and Moggridge, to attend a limited competition over the weekend of 22-23 January 1993 to design an external environment to complement the proposed new orangutan house.

A brief formulated by the JWPT Landscape Committee outlined the requirements for the orangutans, zoo staff and public and detailed several technical requirements which had to be met. The objective of the landscape proposals was to provide an imaginative external environment for a family group of Sumatran orangutans using earthworks, water, timber structures and limited areas of planting (S. Hicks *in litt.*).

The briefing given by the Trust established several basic features of the design: Two separate external spaces were required, a two thirds: one third division of space being preferred; a water-filled moat 5 m wide and 1.25 m deep was recommended as the enclosure barrier, backed up with electrical wiring on the animal side of the moat; and timber structures were required to provide an aerial habitat for the arboreal orangutans.

It was intended that the whole area should provide the visiting public with an enjoyable and informative experience. Facilities for education were, therefore, of paramount importance and it was decided that insufficient educational material could be put into the new building itself. Accordingly, the decision was made to build a small education area on one of the public pathways leading to the new building.

Indoor Accommodation

It was decided that the form of the new building should draw upon the character of Indonesian "long- house" architecture, with flared roof ends and natural timber log cladding. The roof is covered with cedar shingles which, together with the log cladding, will weather to a silvery green color to harmonize with the newly landscaped area and rural surroundings of the Zoological Park (see Fig. 1).

The building in its plan form is simple in concept, comprising three contiguous parts: a public space 15 x 3 x 4.5 m high; the main exhibit area (divided into two) 15 x 9 x 7.5 m high over approximately two-thirds of its area, and 6.5 m high over the remainder; off view holding and service areas 7.275 x 10.55 m with height varying between 5 m in the two interconnecting holding bays to 2.4 m in the food preparation and service and plant room areas (Figs. 2 and 3).

Public Space

The public "walk-through" area affords views of both the internal accommodation and external habitat. Internal viewing is through a centrally placed glass screen 6 m wide x 4 m high together with symmetrically placed flanking screens 1.5 m wide x 4 m high. These screens are constructed with mild steel angle framework and are glazed with 38 mm multi-laminated safety glass. Screens commence 600 mm above floor level and give clear views to all parts of the interior. Viewing of the external habitat from the public space is afforded by a centrally placed window unit 6 m wide x 1.5 m high equipped with a pair of horizontally sliding windows to give a clear opening of 3 m. This window commences at a height of 600 mm above floor level and directly overlooks the moat, two islands and the central peninsula. Flanking the main window are two fixed windows each 1.5 m square symmetrically arranged. At high level, two tunnels span the 3 m width of the public space from the interior accommodation to the outside habitat. These glass-sided tunnels provide a means of access/egress for the orangutans. Use of glass permits visitors to view the orangutans from the public space.

Main Exhibit Area

The main public exhibition area (areas 1 and 2, Fig. 2) is centrally divided with metal grillwork and a keepers' lobby to create two spaces each 7.5 m wide x 9 m long. The grillwork dividing the space is 75 x 75 x 10 mm diameter mesh. This size of mesh prevents an adult male orangutan from reaching through the bars to grab an animal in the adjacent area. At high level in the main section a 900 x 900 mm slide operated manually from the keepers' service area above the public space permits transfer of the orangutans. The keepers' lobby is 2 m wide x 1.25 m deep x 6.5 m high, is constructed with 50 x 50 x 8 mm bar work and contains two access doors, one into each habitat, together with facilities manually to operate two slides, one at high level and one at floor level. This smaller size of mesh is used in all barriers between orangutans and keepers as this mesh size prevents all but the smallest animals from putting a hand through to the keepers.

Centrally placed in the wall dividing the habitat from the public area and at a level of 3.5 m above floor level, two openings 1 x 1 m give access to the transfer tunnels. Access to the tunnels is remotely controlled from the keepers' area and public space by electrically-operated mild steel horizontal slides which have step controllers to adjust speed of travel together with safety cut-out devices. The floor of each tunnel projects into the habitat 1 m to form a "landing pad" for the orangutans. The roof level of the tunnels is extended over the whole area of the public space to provide a keepers' service area. Within this area there

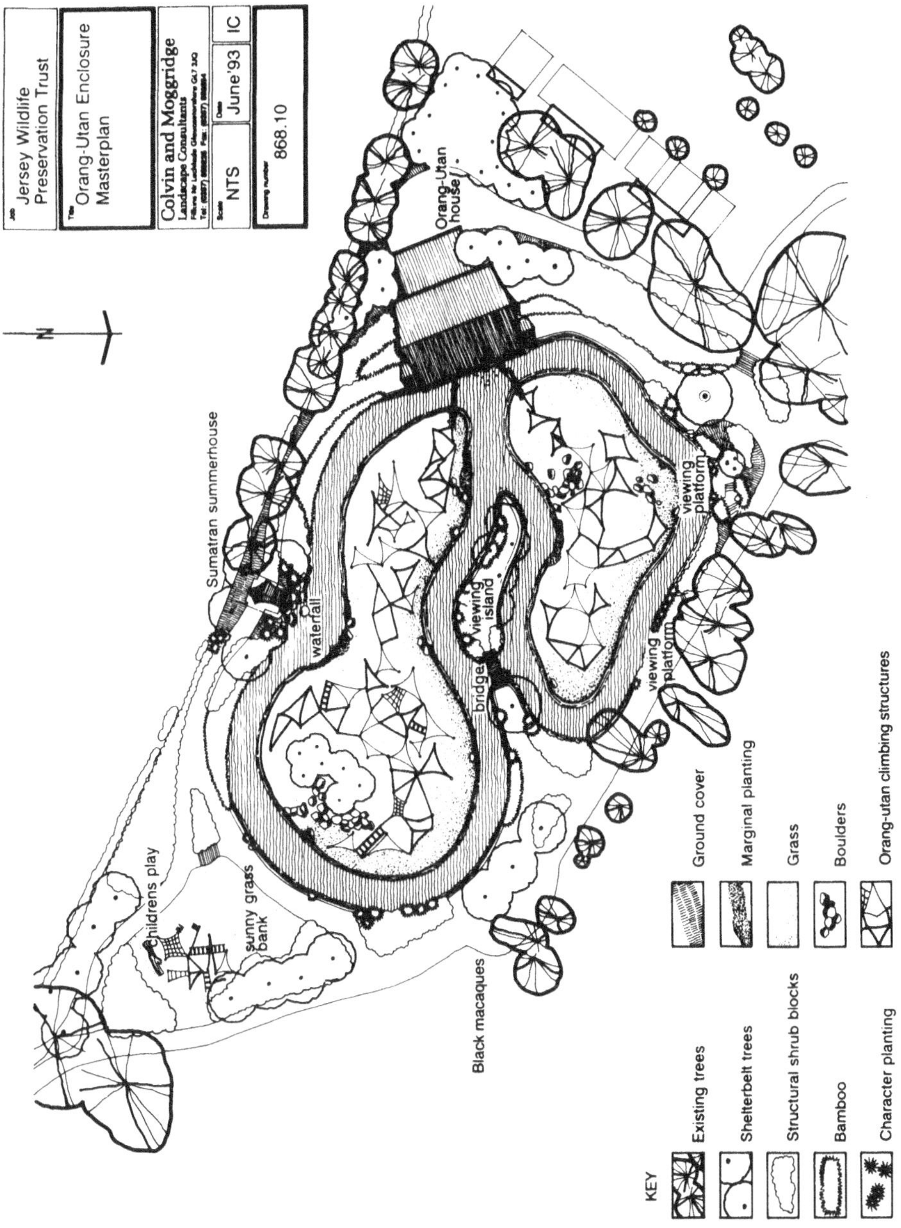

Figure 1. Outside habitat of the Sumatran orangutans at JWPT. Colvin and Moggridge, Landscape Consultants.

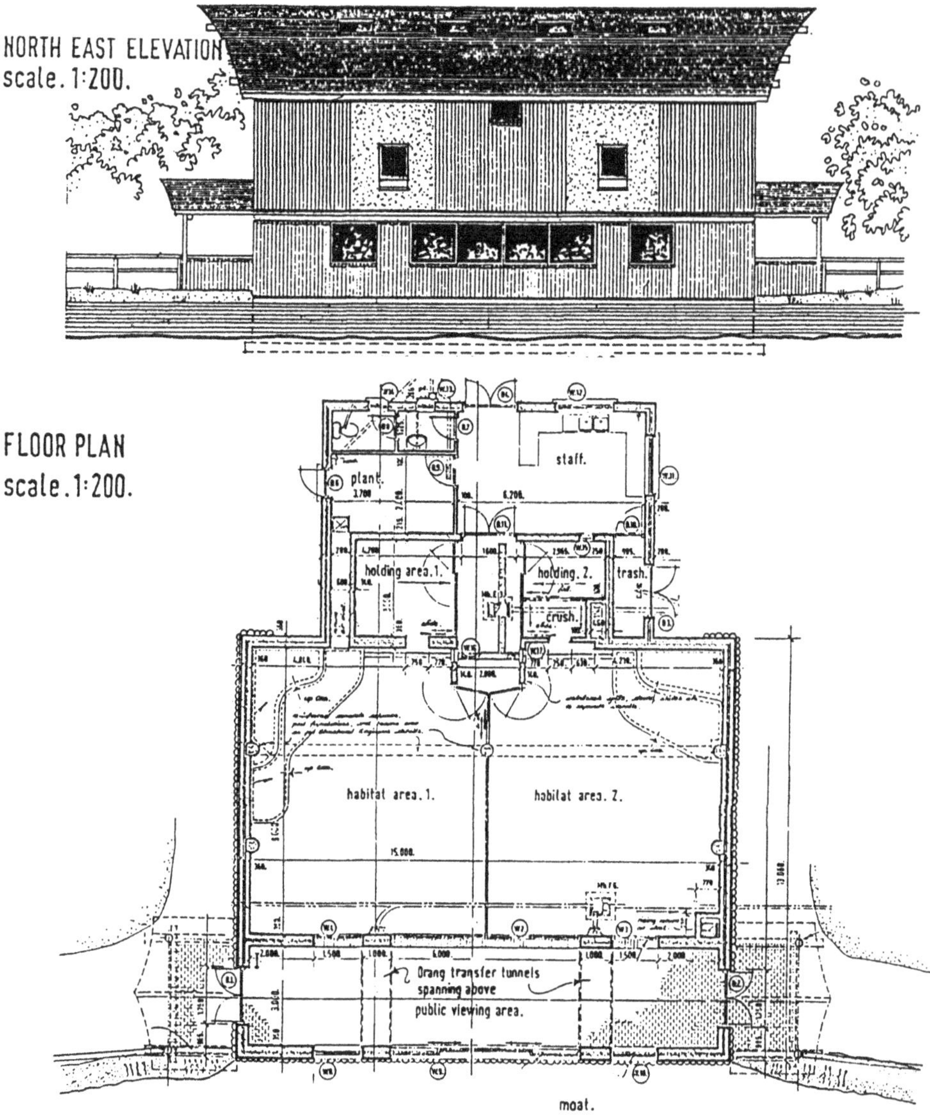

Figure 2. North-east elevation and floor plan of indoor facilities. J. Douglas Smith, Architect.

are openings at a high level facing the inner enclosure and one viewing window facing the external habitat to give the keepers viewing facilities together with facilities for high level feeding. In the floor of this area and immediately above each tunnel is a trap door to allow keepers access to the tunnels for cleaning and maintenance. Access to the whole of the keepers' area is by means of a metal wall ladder leading from the public viewing space.

In the south and west corners of each exhibit area, there are curved raised terraces arranged on two levels 600 mm high. These terraces are equipped with electric heat pads to provide "warm spots" for orangutan comfort. Each exhibition area is also equipped with a 200 mm diameter climbing post, approximately 7 m high, to which are welded two mild steel nesting baskets at intermediate and high level, and an alternating arrangement of 50

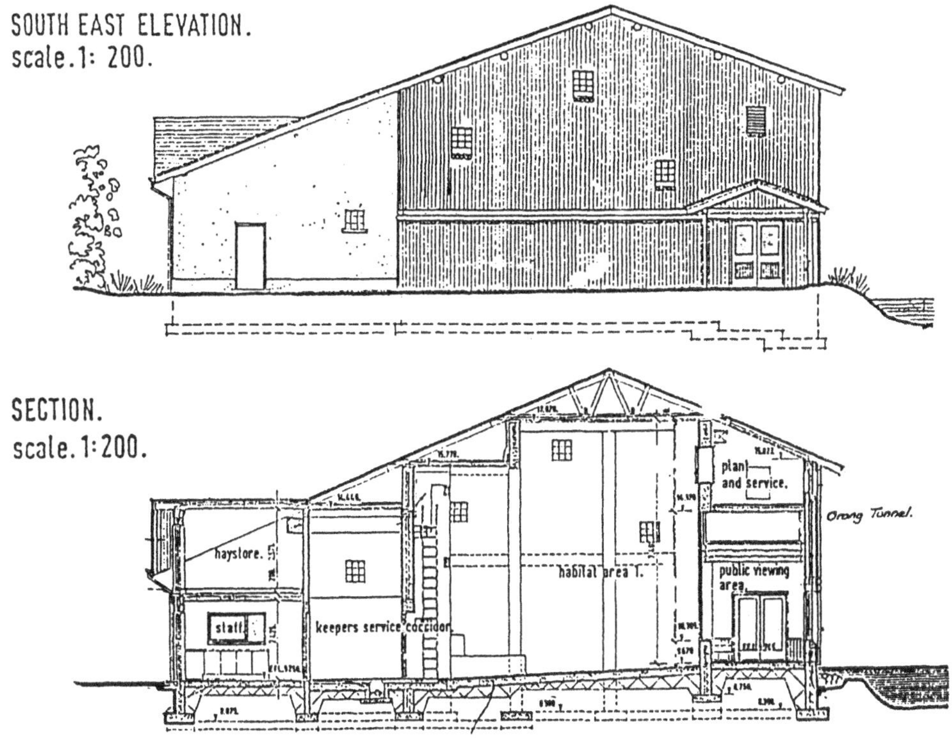

Figure 3. South-east elevation and section of indoor facilities. J. Douglas Smith, Architect.

mm pegs and 100 mm diameter rings throughout its height. There is a network of 50 mm diameter brachiation bars welded from these poles to various parts of the area and an intricate system of ropework extending from random points around the walls. The walls of the area are finished in a sand/cement render and painted with an acrylated rubber paint. Walls are pale green in color. The floor of the exhibit area has been finished with a polymer-modified cementitious screen, 25 mm thick, followed by three coats of a durable two-part epoxy resin-based coating in dark green. All junctions between walls and floors are covered for ease of maintenance. The ceiling is covered with checkered sheet aluminum panels 2.5 mm thick fixed at 200 mm centers with countersunk screws. Each exhibition area is naturally lit by four Velux roof windows set into deep light wells, each protected by openable barwork fitted in the same plane as the main ceiling level. These windows are double-glazed and can be remotely controlled to provide ventilation. The areas are also artificially lit by two Sonn Lamps situated in the keeper service passage.

Off-view Holding And Service Areas

Immediately adjoining the rear of the two main indoor exhibit areas are two holding spaces 3 x 3 m separated by a 2 m wide keepers' corridor. The holding spaces are 5 m high and the floor, walls and ceiling are finished in the same manner as the main exhibit areas. All barwork in the spaces is 50 x 50 x 8 mm. There is a 2 m wide transfer cage at high level, fitted with sheet steel manually-operated slides to permit the transfer of animals from one space to the other. Raised ledges are provided in both spaces and these have inbuilt electric heat pads. Manually-operated slides allow animals to transfer from the exhibit areas to the holding spaces.

One of the holding spaces is fitted with a restraint cage 2 x 1.5 x 1.5 m, constructed in 50 x 50 x 8 mm barwork. This cage is positioned centrally behind a 900 x 900 mm opening into the exhibit area and will be used mainly when crating animals to move them. In the front wall of the cage is a manually-operated slide and the orangutan to be transferred is introduced into the space, the slide is closed and the whole of the front wall of the cage can be manually moved forward towards the opening in the exhibit area wall by a geared cranking handle. The space available to the orangutan continues to be diminished until such time as it has no option but to pass through the opening into a crate placed in the exhibit area. The holding spaces are lit with remotely controllable roof lights as in the main exhibition area and also have artificial lighting.

Immediately adjoining the holding areas is a food preparation kitchen, a plant room and other keeper facilities. Above the food preparation and plant room is a hay-loft accessed externally by a pair of doors at high level.

Environmental Control

Temperature and ventilation control has been achieved by the installation of two split package ducted air-to-air heat pump systems, manufactured by Hitachi. The system provides a design temperature of 24°C against an ambient temperature of -1°C. The accommodation area is maintained at 20°C in general use.

ISLAND HABITATS

The Site

The site, previously an agricultural field, occupied an area of approx. 9,625m^2 in an open elevated position to the south of the Manor House, at the zoo's perimeter. The remnants of hedgerow trees provided some shelter on three sides, but the site was exposed, particularly to the prevailing south-westerly winds. A footpath along the northern perimeter of the site provided access from the gorilla enclosure alongside the black macaque (*Macaca nigra*) and Przewalski horse (*Equus przewalskii*) enclosures to an area west of the site where golden lion tamarins (*Leontopithecus rosalia*) are located in a small wood.

Design Logic

During their first site appraisal at the zoo, the landscape consultants were struck by the inquisitive nature of many of the animals, particularly the smaller monkeys. As they approached enclosures, the occupants came out, inspected them briefly, and retired again. The design aimed to build on this role reversal by providing a small viewing island at the center of the orangutan habitat where the public are "on view" to the surrounding animals. This key feature of the design also provides an area at the center of the enclosure where vegetation can be established safe from damage by the orangutans.

The use of water as a barrier and the lack of significant fencing allows an intimate relationship between the visitor and the animal. This is heightened by the careful location of viewpoints, mounding and planting to prevent views of other visitors across the enclosure. It was possible to achieve a single water level for the moat, although this entailed significant excavation at the south-east corner, and mounding and dam construction at the north-west corner. Surplus excavated soil was used to construct the animal islands and to build up levels to the south-west of the enclosure to provide shelter.

A footpath around the perimeter of the enclosure provides a sequence of views into the animal habitat. This primary route is designed to be accessible to handicapped persons. The orangutans occupy an elevated position from most viewpoints, although a high point above a waterfall puts the visitor level with the animals. Smaller viewpoints off the main footpath allow a more private vantage for quiet study. Access to the servicing and maintenance areas of the new orangutan house is by a private pathway, branching off from the visitor circulation route.

In order to assimilate the new enclosure into the existing zoo landscape, much use has been made of features already used for other enclosures. Simple timber fencing, "hoggin" (a pink granite substrate) footpaths, boulders and palisade edging ensure this continuity. The islands were designed as relatively steep-sided mounds rising from a small shelf at the water's edge. Boulders and ledges cut into slopes provide a variety of microclimates with sun-traps and wind shelters available. The orangutans have access to the water's edge, but hotwiring 1 m from the bank prevents entry into deeper water. The larger island has an area of about 1,500 m^2 while the smaller is about 850 m^2.

The design of the climbing structures seeks to provide a network of arboreal "walkways" so that the animals do not need to walk on the ground. Irregular posts, positioned obliquely, are connected with ropes and webbing nets to provide shade and shelter at height and to allow the animals to move in their natural manner. The posts are arranged as tripods bolted into mild steel brackets at ground level and tied with stainless steel cable where they meet. This maximizes stability. It is intended that the structures should be extended and modified as necessary over a period of time to develop a complete arboreal habitat for the orangutans. The bridge link over the moat from the new building allows the orangutans to access this habitat without coming to the ground.

Power Fencing

The survey by Ogden, *et al.* (1992) on the effectiveness of electric fencing in the control of apes found that it worked well for gorillas and relatively well for chimpanzees (*Pan troglodytes*), whereas with orangutans it was ineffective. However, from observations of electric fence installations, A. Venables (*in litt.*) concluded that in many cases they were so poorly thought out, applied, installed and maintained that it was surprising that they worked at all. In the case of the JWPT orangutan environment it was decided to use power fencing as a secondary barrier to the primary barriers of water and the vertical face of the building around the tunnel openings that provide access to the islands via the bridges.

Island Fencing. The installation of three powered wires around the islands uses fiberglass rods, set in metal brackets fixed to the bank in concrete. The use of flexible, non-conductive fiberglass ensures even wire tension and is visually acceptable. It avoids the use of easily recognized plastic insulators and steel rods. These are less vandal-proof than fiberglass and clips, an important consideration for orangutans (Fig. 4).

The use of high visibility tape on initial installation serves as a visual deterrent to the orangutans. The same material was used on training and introduction panels in the night dens. The orangutans therefore entered the outside environment already sensitized to the fence materials. This type of controlled introduction is vital in ensuring that power fencing is an effective barrier.

The fences around the islands are connected by separate switches to the main control panel in the keeper area. The system is powered by a Mains/Battery energizer producing 7,000 volts at 1.3 joules of expressed energy. The system has a voltage sensor and alarm siren. If a short-circuit occurs, such as by an orangutan falling into the water and touching the fence, staff would be alerted immediately.

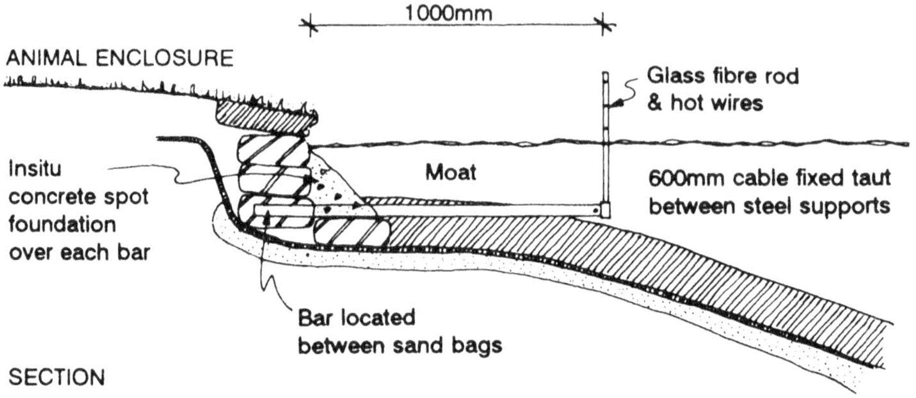

Figure 4. Design of steel supports for power fencing on the island side of the moat in the outdoor habitat for the Sumatran orangutans. Colvin and Moggridge, Landscape Consultants.

The Building and Facade. In order to ensure that the orangutans did not create hand-holds in the cement-rendered face of the building or reach the log-cladding close to the tunnels, electrified weld mesh panels is used to protect those areas. These panels are powered by the energizer mentioned above and are highlighted with white conductive electrical tape. The mesh is painted green to reduce its public visual impact. The areas the orangutans would naturally sit on in the tunnel entrance and on the bridge were tested for electrical conductivity and grounding. Two meter long strips of flat iron were bolted to the wooden bridge and grounded via the building to improve conductivity. Adequate grounding such as this is vital and often overlooked (A. Venables *in litt.*).

Planting Design

The planting is divided into three categories. First, shelterbelt planting at the perimeter of the enclosure, together with the strong ground relief will, in time, create a sheltered environment within the enclosure. Plants are selected for their tolerance to exposure and speed of growth. Second, future structure planting within the enclosure will create spatial composition and define views. Plants selected are predominantly evergreen with contrasting foliage colors and textures. Particular use will be made of scented plants at key positions (e.g. vanilla scented *Azara microphylla* at the sweet kiosk!). Third, Jersey's climate prohibits the planting of tropical species outdoors, but the careful use of character plants within this sheltered area evokes thoughts of the jungle. Plant selection favors bamboos, foliage contrasts, evergreens, strong colors and, in particular, large leaved plants.

The majority of plants are planted at normal stock sizes, with a few specimen plants at key locations to provide immediate impact. Herbaceous perennials, particularly *Geranium sp.* are planted as short-term ground cover between shrubs. These will be removed as the shrubs grow together to fill beds.

Play Area

A children's play area is built on a grassy mound to the east of the enclosure. Play equipment is modelled closely on the orangutan climbing structures in order to encourage mimicry and interactive play. The construction of the play area has been sponsored by Uniroyal.

"LONG-HOUSE" INTERPRETIVE CENTER

The "Long-House" is situated above the waterfall with commanding views across the larger of the two islands, over the island peninsula to the second island, and beyond to the perimeter treeline of the Zoological Park. Educational material associated with the "Long-House" is divided into two parts. Seven panels in the public viewing area of the main house focus on the biology of orangutans including adaptation to arboreal life, diet, family life etc., as well as providing personal details of the "Jersey" orangutans (Fig. 5). In the "Long-House" itself, 12 panels outline conservation issues. We want visitors to understand why we are keeping orangutans and that, along with other zoos in different parts of the world, we are working to save species, whilst at the same time efforts are being made to save them in Sumatra and Borneo. We want them to know where orangutans come from and that their range has been reduced in historical times and is continuing to shrink. We wish to highlight the problems caused by forest fragmentation and the subsequent need to manage dwindling populations of endangered species. In addition to the fixed panels with text and hand-painted illustrations, a video monitor shows a short film about orangutan rehabilitation programs in Indonesia, and a display unit shows photographs of wild orangutans, natural forest and forest clearance by burning. It is intended to use the orangutans as a "flagship" species, as Sumatra represents a key region to highlight the conservation of biodiversity (Coffey and McDonnell, unpublished).

DISCUSSION

It is often felt that the needs of zoo animals are compromised by the need to display them to the visiting public. This is, however, not necessarily the case. In well-designed accommodations those needs are carefully integrated without conflict.

The orangutan environment at JWPT is designed to accommodate the needs of the animals for space, height from the ground, suspensory locomotion and high-level nesting areas. Use of ropes and platforms ensures that the animals do not have to come to ground level unless they choose to do so.

The design of the areas ensures that keepers can work safely and effectively. There are no "blind spots" in the cages where animals cannot be seen, ensuring effective observation of the animals and eliminating the risk of misidentification. The thoughtful and careful use of vegetation and observation areas gives the public the impression that they are "entering" the realm of the orangutans rather than viewing them from outside. The intimacy provided by this feeling enhances the visitor experience. This helps to create receptivity towards the educational materials available both in the orangutan house and in the Long House Interpretive Center.

There are several goals of this environment with respect to the animals and the public. First, there is the goal of continuing to breed orangutans to contribute further to the long-term program for this endangered species. Second, the display is designed to engender empathy for orangutans and sympathy for their plight. Last, there is the goal of informing the public of the dire need for conservation action in Indonesia to help not only the orangutan but also their habitat and the thousands of species it contains.

The results of forest exploitation and depletion in Sumatra and Borneo make the orangutan a classic example of what is happening increasingly to other animals at high risk in greatly reduced, highly fragmented habitats. Initiatives by which the global zoo network can help to achieve its conservation goals can be greatly aided by the collaboration of scientists and managers, and by their active support of both *ex-situ* and *in-situ* management

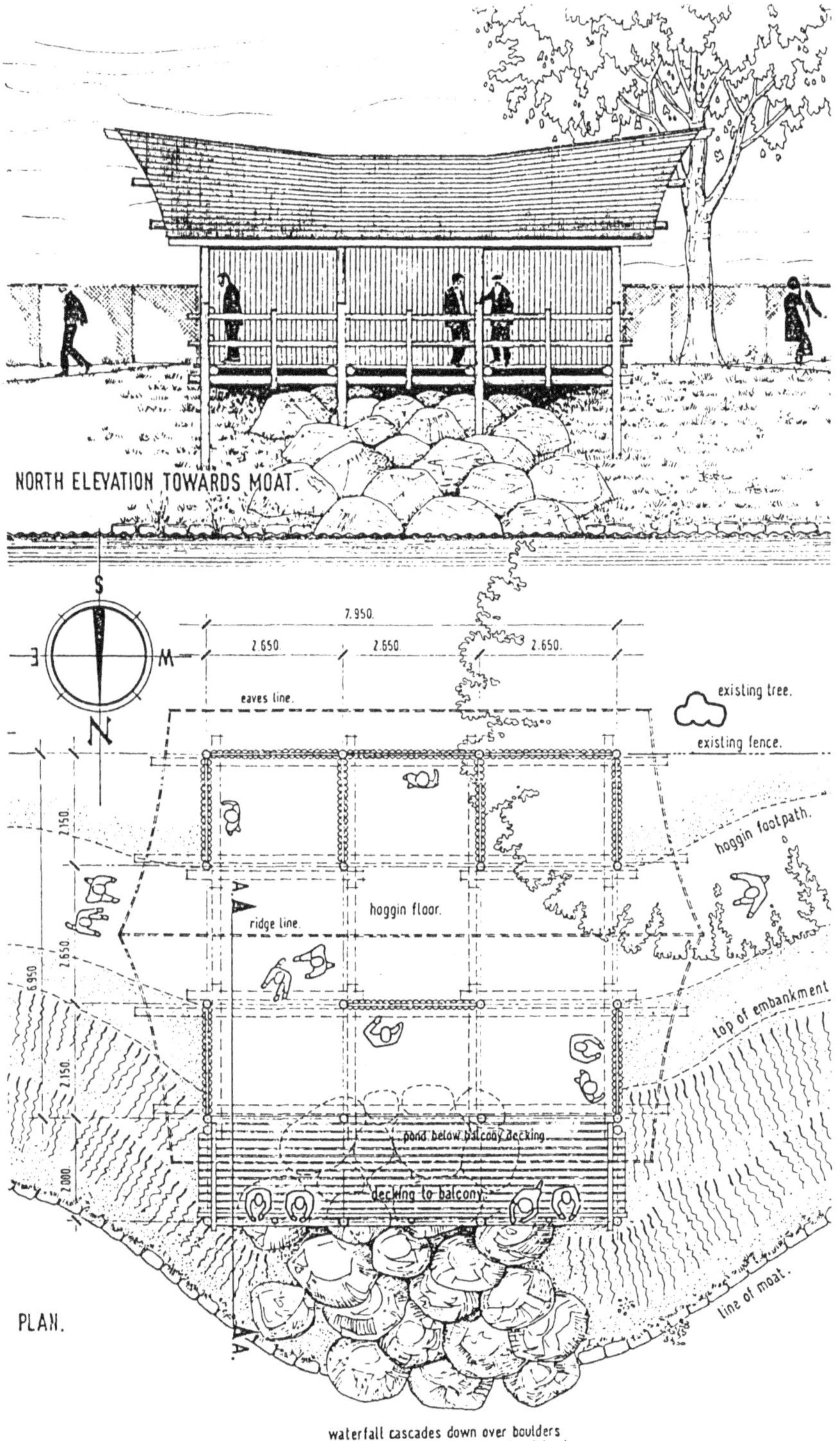

Figure 5. North elevation and plan of conservation education "Long-House". J. Douglas Smith, Architect.

programs for endangered species. As an example of such zoo support, JWPT, Zoo Atlanta, USA and Taronga Zoo, Australia, were co-sponsors with the Survival Service Commission of the International Union for the Conservation of Nature/Captive Breeding Specialist Group (CBSG) of the Orang-utan Population and Habitat Viability Analysis Workshop held in Medan, North Sumatra, 18 - 20 January, 1993, chaired by CBSG Chairman, Dr Ulysses Seal. This meeting arrived at goals and strategies aimed at conservation of the orangutan both in the wild and in captivity and highlights the contribution that zoos may make to an integrated approach to wildlife conservation.

ACKNOWLEDGMENTS

"The Jim Scriven Orang-Utan Home Habitat" was named in memory of the late Jim Scriven by his son and daughter-in-law, Jerry and Mandy. In addition to the major donation from the Scrivens, the Trust is grateful to its Life Members, Gardens, members of the WPTs (JWPT/WPTI/WPTC), supporters of SAFE, the many zoo visitors, and to the State of Jersey Tourism Committee, who have all given most generously to this major development as well as providing their valued support; also to Uniroyal for funding the children's play area. Thanks are due particularly to Richard Johnstone-Scott and Ian Singleton, also to Phillip Coffey, Anne Binney, Simon Hicks and all members of JWPT's Landscape Committee for their considerable input into the design criteria of both the inside and outside environments; to Christopher Carter of Colvin and Moggridge (Landscape Consultants), to Robert Orford and Roger Orford of G. Miles and Son Limited (Waterscape and Landscape Contractors), to David Steer of J. Douglas Smith (Architect), to Andrew Venables of Wildlife Management Services (power fencing), and to John D'Abbot Doyle and Joe Cook of J.P. Mauger Limited, for their collective work on design, landscaping and building construction. The Jersey Field Squadron Royal Engineers, supported by Raleigh International and the British Trust for Conservation Volunteers constructed the climbing "pods" and bridges. Thanks to Dr Simon Slaffer, who coordinated the team of specialists (consultant anaesthetists, gynecologists, general practitioners, radiologists, dentists, veterinarian and theater sister) who provided the orangutan family with a thorough medical check-up both prior to and during their transfer to the new accommodation. To Rachael Moore for typing from some untidy drafts of the paper presented by JM at the international orangutan conference The Neglected Ape, to Bronwen Garth-Thornton for organizing the type-script of the final draft, and to Anna Feistner for contributing some valuable comments on the various drafts of the overall contents of this paper. Finally, the Trust's gratitude to Sir David Attenborough who officially opened "The Jim Scriven Orang-Utan Home Habitat" on Sunday, 15th May, 1994.

REFERENCES

Bruner, G. and Meller, L., 1992, Convergent evolution in design philosophy of gorilla habitats. *Int. Zoo Yb.* 31:213-221.

Durrell, G.M., 1992, I want to be... Liked ... Secure ... Here! *A fundraising booklet, Jersey Wildlife Preservation Trust.* pp. 15.

Embury, A.S., 1992, Gorilla rainforest at Melbourne Zoo. *Int. Zoo Yb.* 31:203-213.

Galdikas, B.M.F., 1988, Orangutan diet, range, and activity at Tanjung Puting, central Borneo. *International Journal of Primatology* 9 (1):1-35.

Mackinnon, J., 1974, *In Search of the Red Ape.* Collins, London.

Mager, W.B. and Griede, T., 1986, Using outside areas for tropical primates in the northern hemisphere: Callitrichidae, Saimiri and Gorilla. In: *Primates the Road to Self-Sustaining Populations.* Benirschke, K. (Ed.), Springer-Verlag, New York. pp. 471 - 477.

Mallinson, J.J.C., 1980, The concept behind the design of the new gorilla environment at the Jersey Wildlife Preservation Trust. *Dodo, J. Jersey Wildl. Preserv. Trust* 17: 79-85.

Mallinson, J.J.C., 1982, Cage furnishings and environments for primates with special reference to marmosets and anthropoid apes. National Federation of Zoological Gardens of Great Britain and Ireland. *Pull-out Supplement No 12 pp. 6.*

Mallinson, J.J.C., 1986, The importance of an interdisciplinary approach: getting the conservation act together. In: *Primates the Road to Self-Sustaining Populations.* Benirschke, K. (Ed.), Springer-Verlag, New York. pp. 995-1003.

Mallinson, J.J.C., 1994, Concept behind the Jersey Wildlife Preservation Trust's new orang-utan accommodation. In: *Int. Conf. Orang utans: the Neglected Ape.* California State University, Fullerton (5 - 7 Mar). Abstract. p. 18.

Mallinson, J.J.C. and Redshaw, M.E., 1992, Maximising natural behaviour for primates in captivity. In: *Proc. Fourth Int. Symp. Zoo Design and Construction.* Stevens, E. M. C. (Ed.), Whitley Wildlife Conservation Trust. pp. 179 - 185.

Mallinson, J.J.C., Smith, J.D., Darwent, M. and Carroll, J.B., 1994, The design of the Sumatran orang-utan *Pongo pygmaeus abelii* 'Home Habitat' at the Jersey Wildlife Preservation Trust, *Dodo, J. Wildl. Preserv. Trust* 30:15-32.

Maple, T.L. and Finlay, T.W., 1986, Evaluating the environments of captive nonhuman primates. In: *Primates the Road to Self-Sustaining Populations.* Benirschke, K. (Ed.), Springer-Verlag, New York. pp. 479 - 488.

Ogden, J.J., Bruner, G. and Maple, T.L., 1992, A survey of the use of electric fencing with captive great apes. *Int. Zoo Yb.* 31:229-236.

Redshaw, M.E. and Mallinson, J.J.C., 1991, Stimulation of natural patterns of behaviour studies with golden lion tamarins and gorillas. In: *Primates Responses to Environmental Change.* Box, H. (Ed.), Chapman and Hall. pp. 217 -238.

Rijksen, H.D., 1978, *A Field Study on Sumatran Orang Utans (Pongo pygmaeus abelii* Lesson, 1827), *Ecology, Behaviour and Conservation*, 78-2. Wageningen, H. Veenman and Zonen B.V.

INDEX

9 780306 452130

GPSR Compliance

The European Union's (EU) General Product Safety Regulation (GPSR) is a set of rules that requires consumer products to be safe and our obligations to ensure this.

If you have any concerns about our products, you can contact us on ProductSafety@springernature.com

In case Publisher is established outside the EU, the EU authorized representative is:

Springer Nature Customer Service Center GmbH
Europaplatz 3
69115 Heidelberg, Germany

Batch number: 10755805

Printed by Printforce, the Netherlands